S0-EPC-058

Recent Titles in This Series

164 **N. N. Uraltseva, Editor,** Nonlinear Evolution Equations

163 **L. A. Bokut′, M. Hazewinkel, and Yu. G. Reshetnyak, Editors,** Third Siberian School "Algebra and Analysis"

162 **S. G. Gindikin, Editor,** Applied Problems of Radon Transform

161 **Katsumi Nomizu, Editor,** Selected Papers on Analysis, Probability, and Statistics

160 **K. Nomizu, Editor,** Selected Papers on Number Theory, Algebraic Geometry, and Differential Geometry

159 **O. A. Ladyzhenskaya, Editor,** Proceedings of the St. Petersburg Mathematical Society, Volume II

158 **A. K. Kelmans, Editor,** Selected Topics in Discrete Mathematics: Proceedings of the Moscow Discrete Mathematics Seminar, 1972–1990

157 **M. Sh. Birman, Editor,** Wave Propagation. Scattering Theory

156 **V. N. Gerasimov, N. G. Nesterenko, and A. I. Valitskas,** Three Papers on Algebras and Their Representations

155 **O. A. Ladyzhenskaya and A. M. Vershik, Editors,** Proceedings of the St. Petersburg Mathematical Society, Volume I

154 **V. A. Artamonov et al.,** Selected Papers in K-Theory

153 **S. G. Gindikin, Editor,** Singularity Theory and Some Problems of Functional Analysis

152 **H. Draškovičová et al.,** Ordered Sets and Lattices II

151 **I. A. Aleksandrov, L. A. Bokut′, and Yu. G. Reshetnyak, Editors,** Second Siberian Winter School "Algebra and Analysis"

150 **S. G. Gindikin, Editor,** Spectral Theory of Operators

149 **V. S. Afraĭmovich et al.,** Thirteen Papers in Algebra, Functional Analysis, Topology, and Probability, Translated from the Russian

148 **A. D. Aleksandrov, O. V. Belegradek, L. A. Bokut′, and Yu. L. Ershov, Editors,** First Siberian Winter School "Algebra and Analysis"

147 **I. G. Bashmakova et al.,** Nine Papers from the International Congress of Mathematicians, 1986

146 **L. A. Aĭzenberg et al.,** Fifteen Papers in Complex Analysis

145 **S. G. Dalalyan et al.,** Eight Papers Translated from the Russian

144 **S. D. Berman et al.,** Thirteen Papers Translated from the Russian

143 **V. A. Belonogov et al.,** Eight Papers Translated from the Russian

142 **M. B. Abalovich et al.,** Ten Papers Translated from the Russian

141 **H. Draškovičová et al.,** Ordered Sets and Lattices

140 **V. I. Bernik et al.,** Eleven Papers Translated from the Russian

139 **A. Ya. Aĭzenshtat et al.,** Nineteen Papers on Algebraic Semigroups

138 **I. V. Kovalishina and V. P. Potapov,** Seven Papers Translated from the Russian

137 **V. I. Arnol′d et al.,** Fourteen Papers Translated from the Russian

136 **L. A. Aksent′ev et al.,** Fourteen Papers Translated from the Russian

135 **S. N. Artemov et al.,** Six Papers in Logic

134 **A. Ya. Aĭzenshtat et al.,** Fourteen Papers Translated from the Russian

133 **R. R. Suncheleev et al.,** Thirteen Papers in Analysis

132 **I. G. Dmitriev et al.,** Thirteen Papers in Algebra

131 **V. A. Zmorovich et al.,** Ten Papers in Analysis

130 **M. M. Lavrent′ev, K. G. Reznitskaya, and V. G. Yakhno,** One-dimensional Inverse Problems of Mathematical Physics

129 **S. Ya. Khavinson,** Two Papers on Extremal Problems in Complex Analysis

128 **I. K. Zhuk et al.,** Thirteen Papers in Algebra and Number Theory

127 **P. L. Shabalin et al.,** Eleven Papers in Analysis

126 **S. A. Akhmedov et al.,** Eleven Papers on Differential Equations

(*Continued in the back of this publication*)

Nonlinear Evolution Equations

American Mathematical Society
TRANSLATIONS
Series 2 • Volume 164

Advances in the Mathematical Sciences – 22

(*Formerly Advances in Soviet Mathematics*)

Nonlinear Evolution Equations

N. N. Uraltseva
Editor

American Mathematical Society
Providence, Rhode Island

Translation edited by A. B. SOSSINSKY

ABSTRACT. This book contains papers on nonlinear problems in partial differential equations. Most of them resulted from talks given at the seminar on differential equations and mathematical physics at St. Petersburg State University.

Among the topics discussed in these papers are the existence and properties of solutions of various classes of nonlinear evolution equations, nonlinear embedding theorems, bifurcation of solutions, and equations of mathematical physics (Navier-Stokes type equations and the nonlinear Schrödinger equation).

The book will be useful for researchers and graduate students working in partial differential equations and mathematical physics.

1991 *Mathematics Subject Classification.* Primary 35G; Secondary 35J, 35K, 35L, 35Q, 46E.

Library of Congress Cataloging-in-Publication Data

Nonlinear evolution equations / N. N. Ural′tseva, editor.
p. cm. — (American Mathematical Society translations, ISSN 0065-9290; ser. 2, v. 164)
Includes bibliographical references.
ISBN 0-8218-4123-8 (alk. paper)
1. Evolution equations, Nonlinear. I. Ural′t͡seva, N. N. (Nina Nikolaevna) II. Series.
QA3.A572 ser. 2, vol. 164
[QA377]
510 s—dc20
[515′.353]
95-2339
CIP

♾ The paper used in this book is acid-free and falls within the guidelines
established to ensure permanence and durability.
♲ Printed on recycled paper.
This volume was typeset by the authors using AMS-TEX,
the American Mathematical Society's TEX macro system.

10 9 8 7 6 5 4 3 2 1 00 99 98 97 96 95

Contents

Foreword ix

Hölder Estimates of Solutions to Initial-Boundary Value Problems for Parabolic Equations of Nondivergent Form with Wentzel Boundary Condition
D. E. Apushkinskaya and A. I. Nazarov 1

Reverse Hölder Inequalities with Boundary Integrals and L_p-Estimates for Solutions of Nonlinear Elliptic and Parabolic Boundary-Value Problems
A. A. Arkhipova 15

Quasilinear Parabolic Equations with Small Parameter in a Hilbert Space
Ya. Belopol'skaya 43

On the Stability of Solitary Waves for Nonlinear Schrödinger Equations
V. S. Buslaev and G. S. Perelman 75

On Semigroups Generated by Initial-Boundary Value Problems Describing Two-Dimensional Visco-Plastic Flows
O. Ladyzhenskaya and G. Serëgin 99

Elliptic Differential Inequalities, Embedding Theorems, and Variational Problems
V. A. Malyshev 125

Long Time Behavior of Flows Moving by Mean Curvature
V. I. Oliker and N. N. Uraltseva 163

Bifurcation Problem for Nonlinear Second Order Equations in Variable Regions
V. G. Osmolovskiĭ and A. V. Sidorov 171

Existence of a Weak Solution of the Minimax Problem Arising in Coulomb-Mohr Plasticity
S. Repin and G. Seregin 189

Foreword

The Mathematical Physics Seminar has been working in Leningrad (St. Petersburg) since September, 1947. Many experts in various areas of mathematics, theoretical mechanics, hydrodynamics and gasdynamics, diffraction theory, and mathematical problems in theoretical physics have participated in it. The seminar is devoted to the study of differential equations (mainly partial differential equations) and their applications both in mathematics itself and in the other disciplines mentioned above.

At the beginning linear problems dominated the seminar discussions, but since the mid-fifties results on nonlinear problems have often been reported as well. A number of mathematicians of our city and the rest of the country, as well as many well-known mathematicians from America, Asia, and Europe, have given talks at the seminar.

From its beginning until 1974 the seminar worked under the supervision of V. I. Smirnov, and after his death it was named the Smirnov Seminar.

Most of the papers in this book are devoted to nonlinear problems. The article by Ladyzhenskaya and Seregin deals with the first initial-boundary value problem for nonlinear systems describing visco-elastic flows. It is proved that their resolving operators form a continuous compact semigroup possessing a compact minimal B-attractor.

The existence of weak solutions to stationary variational problems in plasticity theory is proved under the Coulomb–Mohr fluidity conditions in the paper of Repin and Seregin. The Cauchy problem for an infinite-dimensional quasilinear parabolic equation with a small parameter is studied in the article of Belopol′skaya. It is proved that the solutions are of wave character, and the wave front velocity is computed. The probabilistic approach is used to obtain results in the infinite-dimensional case.

Arkhipova's article is a review of her results on inverse Hölder inequalities and their applications to the study of the regularity of generalized solutions to the Neumann problem for systems of quasilinear elliptic and parabolic equations.

Nonlinear Schrödinger equations with initial data located near a soliton solution are studied by Buslaev and Perelman. They prove that if the spectrum of the problem, linearized on this soliton, is real, then the large-time asymptotics of these solutions is described by moving solitons and scattering states that obey the free Schrödinger equation.

Bifurcations of solutions of fully nonlinear elliptic equations in variable regions are investigated by Osmolovskiĭ and Sidorov. Necessary and sufficient conditions are proved to ensure that the region is a bifurcation region for the equation under consideration.

The Wentzell problem for quasilinear parabolic equations is studied by Apushkinskaya and Nazarov. They obtain a priori estimates for Hölder norms of solutions under optimal restrictions on the data.

A new concept of nonlinear imbedding theorems is developed in Malyshev's article. Unlike the standard imbedding theorems, nonlinear ones are valid for some convex sets in Banach spaces, but not for the spans of those sets. Classical theorems about the convergence of convex, monotone, or subharmonic functions may be regarded as examples of nonlinear imbedding theorems.

N. N. Uraltseva

Amer. Math. Soc. Transl.
(2) Vol. **164**, 1995

Hölder Estimates of Solutions to Initial-Boundary Value Problems for Parabolic Equations of Nondivergent Form with Wentzel Boundary Condition

D. E. Apushkinskaya and A. I. Nazarov

The purpose of this paper is to obtain a majorant of the Hölder norms for solutions of quasilinear parabolic equations of nondivergent structure with boundary condition such as the Wentzel boundary condition. Analogous results for solutions of elliptic boundary value problems were established in **[L1]** and **[L2]**.

This paper is organized as follows. We describe our problem and formulate the main result in §1. The oscillation estimate of the solution of the linear problem near the boundary is presented in §2, and is used to estimate the Hölder constant in §3. Corresponding results for solutions of stationary boundary value problem are formulated in the Appendix (§4).

We would like to thank Professor N. N. Uraltseva for her constant encouragement.

Notation. $x = (x', x_n) = (x_1, \dots, x_{n-1}, x_n)$ is a vector in $\mathbb{R}^n$ with Euclidean norm $\|x\|$, and $|x|$ is the sup-norm of x. (x, t) is a point in $\mathbb{R}^{n+1}$. The indices l, m always vary from 1 to $n-1$, whereas i, j vary from 1 to n. Repeated indices indicate summation.

Ω is a bounded domain in $\mathbb{R}^n$ and $\partial\Omega$ is its boundary. $\mathbf{n}(x) = (\mathbf{n}_i(x))$ is the unit vector of the outward normal to $\partial\Omega$ at the point x.

Q is a domain in $\mathbb{R}^{n+1}$. By $|\Omega|$ we denote the n-dimensional Lebesgue measure of the set Ω, while the $(n+1)$-dimensional Lebesgue measure of the set Q is denoted by $|Q|$. If $Q = \Omega \times (t^1; t^2)$ is a cylinder in $\mathbb{R}^{n+1}$, then $\partial''Q = \partial\Omega \times (t^1; t^2)$ is the lateral surface of Q, and $\partial'Q = \partial''Q \cup \{\Omega \times \{t^1\}\}$ is the parabolic boundary of Q.

We denote by Ω^+ (Q^+) the part of Ω (Q) in the halfspace $x_n > 0$. We shall generally suppose that the part $\Gamma(\Omega^+)$ of $\partial\Omega^+$ (the part $\Gamma(Q^+)$ of $\partial''Q^+$, respectively) lies in the plane $x_n = 0$.

1991 *Mathematics Subject Classification*. Primary 35K60.
This work was partially supported by Russian Fund for Fundamental Research, grant no. 93-011-1696.

We define

$$K_R(x^0) = \{ x \in \mathbb{R}^n : |x - x^0| < R \}, \qquad K_R = K_R(0);$$
$$Q_R(x^0, t^0) = K_R(x^0) \times (t^0 - R^2; t^0), \qquad Q_R = Q_R(0;0);$$
$$\Gamma_R(x^0, t^0) = \Gamma(Q_R^+(x^0, t^0)), \qquad \Gamma_R = \Gamma(Q_R^+).$$

If u is a continuous function, then $A_k^u = \{ (x,t) : u(x,t) \geqslant k \}$. We set

$$\operatorname*{osc}_Q u = \sup_Q u - \inf_Q u.$$

Below we use D_i to denote the operator of differentiaion with respect to x_i; in particular, $Du = (D'u, D_n u)$ is the gradient of u. Let δ_i be the tangential differential operator on $\partial\Omega$, i.e., $\delta_i = D_i - \mathbf{n}_i \mathbf{n}_k D_k$. Then $\delta u = (\delta_i u)$ is the tangential gradient of u on $\partial\Omega$; in particular, $\delta u = (D'u, 0)$ on $\Gamma(Q^+)$. Also, $u_t = \partial u/\partial t$.

We denote by $\|\cdot; Q\|_p$ the norm in $L_p(Q)$, and by $W_p^{2,1}(Q)$ the Sobolev space with the norm

$$\|u\|_{W_p^{2,1}(Q)} = \|u_t, Q\|_p + \|D(Du), Q\|_p + \|u, Q\|_p.$$

$C^\gamma(\overline{Q})$ is the Hölder space with exponent γ in the parabolic distance

$$d_{\text{par}}((x^1, t^1), (x^2, t^2)) = \|x^1 - x^2\| + |t^1 - t^2|^{1/2}.$$

$[u]_{\gamma,Q}$ is the corresponding Hölder constant.

We set $f_+ = \max\{f; 0\}$ and $f_- = \max\{-f; 0\}$.

We denote by $\operatorname{tr}(a)$ the trace of matrix (a). We use M, N, and C with or without indices, to denote various constants. To indicate the quantities on which they depend we write $N(\cdots)$.

§1. Main results

In the cylinder $Q = \Omega \times (0; T)$ we consider the equation

$$u_t - a^{ij}(x, t, u, Du) D_i D_j u = a(x, t, u, Du) \tag{1.1}$$

and suppose that

$$u\big|_{t=0} = \varphi(x), \tag{1.2}$$

$$u_t - \alpha^{ij}(x, t, u, Du)\delta_i\delta_j u = \alpha(x, t, u, Du) \quad \text{on } \partial''Q. \tag{1.3}$$

Assume that equation (1.1) is uniformly parabolic, i.e., $a = (a^{ij})$ is a symmetric matrix and the inequalities

$$\nu\|\xi\|^2 \leqslant a^{ij}(x, t, z, p)\xi_i\xi_j \leqslant \nu^{-1}\|\xi\|^2, \qquad \nu = \text{const} > 0, \tag{1.4}$$

hold for any $(x,t) \in Q$, $z \in \mathbb{R}^1$, $p \in \mathbb{R}^n$, $\xi \in \mathbb{R}^n$. Assume also that the function $a(x, t, z, p)$ on the right-hand side of (1.1) satisfies

$$|a(x, t, z, p)| \leqslant \mu\|p\|^2 + b(x, t)\|p\| + \Phi(x, t) \tag{1.5}$$

with some positive constant μ and with $b \in L_{n+2}(Q)$, $\Phi \in L_{n+1}(Q)$.

Let the operator on the left-hand side of (1.3) be a uniformly parabolic Wentzel boundary operator, i.e., for $(x,t)\in\partial''Q$, $z\in\mathbb{R}^1$, and $p\in\mathbb{R}^n$ the following conditions are satisfied: $\alpha=(\alpha^{ij})$ is a symmetric matrix;

$$\begin{gathered}\nu_1\|\xi\|^2\leqslant\alpha^{ij}(x,t,z,p)\xi_i\xi_j\leqslant\nu_1^{-1}\|\xi\|^2,\quad \nu_1=\mathrm{const}>0,\\ \text{for all } \xi\in\mathbb{R}^{\mathbf{n}} \text{ such that } \xi\perp n(x);\end{gathered}\tag{1.6}$$

the function $\alpha(x,t,z,p)$ is differentiable with respect to the variables p_i, and for $(x,t,)\in\partial''Q$, $z\in\mathbb{R}^1$ we have

$$0\leqslant-\alpha_{p_i}(x,t,z,p)\mathbf{n}_i(x)\leqslant\beta(x,t)\quad\forall p\in\mathbb{R}^n\tag{1.7}$$

$$|\alpha(x,t,z,p)|\leqslant\mu_1\|p\|^2+\beta(x,t)\|p\|+\Theta(x,t)\quad\forall p\in\mathbb{R}^n,\ p\perp\mathbf{n}(x)\tag{1.8}$$

Here $\mu_1=\mathrm{const}>0$, $\beta\in L_{n+1}(\partial''Q)$, $\Theta\in L_n(\partial''Q)$.

THEOREM 1. *Let $\partial\Omega\in W^2_{n+2}$, and let $u\in W^{2,1}_{n+1}(Q)\cap W^{2,1}_n(\partial''Q)\cap C(\overline{Q})$ be a solution of (1.1)–(1.3). Assume also that $|u|\leqslant M_0$ in $\overline{Q}$ and $[\varphi]_{\gamma,\Omega}\leqslant M_\gamma$ with $\gamma=\mathrm{const}>0$. Then there exists $\gamma_1>0$, completely determined by n, ν, ν_1, γ, and $\partial\Omega$, such that*

$$[u]_{\gamma_1,Q}\leqslant M_{\gamma_1},\tag{1.9}$$

where M_{γ_1} depends on the same arguments as γ_1 and, in addition, on μ, μ_1, M_0, M_γ, $\|\Phi;Q\|_{n+1}$, $\|\Theta;\partial''Q\|_n$ and on the moduli of continuity of $b(x,t)$ in $L_{n+2}(Q)$ and of $\beta(x,t)$ in $L_{n+1}(\partial''Q)$.

REMARK. It is sufficient to assume that conditions (1.4)–(1.8) hold for $|z|\leqslant M_0$ only. Without loss of generality we may assume $\mu=\mu_1$, $\nu=\nu_1$.

§2. Estimate of nonnegative solutions of the Wentzel boundary value problem near the boundary

Suppose L is a linear parabolic operator:

$$Lu=u_t(x,t)-a^{ij}(x,t)D_iD_ju(x,t)+b^i(x,t)D_iu(x,t),$$

$a=(a^{ij})$ is a symmetric matrix, and the inequalities

$$\nu\|\xi\|^2\leqslant a^{ij}\xi_i\xi_j\leqslant\nu^{-1}\|\xi\|^2,\qquad \nu=\mathrm{const}>0,$$

hold for any $(x,t)\in Q_1^+$, $z\in\mathbb{R}^n$, $b^i\in L_{n+2}(Q_1^+)$.

Suppose B is a linear parabolic boundary operator:

$$Bu=u_t(x,t)-\alpha^{lm}(x,t)D_lD_mu(x,t)+\beta^i(x,t)D_iu(x,t),$$

$\alpha=(\alpha^{lm})$ is a symmetric matrix, and the inequalities

$$\nu\|\eta\|^2\leqslant\alpha^{lm}\eta_l\eta_m\leqslant\nu^{-1}\|\eta\|^2,\quad \nu=\mathrm{const}>0;\qquad \beta^n(x,t)\leqslant 0$$

hold for any $(x,t)\in\Gamma_1$, $\eta\in\mathbb{R}^{n-1}$, $\beta^i\in L_{n+1}(\Gamma_1)$.

Let us denote $b_*(x,t) = \big(b^i(x,t)\big)$, $\beta_*(x,t) = \big(\beta^i(x,t)\big)$.

LEMMA 2.1. *Suppose $\rho \leqslant 1$ and u is a function such that*

$$u \in W^{2,1}_{n+1}(Q^+_\rho) \cap W^{2,1}_n(\Gamma_\rho) \cap C(\overline{Q}^+_\rho); \qquad u \geqslant 0 \quad in\ Q^+_\rho.$$

Then for any $\delta_1 \in (0;1)$ there exist constants $\sigma_1 > 0$, $\xi < 1$, depending only on n, ν, δ_1, such that the inequalities

$$\|b_*(x,t); Q^+_\rho\|_{n+2} \leqslant \sigma_1, \tag{2.1}$$

$$\|\beta_*(x,t); \Gamma_\rho\|_{n+1} \leqslant \sigma_1, \tag{2.2}$$

$$|A^u_k \cap \Gamma_\rho| \geqslant \xi|\Gamma_\rho| \quad \textit{for some } k > 0 \tag{2.3}$$

imply

$$\begin{aligned} u \geqslant (1-\delta_1)k - N_1(n,\nu)\Big[\rho^{n/(n+1)}\big\|(Lu)_-; Q^+_\rho\big\|_{n+1} \\ + \rho^{(n-1)/n}\big\|(Bu)_-; \Gamma_\rho\big\|_n\Big] \quad on\ \Gamma_{\rho/2}. \end{aligned} \tag{2.4}$$

PROOF. It is obvious that the constant k may be assumed to be 1. By means of the change of coordinates of $(x,t) \in Q^+_\rho$ to the point $(\rho^{-1}x, \rho^{-2}t)$, we take Q^+_1 instead of Q^+_ρ. After such a coordinate transformation the function $\widetilde{u}(x,t) = u(\rho x, \rho^2 t)$ satisfies

$$\widetilde{L}\widetilde{u} = \widetilde{u}_t - \widetilde{a}^{ij} D_i D_j \widetilde{u} + \rho \widetilde{b}^i D_i \widetilde{u} = \rho^2 (Lu) \quad \text{in } Q^+_1, \tag{2.5}$$

$$\widetilde{B}\widetilde{u} = \widetilde{u}_t - \widetilde{\alpha}^{lm} D_l D_m \widetilde{u} + \rho \widetilde{\beta}^i D_i \widetilde{u} = \rho^2 (Bu) \quad \text{on } \Gamma_1. \tag{2.6}$$

Here $\widetilde{a}^{ij}(x,t) = a^{ij}(\rho x, \rho^2 t)$, $\widetilde{\alpha}^{lm}(x,t) = \alpha^{lm}(\rho x, \rho^2 t)$, etc.

Let $\underleftarrow{\min}_{\Gamma_{1/2}} \widetilde{u} = \widetilde{u}(x^0, t^0)$. Obviously, $Q^+_{1/2}(x^0,t^0) \subseteq Q^+_1$. We introduce the auxiliary function

$$\psi(x,t) = 1 - 4\big(\|x' - x^{0\prime}\|^2 - (t - t^0)\big) + Cx_n^2 - C_1 x_n, \tag{2.7}$$

where C, C_1 are the positive constants to be determined later.

Let Q denote the set $Q = \{\,(x,t) \in Q^+_{1/2}(x^0,t^0) : x_n < C^{-1/2}\,\}$.

If we take $C_1 = 2C^{1/2}$, then $\psi \leqslant 0$ on $\partial' Q \setminus \Gamma(Q)$, and therefore also $(\psi - \widetilde{u}) \leqslant 0$ on $\partial' Q \setminus \Gamma(Q)$. Applying the maximum principle (see Theorem 1 from **[A]**) to the difference $\psi - \widetilde{u}$ for $(x,t) \in Q$, we obtain

$$\begin{aligned} \psi - \widetilde{u} \leqslant \widetilde{N}\Big\{\big\|\big(\widetilde{L}(\psi - \widetilde{u})\big)_+; A_0^{\psi-\widetilde{u}} \cap Q\big\|_{n+1} \\ + \big\|\big(\widetilde{B}(\psi - \widetilde{u})\big)_+; A_0^{\psi-\widetilde{u}} \cap \Gamma(Q)\big\|_n\Big\} \quad \text{in } Q, \end{aligned}$$

where $\widetilde{N}$ depends only on n, ν, $\big\|\rho\widetilde{b}_*; Q^+_1\big\|_{n+1}$, $\big\|\rho\widetilde{\beta}_*; \Gamma_1\big\|_n$. We remark that conditions (2.1) and (2.2) imply

$$\big\|\rho\widetilde{b}_*; Q^+_1\big\|_{n+1} = \rho^{-1/(n+1)}\big\|b_*; Q^+_\rho\big\|_{n+1} \leqslant 2\big\|b_*; Q^+_\rho\big\|_{n+2} \leqslant 2\sigma_1, \tag{2.8}$$

$$\big\|\rho\widetilde{\beta}_*; \Gamma_1\big\|_n = \rho^{-1/n}\big\|\beta_*; \Gamma_\rho\big\|_n \leqslant 2\big\|\beta_*; \Gamma_\rho\big\|_{n+1} \leqslant 2\sigma_1. \tag{2.9}$$

Without loss of generality we shall assume $\sigma_1 \leqslant \nu$. Hence we get

$$\begin{aligned}\psi - \widetilde{u} \leqslant N_1(n;\nu)\Big\{\big\|(\widetilde{L}\psi)_+; A_0^{\psi-\widetilde{u}} \cap Q\big\|_{n+1} + \big\|(\widetilde{L}\widetilde{u})_-; A_0^{\psi-\widetilde{u}} \cap Q\big\|_{n+1} \\ + \big\|(\widetilde{B}\psi)_+; A_0^{\psi-\widetilde{u}} \cap \Gamma(Q)\big\|_n + \big\|(\widetilde{B}\widetilde{u})_-; A_0^{\psi-\widetilde{u}} \cap \Gamma(Q)\big\|_n\Big\} \quad \text{in } Q.\end{aligned} \tag{2.10}$$

An elementary computation gives

$$\begin{aligned}\widetilde{L}\psi &= 4 + 8\,\mathrm{tr}(a) - 2(C+4)a^{nn} - 8\rho\widetilde{b}^l(x_l - x_l^0) + 2C\rho\widetilde{b}^n x_n - 2C^{1/2}\rho\widetilde{b}^n \\ &\leqslant 4 + 8(n-1)\nu^{-1} + 8\rho\|\widetilde{b}_*\|\|x' - x^{0\prime}\| + 2C^{1/2}\rho|\widetilde{b}^n| \quad \text{in } Q.\end{aligned}$$

Let $C = \big[2(2(n-1)+\nu)\big]/\nu^2$. It is easy to verify that

$$\widetilde{L}\psi \leqslant C_2(n,\nu)\rho\|\widetilde{b}_*\| \quad \text{in } Q, \tag{2.11}$$

$$\begin{aligned}\widetilde{B}\psi &= 4 + 8\,\mathrm{tr}(\alpha) - 8\rho\widetilde{\beta}^l(x_l - x_l^0) - 2C^{1/2}\rho\widetilde{\beta}^n \\ &\leqslant C_3(n,\nu)\big(1 + \rho\|\widetilde{\beta}_*\|\big) \quad \text{on } \Gamma(Q).\end{aligned} \tag{2.12}$$

Now we observe that $\widetilde{u} \leqslant 1$ on the set $A_0^{\psi-\widetilde{u}}$. Then combining estimate (2.10) with (2.3), (2.8), (2.9), (2.11), and (2.12), we conclude that

$$\begin{aligned}\psi - \widetilde{u} \leqslant N_1\big\{\big\|(\widetilde{L}\widetilde{u})_-; Q_1^+\big\|_{n+1} + \big\|(\widetilde{B}\widetilde{u})_-; \Gamma_1\big\|_n\big\} \\ + 2N_1(C_2 + C_3)\sigma_1 + N_1C_3(1-\xi) \quad \text{in } Q.\end{aligned} \tag{2.13}$$

Relations (2.5) and (2.6) give

$$\begin{aligned}\big\|(\widetilde{L}\widetilde{u})_-; Q_1^+\big\|_{n+1} &= \rho^{n/(n+1)}\big\|(Lu)_-; Q_\rho^+\big\|_{n+1}, \\ \big\|(\widetilde{B}\widetilde{u})_-; \Gamma_1\big\|_n &= \rho^{(n-1)/n}\big\|(Bu)_-; \Gamma_\rho\big\|_n.\end{aligned}$$

Choosing σ_1, ξ so that $N_1\{2(C_2 + C_3)\sigma_1 + C_3(1-\xi)\} < \delta_1$, we obtain

$$\begin{aligned}\widetilde{u}(x,t) \geqslant \psi(x,t) - \delta_1 - N_1(n,\nu)\{\rho^{n/(n+1)}\big\|(Lu)_-; Q_\rho^+\big\|_{n+1} \\ + \rho^{(n-1)/n}\big\|(Bu)_-; \Gamma_\rho\big\|_n\}.\end{aligned}$$

Taking this inequality at the point (x^0, t^0) and noting that $\psi(x^0, t^0) = 1$, we get estimate (2.4). □

LEMMA 2.2. *Let $\rho \leqslant 1$, $u \in W_{n+1}^{2,1}(Q_\rho^+) \cap C(\overline{Q}_\rho^+)$, $u \geqslant 0$ in $\overline{Q}_\rho^+$, $u \geqslant k > 0$ on Γ_ρ. Then for any $\delta_2 \in (0;1)$ there exist positive constants ε and σ_2, depending only on n, ν, δ_2, such that $\big\|b_*; Q_\rho^+\big\|_{n+2} \leqslant \sigma_2$ implies*

$$u \geqslant (1-\delta_2)k - N_2(n,\nu)\rho^{n/(n+1)}\big\|(Lu)_-; Q_\rho^+\big\|_{n+1} \quad \textit{in } Q_{\varepsilon\rho}^+. \tag{2.14}$$

PROOF. By the same coordinate stretching as in the proof of Lemma 2.1 we map Q_ρ^+ into Q_1^+. We shall assume also that $k = 1$ and $\sigma_2 \leqslant \nu$. We introduce the barrier function

$$\psi_1(x,t) = 1 - \|x'\|^2 + t + (C/4)x_n^2 - C^{1/2}x_n,$$

where $C = C(n,\nu)$ is the same as in Lemma 2.1.

Let $Q = Q_1^+ \cap \{x_n < 2C^{-1/2}\}$. It follows from the assumptions of Lemma 2.2 that $\psi_1 - \widetilde{u} \leqslant 0$ on $\partial' Q$, and $\widetilde{L}\psi_1 \leqslant C_4(n,\nu)\rho\|\widetilde{b}_*\|$ in Q. Applying the inequality of Nazarov and Ural′tseva (see Theorem 1 in **[NU]**) in Q to $\psi_1 - \widetilde{u}$, and using (2.5) and (2.8), we obtain

$$\widetilde{u}(x,t) \geqslant \psi_1(x,t) - N_2(n,\nu)\rho^{n/(n+1)}\big\|(Lu)_-; Q_\rho^+\big\|_{n+1} - 2N_2C_4\sigma_2 \quad \text{in } Q. \tag{2.15}$$

Since $\psi_1(0,0) = 1$, we may choose $\varepsilon = \varepsilon(n,\nu,\delta_2)$ such that $Q_\varepsilon^+ \subset Q$ and $\psi_1 \geqslant 1 - \delta_2/2$ in Q_ε^+. Let us take $\sigma_2 = \delta_2/(4N_2C_4)$ in (2.15). Then, returning to the original variables, we get (2.14). □

LEMMA 2.3. *Let $\varkappa \in (0;1)$, $\rho \leqslant 1$, $\varepsilon_1 \in (0;1)$,*

$$\varkappa\rho^2 \leqslant \tau \leqslant \varkappa^{-1}\rho^2, \qquad 2\varkappa\rho \leqslant x_l^{(2)} - x_l^{(1)} \leqslant 2\varkappa^{-1}\rho$$
$$x_l^{(1)} + \varkappa\rho \leqslant x_l^{(0)} \leqslant x_l^{(2)} - \varkappa\rho, \qquad l = 1,2,\dots,n-1,$$
$$2\varkappa\rho \leqslant x_n^{(2)} \leqslant 2\varkappa^{-1}\rho, \qquad x_n^{(0)} = 0,$$
$$Q = \big\{\,(x,t) : x_l \in (x_l^{(1)}; x_l^{(2)}),\ x_n \in (0; x_n^{(2)}),\ t \in (t^1; t^1+\tau)\,\big\}.$$

Assume also that $u \in W_{n+1}^{2,1}(Q) \cap W_n^{2,1}(\Gamma(Q)) \cap C(\overline{Q})$, $u \geqslant 0$ in Q, and that $u(x,t^1) \geqslant k > 0$ for $|x - x^{(0)}| \leqslant \varepsilon_1\rho$, $x_n > 0$.

Then there exist constants δ_3 and p, depending on n, ν, $\varkappa$, $\sigma_3 = \sigma_3(n,\nu,\varkappa,\varepsilon_1)$ only, such that the inequalities

$$\|b_*; Q\|_{n+2} \leqslant \sigma_3, \qquad \|\beta_*; \Gamma(Q)\|_{n+1} \leqslant \sigma_3,$$

imply

$$\begin{aligned} u \geqslant \delta_3\varepsilon_1^{2p+1}k - N_1\big\{&\rho^{n/(n+1)}\big\|(Lu)_-; Q\big\|_{n+1} \\ &+ \rho^{(n-1)/n}\big\|(Bu)_-; \Gamma(Q)\big\|_n\big\} \quad \text{in } Q', \end{aligned} \tag{2.16}$$

where N_1 is the constant from Lemma 2.1 *and*

$$\begin{aligned} Q' = \big\{\,(x,t) : x_l^{(1)} + \varkappa\rho \leqslant x_l \leqslant x_l^{(2)} - \varkappa\rho,& \\ 0 \leqslant x_n \leqslant x_n^{(2)} - \varkappa\rho,\quad t^1 + \varkappa\rho^2 \leqslant t \leqslant t^1 + \tau\,\big\}.& \end{aligned}$$

PROOF. Without loss of generality we may assume $x^{(0)} = 0$, $t^1 = 0$, $k = 1$, $\sigma_3 \leqslant \nu$, $\varepsilon_1 \leqslant \varkappa 2^{-1/2}$. As in the proofs of Lemmas 2.1 and 2.2 we map Q_ρ^+ by means of a coordinate transformation into Q_1^+. Let

$$\min_{Q'} \widetilde{u} = u(x^*, t^*).$$

Without loss of generality we also assume that $t^* = \tau$. Denote

$$\gamma = \varkappa^3/2, \quad r = \|x - \tau^{-1}tx^*\|/(\gamma t + \varepsilon_1^2), \quad D = \{\,(x,t) : r < 1,\ 0 < t < \tau\,\}.$$

Obviously $D^+ \subset Q$.

We introduce the function (see **[KS]**)

$$\varphi(x,t) = \begin{cases} (1-r)^2/(\gamma t + \varepsilon_1^2)^p & \text{in } D, \\ 0 & \text{in } \mathbb{R}^{n+1} \setminus D. \end{cases}$$

As is easily seen, $\varepsilon_1^{2p+1}\varphi - u \leqslant 0$ on $\partial' D$, and therefore it will be true on $\partial' D^+ \setminus \Gamma(D^+)$.

Hence we can use Theorem 1 from **[A]** and apply it to $\varepsilon_1^{2p+1}\varphi - \widetilde{u}$ in D^+. This gives the estimate

$$\varepsilon_1^{2p+1}\varphi - \widetilde{u} \leqslant \widetilde{N}\Big\{\big\|\big(\widetilde{L}(\varepsilon_1^{2p+1}\varphi - \widetilde{u})\big)_+; D^+\big\|_{n+1} + \big\|\big(\widetilde{B}(\varepsilon_1^{2p+1}\varphi - \widetilde{u})\big)_+; \Gamma(D^+)\big\|_n\Big\} \quad \text{in } D^+.$$

Further, using the same argument as in the proof of Lemma 2.1, we have

$$\begin{aligned}\varepsilon_1^{2p+1}\varphi - \widetilde{u} \leqslant N_1(n,\nu)\Big\{&\varepsilon_1^{2p+1}\big\|(\widetilde{L}\varphi)_+; D^+\big\|_{n+1} + \big\|(\widetilde{L}\widetilde{u})_-; Q\big\|_{n+1} \\ &+ \varepsilon_1^{2p+1}\big\|(\widetilde{B}\varphi)_+; \Gamma(D^+)\big\|_n + \big\|(\widetilde{B}\widetilde{u})_-; \Gamma(Q)\big\|_n\Big\} \quad \text{in } D^+.\end{aligned} \tag{2.17}$$

Calculations show that $\|D\varphi\| \leqslant 4/\varepsilon_1^{2p+1}$. Moreover, it is known **[KS,** Lemma 1.3] that if $p = p(n, \varkappa, \nu)$ is sufficiently large, then we have

$$\begin{aligned}\varphi_t - \widetilde{a}^{ij}D_iD_j\varphi \leqslant 0 \quad &\text{in } D^+, \\ \varphi_t - \widetilde{\alpha}^{lm}D_lD_m\varphi \leqslant 0 \quad &\text{on } \Gamma(D^+),\end{aligned}$$

respectively.

Hence

$$\begin{aligned}\widetilde{L}\varphi &= \varphi_t - \widetilde{a}^{ij}D_iD_j\varphi + \rho\widetilde{b}^iD_i\varphi \\ &\leqslant \rho\|\widetilde{b}_*\|\|D\varphi\| \leqslant \big(4\rho/\varepsilon_1^{2p+1}\big)\|\widetilde{b}_*\| \quad \text{in } D^+,\end{aligned} \tag{2.18}$$

$$\begin{aligned}\widetilde{B}\varphi &= \varphi_t - \widetilde{\alpha}^{lm}D_lD_m\varphi + \rho\widetilde{\beta}^iD_i\varphi \\ &\leqslant \rho\|\widetilde{\beta}_*\|\|D\varphi\| \leqslant \big(4\rho/\varepsilon_1^{2p+1}\big)\|\widetilde{\beta}_*\| \quad \text{on } \Gamma(D^+).\end{aligned} \tag{2.19}$$

We remark that $\varphi(x^*, \tau) \geqslant \varkappa^{-2p}$. Then using (2.5), (2.6), (2.8), (2.9), (2.18), (2.19), and returning to the original variables, we obtain from (2.17) that

$$\begin{aligned}u \geqslant\ &\varepsilon_1^{2p+1}/\varkappa^{2p} - 16N_1\sigma_3 \\ &- N_1\Big\{\rho^{n/(n+1)}\big\|(Lu)_-; Q\big\|_{n+1} + \rho^{(n-1)/n}\big\|(Bu)_-; \Gamma(Q)\big\|_n\Big\} \quad \text{in } Q'.\end{aligned} \tag{2.20}$$

By taking $\sigma_3 = \varepsilon_1^{2p+1}/(32N_1\varkappa^{2p})$, $\delta_3 = 1/(2\varkappa^{2p})$ in (2.20) we get (2.16). □

LEMMA 2.4. *Let $\rho \leqslant 1$, and let*

$$u \in W_{n+1}^{2,1}(Q_\rho^+) \cap W_n^{2,1}(\Gamma_\rho) \cap C(\overline{Q}_\rho^+), \qquad u \geqslant 0 \quad \text{in } \overline{Q}_\rho^+.$$

Then there exist constants ξ, δ_4, σ_4, depending only on n and ν, such that the inequalities

$$\big\|b_*; Q_\rho^+\big\|_{n+2} \leqslant \sigma_4, \quad \big\|\beta_+; \Gamma_\rho\big\|_{n+1} \leqslant \sigma_4, \quad \big|A_k^u \cap \Gamma_{\rho/2}(0; -3\rho^2/4)\big| \geqslant \xi\big|\Gamma_{\rho/2}\big|,$$

imply

$$\begin{aligned}u \geqslant \delta_4 k - N_3(n,\nu)\Big\{&\rho^{n/(n+1)}\big\|(Lu)_-; Q_\rho^+\big\|_{n+1} \\ &+ \rho^{(n-1)/n}\big\|(Bu)_-; \Gamma_\rho\big\|_n\Big\} \quad \text{in } Q_{\rho/2}^+.\end{aligned} \tag{2.21}$$

PROOF. Suppose that $\sigma_4 \leqslant \min\{\sigma_1; \sigma_2\}$, where $\sigma_1 = \sigma_1(n, \nu, 1/2)$ and $\sigma_2 = \sigma_2(n, \nu, 1/2)$ are the constants from Lemmas 2.1 and 2.2, respectively.

Applying Lemma 2.1 with $\delta_1 = 1/2$ in the cylinder $Q^+_{\rho/2}(0; -3\rho^2/4)$ to u and then using Lemma 2.2 with $\delta_2 = 1/2$ in the cylinder $Q^+_{\rho/4}(0; -3\rho^2/4)$, we obtain

$$u \geqslant k_1 = \Big[k/4 - C_5(n,\nu)\{\rho^{n/(n+1)}\|(Lu)_-; Q^+_\rho\|_{n+1} + \rho^{(n-1)/n}\|(Bu)_-; \Gamma_\rho\|_n\}\Big]_+ \quad \text{in } Q^+_{\varepsilon\rho/4}(0; -3\rho^2/4). \tag{2.22}$$

Now, by Lemma 2.3 (with $\varkappa = 1/2$, $\varepsilon_1 = \varepsilon/4$, $\tau = 3\rho^2/4$, $t = -3\rho^2/4$, $x^{(0)} = 0$), assuming that $\sigma_4 \leqslant \sigma_3 = \sigma_3(n, \nu, 1/2, \varepsilon/4)$, we have the estimate

$$u \geqslant \delta_3 \varepsilon_1^{2p+1} k_1 - N_1\Big\{\rho^{n/(n+1)}\|(Lu)_-; Q^+_\rho\|_{n+1} + \rho^{(n-1)/n}\|(Bu)_-; \Gamma_\rho\|_n\Big\} \quad \text{in } Q^+_{\rho/2}.$$

Taking $\delta_4 = \delta_3\varepsilon_1^{2p+1}/4$, $N_3 = 4\delta_4 C_5 + N_1$ and using (2.22), we get (2.21). □

Lemma 2.5. *Let $\rho \leqslant 1$ and*

$$u \in W^{2,1}_{n+1}(Q^+_\rho) \cap W^{2,1}_n(\Gamma_\rho) \cap C(\overline{Q}^+_\rho), \qquad u \geqslant 0 \quad in\ \overline{Q}^+_\rho.$$

Then for any $\zeta \in (0;1)$ there exist constants δ_5, σ_5, $\eta > 0$, depending only on n, ν, ξ, such that if

$$\|b_*; Q^+_\rho\|_{n+2} \leqslant \sigma_5, \qquad \|\beta_*; \Gamma_\rho\|_{n+1} \leqslant \sigma_5,$$
$$\zeta|\Gamma_{\rho/2}| \leqslant |A^u_k \cap \Gamma_{\rho/2}(0; -3\rho^2/4)| \leqslant \xi|\Gamma_{\rho/2}|$$

(here $\xi = \xi(n,\nu)$ is the same as in Lemma 2.4), then at least one of the two following inequalities holds:

$$u \geqslant \delta_5 k - N_4(n,\nu,\zeta)\Big\{\rho^{n/(n+1)}\|(Lu)_-; Q^+_\rho\|_{n+1} + \rho^{(n-1)/n}\|(Bu)_-; \Gamma_\rho\|_n\Big\} \quad in\ Q^+_{\rho/2}, \tag{2.23}$$

$$|A^u_{k_1} \cap \Gamma_{\rho/2}(0; -3\rho^2/4)| > (1+\eta)|A^u_k \cap \Gamma_{\rho/2}(0; -3\rho^2/4)|. \tag{2.24}$$

Here k_1 is some constant, completely determined by k, n, ν, ζ.

Proof. Applying the same arguments as in the proofs of Lemmas 3.4–3.6 from [**LU**] to the set $A^u_k \cap \Gamma_{\rho/2}(0; -3\rho^2/4)$, which lies in Γ_ρ, and using Lemmas 2.1–2.3 instead of Corollary 3.1 from [**LU**], we get the desired estimates. □

Corollary 2.6. *Let $\rho \leqslant 1$,*

$$u \in W^{2,1}_{n+1}(Q^+_\rho) \cap W^{2,1}_n(\Gamma_\rho) \cap C(\overline{Q}^+_\rho), \qquad u \geqslant 0 \quad in\ \overline{Q}^+_\rho.$$

Then for any $\zeta_1 \in (0;1)$ there exist constants δ, σ, depending only on n, ν, ζ_1, such that the inequalities

$$\|b_*; Q^+_\rho\|_{n+2} \leqslant \sigma, \qquad \|\beta_*; \Gamma_\rho\|_{n+1} \leqslant \sigma, \qquad |A^u_k \cap \Gamma_{\rho/2}(0; -3\rho^2/4)| \geqslant \zeta_1|\Gamma_{\rho/2}|$$

imply

$$u \geqslant \delta k - N(n,\nu,\zeta_1)\Big\{\rho^{n/(n+1)}\|(Lu)_-; Q^+_\rho\|_{n+1} + \rho^{(n-1)/n}\|(Bu)_-; \Gamma_\rho\|_n\Big\}.$$

PROOF. The proof of this corollary is similar to that of Lemma 2.5 from [LU]. The only difference is that we must use Lemmas 2.4 and 2.5 instead of Lemmas 2.3 and 2.4 from [LU]. □

§3. An estimate of the Hölder constant for u

THEOREM 2. *Suppose $\rho_0 \leqslant 1$, and a function u is given so that*

$$u \in W^{2,1}_{n+1}(Q^+_{\rho_0}) \cap W^{2,1}_n(\Gamma_{\rho_0}) \cap C(\overline{Q}^+_{\rho_0}),$$

and u satisfies the inequalities

$$\begin{aligned} &\left|u_t - a^{ij}(x,t,u,Du)D_iD_ju\right| \\ &\qquad \leqslant \mu\|Du\|^2 + b(x,t)\|Du\| + \Phi(x,t) \quad \text{in } Q^+_{\rho_0}, \end{aligned} \tag{3.1}$$

$$\begin{aligned} &\left|u_t - \alpha^{lm}(x,t,u,Du)D_lD_mu + \beta^n(x,t)D_nu\right| \\ &\qquad \leqslant \mu\|D'u\|^2 + \beta(x,t)\|D'u\| + \Theta(x,t) \quad \text{on } \Gamma_{\rho_0} \end{aligned} \tag{3.2}$$

while the functions $a^{ij}(x,t,z,p)$ for $(x,t) \in \overline{Q}^+_{\rho_0}$, $z \in \mathbb{R}^1$, $p \in \mathbb{R}^n$ satisfy inequality (1.4), $\alpha = (\alpha^{lm})$ is a symmetric matrix,

$$\nu\|\eta\|^2 \leqslant \alpha^{lm}\xi_l\xi_m \leqslant \nu^{-1}\|\eta\|^2 \quad \forall \eta \in \mathbb{R}^{n-1},$$

$$\beta^n, \beta \in L_{n+1}(\Gamma_{\rho_0}), \quad \beta^n \leqslant 0, \qquad \Theta \in L_n(\Gamma_{\rho_0}).$$

Assume that $|u| \leqslant M_0$ in $Q^+_{\rho_0}$, and, in addition, suppose that

$$\left\|b_*; Q^+_{\rho_0}\right\|_{n+2} \leqslant \sigma, \qquad \left\|\beta_*; \Gamma_{\rho_0}\right\|_{n+1} \leqslant \sigma,$$

where $\sigma = \sigma(n, \nu, 1/2)$ is the constant from Corollary 2.6.

Then for any $\rho \leqslant \rho_0$ the estimate

$$\operatorname*{osc}_{Q^+_\rho} u \leqslant C\rho^{\gamma_1}\rho_0^{-\gamma_1} \tag{3.3}$$

is valid with some $\gamma_1 = \gamma_1(n,\nu) \in (0;1)$ and $C > 0$ depending only on n, ν, μ, M_0, $\|\Phi; Q^+_{\rho_0}\|_{n+1}$, $\|\Theta; \Gamma_{\rho_0}\|_n$.

PROOF. We introduce the notation $u_1 = \inf_{Q^+_\rho} u$, $u_2 = \sup_{Q^+_\rho} u$, $\omega = u_2 - u_1$, and assume that $\omega > 0$.

First let us consider the case $\mu > 0$. It is easy to see that the functions

$$v^{(h)}(x,t) = 1 - \exp\{-(\lambda/\omega)|u(x,t) - u_h|\}, \qquad h = 1,2,$$

for $\lambda \geqslant \omega\mu\nu^{-1}$ satisfy the inequalities

$$L_hv^{(h)} = v^{(h)}_t - a^{ij}D_iD_jv^{(h)} + b^i_{(h)}D_iv^{(h)} \geqslant -(\lambda/\omega)\Phi \quad \text{in } Q^+_\rho, \tag{3.4}$$

$$B_hv^{(h)} = v^{(h)}_t - \alpha^{lm}D_lD_mv^{(h)} + \beta^i_{(h)}D_iv^{(h)} \geqslant -(\lambda/\omega)\Theta \quad \text{on } \Gamma_\rho. \tag{3.5}$$

Here $b^i_{(h)} = b(x,t)D_iv^{(h)}(x,t)/\|Dv^{(h)}(x,t)\|$ (at points where $\|Dv^{(h)}\| = 0$ we set $b^i_{(h)} = b(x,t)/n^{1/2}$, so that $\|b_{*(h)}\| = b$ everywhere). The coefficients $\beta^l_{(h)}$ are defined in a similar way as $b^l_{(h)}$, $\beta^n_{(h)} = \beta^n$.

If $u(x,t) \geqslant [u_1 + u_2]/2$, then $v^{(1)}(x,t) \geqslant k = 1 - \exp\{-\lambda/2\}$, otherwise $v^{(2)}(x,t) \geqslant k$. Therefore, at least one of these two functions satisfies the assumptions of Corollary 2.6 (with $\zeta_1 = 1/2$). For this function we have

$$v^{(h)} \geqslant \delta k - \widetilde{C}\omega^{-1}\rho^{(n-1)/n} \quad \text{in } Q^+_{\rho/2} \tag{3.6}$$

where $\widetilde{C} = \lambda N\{\|\Phi; Q^+_{\rho_0}\|_{n+1} + \|\Theta; \Gamma_{\rho_0}\|_n\}$.

If $\widetilde{C}\omega^{-1}\rho^{(n-1)/n} \geqslant \delta k/2$, then

$$\operatorname*{osc}_{Q^+_{\rho/2}} u \leqslant \omega \leqslant \widetilde{C}_1\rho^{(n-1)/n},$$

where $\widetilde{C}_1 = 2\widetilde{C}/(\delta k)$. In the converse case, (3.6) implies $v^{(h)} \geqslant \delta k/2$ in $Q^+_{\rho/2}$, and hence

$$\operatorname*{osc}_{Q^+_{\rho/2}} u \leqslant (1-\varkappa)\omega, \qquad \omega = -\lambda^{-1}\ln(1-\delta k/2) > 0.$$

It is well known that the estimate (3.3) with

$$\gamma_1 = \min\{-\log_2(1-\varkappa); (n-1)/n\} > 0,$$
$$C = 2^{\gamma_1}\max\{\widetilde{C}_1\rho_0^{(n-1)/n}; \operatorname*{osc}_{Q^+_{\rho_0}} u\}$$

follows from this alternative.

Further, by the standard method (see **[LU]**), we can free the constants from their dependence on λ (and hence from their dependence on γ_1), μ, and M. This completes the proof of Theorem 2 in the case $\mu > 0$.

When $\mu = 0$, we replace the functions $v^{(h)}$ by $\widetilde{v}^{(h)}(x,t) = \omega^{-1}|u(x,t) - u_h|$, $h = 1, 2$, respectively. After this we apply the previous arguments to the functions $\widetilde{v}^{(h)}$. □

Proof of Theorem 1. Since the function $\alpha(x,t,z,p)$ for $(x,t) \in \partial''Q$, $z \in \mathbb{R}^1$, $p \in \mathbb{R}^n$ is differentiable with respect to the variables p_i, we can rewrite (1.3) as

$$\begin{aligned} &u_t - \alpha^{ij}(x,t,u,Du)\delta_i\delta_j u \\ &\quad = \alpha(x,t,u,\delta u) + \int_0^1 \alpha_{p_i}(x,t,u,\delta u + \varepsilon\mathbf{n}\partial u/\partial n)\mathbf{n}_i\, d\varepsilon\partial u/\partial n. \end{aligned} \tag{3.7}$$

(Here $\partial u/\partial n$ denotes the normal derivative of u.)

Combining (3.7) with (1.8), we obtain the inequality

$$\begin{aligned} &\left|u_t - \alpha^{ij}(x,t,u,Du)\delta_i\delta_j u - \beta^n(x,t)\partial u/\partial n\right| \\ &\qquad \leqslant \mu\|\delta u\|^2 + \beta(x,t)\|\delta u\| + \Theta(x,t). \end{aligned} \tag{3.8}$$

Here β^n denotes the integral on the right-hand side of (3.7). Condition (1.7) gives us $\beta^n \leqslant 0$ and $|\beta^n(x,t)| \leqslant \beta(x,t)$.

Now we fix an arbitrary point $(x^0, t^0) \in \partial''Q$. By means of a coordinate transformation of the independent variables, we rectify $\partial\Omega$ near a point x^0 and map (x^0, t^0) into the origin. Under this transformation a part of $\partial\Omega$ is mapped into a part of the plane $x_n = 0$ and the vector $\mathbf{n}(x^0)$ is mapped to the unit vector opposite to the x_n axis.

By virtue of (1.1), (1.4)–(1.6), (3.8), and the condition $\partial\Omega \in W^2_{n+2}$, the inequalities (3.1), (3.2) with $\widetilde{\nu}$, $\widetilde{\mu}$, $\widetilde{b}(x,t)$, $\widetilde{\beta}(x,t)$ replacing ν, μ, $b(x,t)$, $\beta(x,t)$ respectively are fulfilled in the new coordinate system. We note that $\widetilde{\nu}$, $\widetilde{\mu}$, $\widetilde{b}(x,t)$, $\widetilde{\beta}(x,t)$ are completely determined by the constants without tildes and by $\partial\Omega$.

In the new coordinate system let us consider the cylinder $Q^+_{\rho_0}$. There exists a constant $\rho_0 \leqslant 1$ depending only on $\partial\Omega$ and the moduli of continuity of $b(x,t)$ in $L_{n+2}(Q)$ and of $\beta(x,t)$ in $L_{n+1}(\partial'' Q)$ such that the inequalities

$$\|\widetilde{b}_*; Q^+_{\rho_0}\|_{n+2} \leqslant \sigma, \qquad \|\widetilde{\beta}_*; \Gamma_{\rho_0}\|_{n+1} \leqslant \sigma$$

hold independently of the particular point (x^0, t^0).

Hence, Theorem 2 is applicable to u in $Q^+_{\rho_0}$ and guarantees the estimate (3.3). Taking into account the condition $u\big|_{t=0} \in C^{\gamma}(\Omega)$, we remark that for interior cylinders and for cylinders adjoining the lower base of Q, the same estimate with some γ_2 instead of γ_1 is established in **[LU]**. Thus the estimate of $M_{\widetilde{\gamma}}$ with $\widetilde{\gamma} = \frac{1}{2}\min\{\gamma_1; \gamma_2\}$ follows. (This fact is also proved in **[LU]**.) □

§4. Appendix. The elliptic case

In this section we formulate the elliptic analogs of Theorem 1 and of Theorem 1 from **[A]**.

THEOREM 1′. *Let $\partial\Omega \in W^2_n$, suppose the function u satisfies*

$$u \in W^2_n(\Omega) \cap W^2_{n-1}(\partial\Omega) \cap C(\overline{\Omega})$$

and

$$-a^{ij}(x,u,Du)D_iD_ju = a(x,u,Du) \quad \text{in } \Omega, \tag{4.1}$$

$$-\alpha^{ij}(x,u,Du)\delta_i\delta_ju = \alpha(x,u,Du) \quad \text{on } \partial\Omega \tag{4.2}$$

while the coefficients in (4.1), (4.2) *satisfy the ellipticity condition, i.e.,*

- $a = (a^{ij})$ *is a symmetric matrix and the inequalities*

$$\nu\|\xi\|^2 \leqslant a^{ij}(x,z,p)\xi_i\xi_j \leqslant \nu^{-1}\|\xi\|^2$$

hold for any $x \in \Omega$, $z \in \mathbb{R}^1$, $\xi \in \mathbb{R}^n$, $p \in \mathbb{R}^n$;

- $\alpha = (\alpha^{ij})$ *is a symmetric matrix and the inequalities*

$$\nu_1\|\xi\|^2 \leqslant \alpha^{ij}(x,z,p)\xi_i\xi_j \leqslant \nu_1^{-1}\|\xi\|^2$$

hold for any $x \in \partial\Omega$, $z \in \mathbb{R}^1$, $p \in \mathbb{R}^n$ and for $\xi \in \mathbb{R}^n$ with $\xi \perp \mathbf{n}(x)$, with some positive constants ν and ν_1.

Let

$$|a(x,u,p)| \leqslant \mu\|p\|^2 + b(x)\|p\| + \Phi(x)$$

with some positive constant μ and with $b, \Phi \in L_n(\Omega)$. Assume also that the function $\alpha(x,z,p)$ is differentiable with respect to the variables p_i and for $x \in \partial\Omega$, $z \in \mathbb{R}^1$ the following inequalities are true:

$$0 \leqslant -\alpha_{p_i}(x,z,p)\mathbf{n}_i(x) \leqslant \beta(x) \qquad \forall p \in \mathbb{R}^n,$$

$$|\alpha(x,z,p)| \leqslant \mu_1\|p\|^2 + \beta(x)\|p\| + \Theta(x) \qquad \forall p \in \mathbb{R}^n,\ p \perp \mathbf{n}(x)$$

with some positive constant μ_1 and with $\beta, \Theta \in L_{n-1}(\partial\Omega)$.

Then there exists a $\gamma > 0$, *depending only on* n, ν, ν_1, *and* $\partial\Omega$, *such that*

$$[u]_{\gamma,\Omega} \leqslant M_\gamma,$$

where M_γ *depends only on the same arguments as* γ *and also on* μ, μ_1, $M_0 = \sup_\Omega |u|$, $\|\Phi;\Omega\|_n$, $\|\Theta;\partial\Omega\|_{n-1}$ *and the moduli of continuity of* $b(x)$ *in* $L_n(\Omega)$ *and of* $\beta(x)$ *in* $L_{n-1}(\partial\Omega)$.

THEOREM 3 (An Aleksandrov type maximum estimate for the solutions of the Wentzel boundary value problem). *Let* $u \in W^2_n(\Omega^+) \cap W^2_{n-1}(\Gamma(\Omega^+)) \cap C(\overline{\Omega}^+)$,

$$-a^{ij}(x)D_iD_ju + b^i(x)D_iu + c(x)u = f(x) \quad \text{in } \Omega^+, \tag{4.3}$$

$$-\alpha^{lm}(x)D_lD_mu + \beta^i(x)D_iu + \gamma(x)u = \varphi(x) \quad \text{on } \Gamma(\Omega^+). \tag{4.4}$$

Assume that the coefficients in (4.3), (4.4) *satisfy the following conditions*:

$$a^{ij} = a^{ji}, \qquad a^{ij}\xi_i\xi_j \geqslant 0 \quad \text{for } \xi \in \mathbb{R}^n,$$
$$c \geqslant 0, \qquad \operatorname{tr}(a) > 0 \quad \text{a.e. in } \Omega^+,$$
$$h = \|b_*\|\big(\det(a)\big)^{-1/n} \in L_n(\Omega^+), \qquad \alpha^{lm} = \alpha^{ml},$$
$$\alpha^{lm}\xi_l\xi_m \geqslant 0 \quad \text{for } \xi \in \mathbb{R}^{n-1}, \qquad \gamma \geqslant 0,$$
$$\beta^n \leqslant 0, \qquad \operatorname{tr}(\alpha) > 0 \quad \text{a.e. on } \Gamma(\Omega^+),$$
$$\eta = \|\beta_*\|\big(\det(\alpha)\big)^{-1/(n-1)} \in L_{n-1}(\Gamma(\Omega^+)).$$

If, in addition, $u \leqslant 0$ *on* $\partial\Omega \setminus \Gamma(\Omega^+)$, *then*

$$u \leqslant C\Big(n, \|h;\Omega^+\|_n, \|\eta;\Gamma(\Omega^+)\|_{n-1}\Big) \cdot \operatorname{diam}(\Omega^+)$$
$$\cdot \Big\{ \|f_+/\Delta_1;\Omega^+\|_n + \|\varphi_+/\Delta_2;\Gamma(\Omega^+)\|_{n-1} \Big\}$$

where $\Delta_1 = \big(\det(a)\big)^{1/n}$, $\Delta_2 = \big(\det(\alpha)\big)^{1/(n-1)}$.

REMARK 1. The statement of Theorem 3 remains true if all the sets are replaced by their intersection with A^u_0.

REMARK 2. The Wentzel boundary value problem for an elliptic equation was considered in **[LT1, 2]** and **[L1, 2]**. However, we note that the definition of boundary operator in **[LT1]** is incorrect because the authors use δ_{ij} instead of $\delta_i\delta_j$, which is possible only for a flat boundary of the domain.

Theorem 3 was proved in **[L1]** only for uniformly elliptic operators; it was assumed also that $\beta' = 0$.

The Hölder estimate for u was obtained in **[L1]** and **[L2]** only for bounded b, β, Φ, and Θ.

References

[A] D. E. Apushkinskaya, *The maximum estimation of the solutions of parabolic equations with Wentzel boundary condition*, Vestnik Leningrad Univ. **1991**, no. 8 (Ser. Mat. Mekh. Astr. vyp. 2), 3–12; English transl. in Vestnik Leningrad Univ. Math. **24** (1991).

[KS] N. V. Krylov and M. V. Safonov, *A certain property of solutions of parabolic equations with measurable coefficients*, Izv. Akad. Nauk SSSR Ser. Mat. **44** (1980), no. 1, 161–175; English transl. in Math. USSR-Izv. **16** (1981).

[LU] O. A. Ladyzhenskaya and N. N. Ural′tseva, *Estimates of the Hölder constant for functions satisfying a uniformly elliptic or uniformly parabolic quasilinear inequality with unbounded coefficients*, Zap. Nauchn. Sem. Leningrad. Otdel. Mat. Inst. Steklov. (LOMI) **147** (1985), 72–94; English transl. in J. Soviet Math. **37** (1987), no. 1.

[L1] You-Song Luo, *An Aleksandrov-Bakel′man type maximum principle and applications*, J. Differential Equations **101** (1993), 213–231.

[L2] ———, *On the quasilinear elliptic Venttsel′ boundary value problem*, Nonlinear Anal. **16** (1991), 761–769.

[LT1] You-Song Luo and Neil S. Trudinger, *Linear second order elliptic equations with Venttsel′ boundary conditions*, Proc. Roy. Soc. Edinburgh Sect. A **118** (1991), 193–207.

[LT2] ———, *Quasilinear second order elliptic equations with Venttsel′ boundary conditions*, Potential Anal. **3** (1994), 219–243.

[NU] A. I. Nazarov and N. N. Ural′tseva, *Convex-monotone hulls and an estimate of the maximum of the solution of a parabolic equation*, Zap. Nauchn. Sem. Leningrad. Otdel. Mat. Inst. Steklov. (LOMI) **147** (1985), 95–109; English transl. in J. Soviet Math. **37** (1987), no. 1.

St. Petersburg Academy of Civil Engineering

Department of Mathematics and Mechanics, St. Petersburg State University

Translated by the authors

Amer. Math. Soc. Transl.
(2) Vol. **164**, 1995

Reverse Hölder Inequalities with Boundary Integrals and L_p-Estimates for Solutions of Nonlinear Elliptic and Parabolic Boundary-Value Problems

A. A. Arkhipova

Contents

INTRODUCTION
§1. Reverse Hölder inequalities and higher integrability theorems (the Gehring lemma and its modifications)
§2. L_p-estimates for solutions of elliptic boundary-value problems
§3. Boundary regularity for solutions of elliptic Neumann-type problems
§4. Parabolic reverse Hölder inequalities with additional anisotropic terms
§5. L_p-estimates for the gradients of solutions of parabolic boundary-value problems
§6. Proof of Theorem 2

Introduction

In 1973 F. W. Gehring **[Ge]** proved the higher integrability property of functions satisfying the so-called reverse Hölder inequalities (the Gehring lemma, see §1 below). Many authors since have used this statement and its modifications to study the regularity of solutions of nonlinear systems of partial differential equations. In particular, it was noticed that some power of the gradient of a solution in the sense of distributions of nonlinear elliptic or parabolic system satisfies the local reverse Hölder inequalities. This implies the higher integrability of the gradient inside the domain. In particular, it was proved that the first-order derivatives of the $W_2^1(\Omega)$-solution in the sense of distributions of quasilinear elliptic second-order systems belong to $L_{p,\mathrm{loc}}(\Omega)$ for some $p > 2$.

The study of the partial regularity of solutions of elliptic and parabolic systems was initiated in the papers **[Mor, G, GM]**, where the indirect method was used. The L_p-estimates of derivatives obtained with the help of a modification of the Gehring lemma **[GiMo1, St]** laid the foundation to the so-called "direct method" of proving partial regularity for solutions of elliptic and parabolic systems. This method made it possible to study wider classes of systems with strong nonlinearities. Details concerning the methods and the results may be found in the monograph by

---

1991 *Mathematics Subject Classification*. Primary 35J65, 35K60.

This work was partially supported by the Russian Fund for Fundamental Research, grant no. 93-011-1696.

0065-9290/95 \$1.00 + \$.25 per page

M. Giaquinta [**Gi3**] and the survey by O. John and J. Stara [**JS**]. The main results dealt with regularity, inside the domain and up to the boundary, of solutions of the Dirichlet problem. The regularity of solutions of the Neumann problem was not studied except by M. Giaquinta and G. Modica [**GiMo2**], where the case of smooth boundary data was considered. In this case the boundary condition can be easily reduced to the condition that the conormal derivative equals zero, which corresponds to the absence of boundary integrals in the integral relation which determines the solution in the sense of distributions.

In [**A1–A5**] the author studied the regularity of solutions of the Neumann problem with nonsmooth boundary data for nonlinear elliptic and parabolic systems. In this case the integral relation for the solution in the sense of distributions includes the boundary integrals. This circumstance causes analytical difficulties in the derivation of L_p-estimates for the gradients of the solutions. For this approach the crucial point was to derive a modification of the Gehring lemma. The higher integrability property of a function satisfying the reverse Hölder inequality with additional integrals over the lower-dimensional manifolds was proved in [**A1**] (see Theorem 2 below). The anisotropic version of the Gehring lemma for the parabolic metric was proved in [**A3**] (see Theorem 8 below). These results yield L_p-estimates for the gradients of solutions of the Neumann problem for a wide class of nonlinear elliptic and parabolic systems [**A1–A4**]. The estimates were used in [**A5**] and [**A6**] to prove the partial regularity of solutions of the Neumann problem for quasilinear elliptic and parabolic systems.

This article is a review of the author's work on to the problem of regularity of solutions of the Neumann problem for nonlinear elliptic and parabolic systems. Some related results of other authors concerning the reverse Hölder inequalities, L_p-estimates and partial regularity of solutions of quasilinear systems are also discussed.

We do not touch on the many papers devoted to regularity problems in the calculus of variations for nonsmooth functionals (see [**Gi1, Gi2**]) and the regularity of weak solutions of nonlinear elliptic and parabolic systems in Morrey and Campanato spaces (see [**C1–C7**]).

It is impossible to present all the proofs here, so we prove only the main theorem on the reverse inequalities with surface integrals (Theorem 2). The proof of the remaining statements can be found in the books and papers cited in the references.

We use the following standard notation.

- $x = (x_1, \dots, x_n)$.
- $B_\rho(x^0)$ is the ball in $\mathbb{R}^n$ centered at x^0 of radius ρ, and

$$B_\rho^+(x_0) = B_\rho(x_0) \cap \{ x_n > x_n^0 \}.$$

- Ω is a bounded domain in $\mathbb{R}^n$, $n \geqslant 2$, with boundary $\partial\Omega$.
- $\mathbf{n}(x)$ is the unit outward normal vector at a point $x \in \partial\Omega$.

The domain Ω is said to be *strongly Lipschitz* if for any point $x^0 \in \partial\Omega$ the local coordinates $(\widetilde{y}, y_n)$, $\widetilde{y} = (y_1, \dots, y_{n-1})$, may be chosen so that for the cylinder

$$C_{R,L} = \{ y \in \mathbb{R}^n : |\widetilde{y}| < R,\ |y_n| < 2LR \}$$

the surface $\partial\Omega \cap \overline{C}_{R,L}$ is given by $y_n = w(\widetilde{y})$, w being a Lipschitz function in $\{ |\widetilde{y}| \leqslant R \}$ with Lipschitz constant not exceeding L and

$$\overline{\Omega} \cap \overline{C}_{R,L} = \{ y \in \mathbb{R}^n : |\widetilde{y}| \leqslant R,\ w(\widetilde{y}) \leqslant y_n \leqslant 2LR \}.$$

The constants L and R are fixed for the whole domain Ω.

For $u\colon \Omega \to \mathbb{R}^N$, $u = (u^1, \dots, u^N)$, we denote

$$|u|^2 = \sum_{k=1}^{N} (u^k)^2, \quad u^k_{x_\alpha} = \frac{\partial u^k}{\partial x_\alpha}, \quad u_x = \{u^k_{x_\alpha}\}^{k\leqslant N}_{\alpha\leqslant n}, \quad |u_x|^2 = \sum_{k\leqslant N}\sum_{\alpha\leqslant n}(u^k_{x_\alpha})^2.$$

We denote the Lebesgue measure of $A \subset \mathbb{R}^n$ by $|A| = \operatorname{meas}_n A$, and we define

$$\fint_A g\,dx = \frac{1}{|A|}\int_A g\,dx, \qquad u_{R,x^0} = \fint_{B_R(x^0)} u\,dx, \qquad \omega_n = |B_1(0)|.$$

The function $d(x, S)$ is the distance from x to S.

For function spaces we follow the notation of the monograph [**LSU**].

§1. Reverse Hölder inequalities and higher integrability theorems (the Gerhing lemma and its modifications)

Let Q, Q_0 be cubes in $\mathbb{R}^n$ with sides parallel to the coordinate axes, and let $Q/2$ be a cube with the same center as Q and with half the side of Q.

The simplest reverse Hölder inequalities for the function $g \in L_q(Q_0)$, $g \geqslant 0$, $q > 1$ are

(1)
$$\fint_Q g^q\,dx \leqslant B\left(\fint_Q g\,dx\right)^q, \qquad B > 1,\ \forall Q \subset Q_0.$$

As we pointed out in the Introduction, the first result on higher integrability of functions g satisfying (1) was established in [**Ge**].

LEMMA (F. W. Gehring [**Ge**]). *Let g be a nonnegative function on Q_0 satisfying the inequalities* (1) *with some constants $q > 1$, $B > 1$ for every $Q \subset Q_0$. Then there exists a constant $p_0 = p_0(n, q, B) > q$ such that $g \in L_p(Q_0)$ for every $p \in [q, p_0)$ and*

(2)
$$\fint_{Q_0} g^p\,dx \leqslant \frac{p_0 - q}{p_0 - p}\left(\fint_{Q_0} g^q\,dx\right)^{p/q}$$

By means of this fundamental result, quasiconformal mappings $f : \Omega \to \mathbb{R}^n$, $f \in W^1_{n,\mathrm{loc}}(\Omega)$, were proved to be from $W^1_{p,\mathrm{loc}}(\Omega)$, $p > n$. A similar fact for quasiregular mappings was obtained in [**Ma**] (see also [**ME**]). It is known that $p_0 \to +\infty$ when $B \searrow 1$ [**I, Sb**]. Unfortunately, the constant B and hence the difference $B - 1$ and p_0 cannot be evaluated in applications.

The concept of maximal function (cf. [**S**]) is essential for the proof of this lemma and its modifications. For $h \in L_{1,\mathrm{loc}}(\mathbb{R}^n)$, $h \geqslant 0$, it is defined by

$$M(h)(x) = \sup_R \fint_{Q_R(x)} h(y)\,dy.$$

The cubes $Q_R(x)$ with center x and side $2R$ may be replaced by the balls $B_R(x)$. If

$h \in L_1(\Omega)$, then h is assumed to be zero in $\mathbb{R}^n \setminus \Omega$. Note one property of maximal functions: if $h \in L_m(\mathbb{R}^n)$, $m \in (1,\infty]$, then $M(h) \in L_m(\mathbb{R}^n)$ and

$$\int_{\mathbb{R}^n} [M(h)]^m \, dx \leqslant C(n,m) \int_{\mathbb{R}^n} h^m \, dx. \tag{3}$$

The Gehring lemma may be formulated in terms of maximal functions. Then the inequalities (1) must be replaced by

$$M(g^q)(x) \leqslant B_1[M(g)(x)]^q \tag{1°}$$

for almost all $x \in Q_0$.

The reverse Hölder inequalities for the gradients of weak solutions of elliptic systems also appear to be valid, but with different integration domains. These inequalities are weaker than (1). They are known as the *reverse Hölder inequalities* with different supports or *weak reverse Hölder inequalities* (W.R.I.):

$$⨍_{Q/2} g^q \, dx \leqslant B\left(⨍_Q g \, dx\right)^q, \qquad \forall Q \subset Q_0, \ B > 1. \tag{4}$$

The first higher integrability theorems based on W.R.I. are due to M. Giaquinta and G. Modica [**GiMo1**], and to E. Stredulinsky [**St**] (see also B. Bojarski and T. Iwaniec [**BI**]). The following theorem is presented as stated in the monograph by M. Giaquinta [**Gi3**].

THEOREM 1. *Suppose that* $g \in L_q(Q_0)$, $F \in L_{q+\delta}(Q_0)$, $g, F \geqslant 0$, $q > 1$, $\delta > 0$ *and*

$$⨍_{Q_R(x)} g^q \, dx \leqslant B\left(⨍_{Q_{2R}(x)} g \, dx\right)^q + ⨍_{Q_{2R}(x)} F^q \, dx + \theta ⨍_{Q_{2R}(x)} g^q \, dx \tag{5}$$

for almost every $x \in Q_0$, $R < \frac{1}{2}\min\{\, d(x,\partial Q_0),\ R_0 \,\}$, *where* $R_0 > 0$, $B > 1$, $\theta \in [0,1)$.

Then $g \in L_{p,\mathrm{loc}}(Q_0)$ *and*

$$\left(⨍_{Q_R} g^p \, dx\right)^{1/p} \leqslant C\left\{\left(⨍_{Q_{2R}} g^q \, dx\right)^{1/q} + \left(⨍_{Q_{2R}} F^p \, dx\right)^{1/p}\right\} \tag{6}$$

for any $Q_{2R} \subset Q_0$, $2R < R_0$, $p \in [q, p_0)$, *where* $C > 0$ *and* $p_0 > q$ *depend only on* B, θ, q, n.

Additional terms on the right-hand side of inequalities (5) are common in applications. Note that instead of Q_{2R} in (5), (6) we can write Q_{aR}, $a > 1$, and replace the integration domain Q_{aR} by B_{aR}.

There are various generalizations of this theorem. So, if we replace W.R.I. (5) with $F = \mathrm{const}$, $\theta = 0$ by the weak reverse Jensen inequalities

$$\begin{gathered} A^{-1}\left(⨍_{Q_R} A(g) \, dx\right) \leqslant B ⨍_{Q_{2R}} g \, dx + T, \\ \forall Q_{2R} \subset Q_0, \ T, B = \mathrm{const}, \ B > 1, \end{gathered} \tag{7}$$

where A is a convex function $A\colon [0,\infty) \to [0,\infty)$ satisfying some additional conditions, then there exists $p > 1$ such that $A_p(g) \in L_{1,\mathrm{loc}}(Q_0)$, $A_p(t) = A^p(t)/t^{p-1}$. Moreover, the estimate

$$A_p^{-1}\left(⨍_{Q_R} A_p(g) \, dx\right) \leqslant b_1 + b_2\left(⨍_{Q_{2R}} g \, dx\right), \qquad \forall Q_{2R} \subset Q_0,$$

holds with constants b_1, $b_2 > 0$ determined by B, T, n and the properties of the function $A(t)$. This result was proved by N. Fusco and C. Sbordone [**SbF**]. We do not formulate the conditions on $A(t)$ precisely, but only note that $A(t) = t^q$, $q > 1$, is permitted. Inequalities (7) arise, for example, in variational problems for functionals $F[u] = \int_\Omega A(|Du|)\,dx$ with convex $A(t)$.

Replacement of the Lebesgue measure by μ-measure, $d\mu = w(x)\,dx$, $w \in L_1(Q_0)$, $w \geqslant 0$, gives another modification of the theorem, which can be proved if the weight function w satisfies the following condition (B. Muckenhoupt [**Mu**]):

$$(8) \qquad \fint_Q w\,dx \left(\fint_Q w^{-\frac{1}{q-1}}\,dx \right)^{q-1} \leqslant A, \qquad \forall Q \subset Q_0,\ A > 1.$$

In particular, condition (8) is sharp for the validity of estimate (3) with $d\mu = w\,dx$ instead of dx.

There are higher integrability theorems coming from the weighted W.R.I.:

$$\fint_{Q/2} g^q\,d\mu \leqslant B \left(\fint_Q g\,dv \right)^q, \qquad \forall Q \subset Q_0,\ B > 1.$$

where μ and v are different measures.

Here is one result from the paper by A. Canale ([**C, Theorem 3**]) with $d\mu = w(x)\,dx$, $dv = dx$ (more references on weighted W.R.I. can be found in [**C**]).

THEOREM 1°. *Suppose* $g \in L_q(Q_0; w\,dx)$, $q > 1$, $g \geqslant 0$, *the weight* w *satisfies condition* (8) *and*

$$\fint_{Q/2} g^q w\,dx \leqslant B \left(\fint_Q g\,dx \right)^q, \qquad \forall Q \subset Q_0,\ B > 1.$$

Then there exists a constant $p > q$ *such that* $g \in L_{p,\mathrm{loc}}(Q_0; w\,dx)$ *and*

$$\fint_{Q/2} g^p w\,dx \leqslant C \left(\fint_Q g\,dx \right)^p, \qquad \forall Q \subset Q_0,$$

where

$$\fint_Q hw\,dx = \frac{1}{\mu(Q)} \int_Q h\,d\mu.$$

The weighted W.R.I. appear in the study of regularity for degenerate variational problems [**Mo**].

It will be shown in §2 that the W.R.I. for gradients of weak solutions of the Neumann problem contain additional boundary integrals. That is why the higher integrability result based on the W.R.I. with additional integrals over lower-dimensional manifolds is neccessary. The result was obtained by the author in [**A1**].

Let us describe the class of admissible manifolds Γ. Suppose $\Gamma \subset Q_0$, $\mathscr{H}_{\bar{k}}(\Gamma) < +\infty$ ($\mathscr{H}_{\bar{k}}$ is the $\bar{k}$-dimensional Hausdorff measure), the dimension $\bar{k} < n$ and $l\bar{k} \geqslant n$, where l is the parameter from (10).

We say that Γ satisfies the *(L, ρ)-condition* if

$$(9) \qquad \mathscr{H}_{\bar{k}}(\gamma_r(x)) \leqslant L r^{\bar{k}}, \qquad \forall x \in \Gamma,\ r \leqslant \rho,$$

where $\gamma_r(x) = \Gamma \cap B_r(x)$.

THEOREM 2. *Let g, F, Φ be nonnegative functions, $g \in L_q(Q)$, $F \in L_{1+\delta}(Q)$, $\Phi \in L_{1+\delta}(\Gamma)$, $q > 1$, $\delta > 0$, and let Γ satisfy the (L, ρ)-condition. Suppose that for almost every $x \in Q$*

$$
\begin{aligned}
\fint_{B_R(x)} g^q\,dx \leqslant B\Big(\fint_{B_{aR}(x)} g\,dx\Big)^q + \theta \fint_{B_{aR}(x)} g^q\,dx \\
+ R^{(m-1)n}\Big(\fint_{B_{aR}(x)} F\,dx\Big)^m + R^{-n}\mathscr{H}^l_{\bar{k}}(\gamma_{aR}(x))\Big(\fint_{\gamma_{aR}(x)} \Phi\,d\mathscr{H}_{\bar{k}}\Big)^l,
\end{aligned}
\tag{10}
$$

where $R < \frac{1}{a}\min\{\,d(x,\partial Q),\ R_0\,\}$, $B > 1$, $\theta \in [0,1)$, $R_0 > 0$, $m \geqslant 1$, $l \geqslant 1$, $a \geqslant 2$, $\bar{k}l \geqslant n$.

Then there exist constants $p_0 > q$ and $C_0 > 0$, depending only on θ, B, q, n, $\bar{k}$, m, l, a, L, ρ, such that $g \in L_p(Q')\ \forall \overline{Q}' \subset Q$, and

$$
\|g\|_{p,Q'} \leqslant C_0\Big\{\|g\|_{q,Q} + \|F\|^{\frac{m}{q}}_{1+\frac{m}{q}(p-q),Q} + \|\Phi\|^{\frac{l}{q}}_{1+\frac{l}{q}(p-q),\Gamma}\Big\}.
\tag{11}
$$

The constant C_0 depends also on $d(Q', \partial Q)$.

Note that the term in (10) containing Φ cannot be roughened by using inequality (9). The value of $\mathscr{H}_{\bar{k}}(\gamma_{aR}(x))$ depends on the distance $d(x,\Gamma)$ and tends to zero as $d(x,\Gamma)$ increases. As a rule in applications we derive W.R.I. in the following form:

$$
\begin{aligned}
\int_{B_R(x)} g^q\,dx \leqslant B_1 R^{n(1-q)}\Big(\int_{B_{aR}(x)} g\,dx\Big)^q + \theta \int_{B_{aR}(x)} g^q\,dx \\
+ \Big(\int_{B_{aR}(x)} F\,dx\Big)^m + \Big(\int_{\gamma_{aR}(x)} \Phi\,d\mathscr{H}_{\bar{k}}\Big)^l, \qquad \theta \ll 1.
\end{aligned}
\tag{10°}
$$

Note that for $\Phi \neq 0$ the inequalities (10) cannot be written in terms of the maximal function, and hence this notation cannot be used in the proof of Theorem 2 (see §6). This notation played an essential role in **[Ge, GiMo1, St, BI]**.

§2. L_p-estimates for solutions of elliptic boundary-value problems

Consider a weak solution $u \in W^1_2(\Omega)$ of the quasilinear system

$$
\begin{gathered}
-(A^{\alpha\beta}_{kl}(x,u)u^l_{x_\beta} + a^\alpha_k(x,u))_{x_\alpha} + b_k(x,u,u_x) = 0, \\
x \in \Omega \subset \mathbb{R}^n, \quad n \geqslant 2, \quad \alpha,\beta \leqslant n, \quad k,l \leqslant N,
\end{gathered}
\tag{12}
$$

with functions $A^{\alpha\beta}_{kl}(x,u)$ bounded in $\Omega \times \mathbb{R}^N$ and

$$
A^{\alpha\beta}_{kl}(x,u)\xi^k_\alpha \xi^l_\beta \geqslant \nu|\xi|^2 \qquad \forall \xi \in \mathbb{R}^{nN},\ \nu = \text{const} > 0,\ x \in \Omega,\ u \in \mathbb{R}^N.
\tag{13}
$$

Different conditions on the coefficients imply the so-called controllable or uncontrollable growth (for the definitions see **[Gi3]**, or below). The local higher integrability theorems for gradients of weak solutions of both classes of systems are described in **[Gi3,** Chapter 5, Theorems 2.2 and 2.3]. The gradients of solutions of system (12) are proved to belong to $L_{p,\text{loc}}$ with some $p > 2$. The L_p-estimates for the gradient are of interest in themselves and are essentially used in the proof of the partial regularity of the solutions. A well-known example due to E. Giusti and M. Miranda shows that the solution of the elliptic system

$$(A_{kl}^{\alpha\beta}(u)u_{x_\beta}^l)_{x_\alpha} = 0, \qquad x \in \Omega \subset \mathbb{R}^n,\ n \geqslant 3,\ N \geqslant 2, \tag{14}$$

with continuous functions $A_{kl}^{\alpha\beta}$ can have a singular set. An L_p-estimate for the gradient of the solution (12) was used to estimate the Hausdorff dimension of the singular set $\Sigma \subset \Omega$: $\dim_{\mathscr{H}} \Sigma \leqslant n - p$ for some $p > 2$ if $A_{kl}^{\alpha\beta}(x,u)$ are continuous in both arguments and the system belongs to one of the classes mentioned above.

Let us consider the regularity of the weak solutions of the boundary value problems for system (12). Let Ω be a strongly Lipschitz domain. For the solution u of the Dirichlet problem for the system (12), $u \restriction_{\partial\Omega} = \varphi$, $\varphi \in W_{2+\delta}^1(\Omega)$, $\delta > 0$, the inclusion $u_x \in L_p(\Omega)$ for some $p > 2$ is proved (for quasilinear systems both under controllable and uncontrollable growth conditions, see **[Gi3,** Chapter 5]). Consider the Neumann problem for (12):

$$(A_{kl}^{\alpha\beta}(x,u)u_{x_\beta}^l + a_k^\alpha(x,u))\cos(\mathbf{n}(x), x_\alpha) \restriction_{\partial\Omega} = \varphi_k(x,u), \qquad k \leqslant N. \tag{15}$$

If $\varphi_k \equiv 0$, then for both types of growth conditions the L_p-estimate for u_x in Ω can be obtained with the help of Theorem 1. In the case of nonsmooth functions $\varphi_k \neq 0$, the L_p-estimate was obtained by the author with the help of Theorem 2. (Clearly, the case of smooth φ_k can be easily reduced to the case $\varphi_k \equiv 0$.) Before going on to the gradient higher integrability theorem, let us consider a simple example of the proof of the L_p-estimate.

Take (12) with

$$A_{kl}^{\alpha\beta} = A_{kl}^{\alpha\beta}(x), \quad a_k^\alpha = 0, \quad b_k = a_0 u^k + f_k(x), \quad a_0 = \text{const} > 0,$$
$$f_k \in L_{\frac{2n}{n+2}+\delta}(\Omega), \quad \delta > 0,$$

and (15) with

$$\varphi_k \in L_{\frac{2(n-1)}{n}+\delta}(\partial\Omega), \qquad \delta > 0.$$

Assume u to be a solution of the linear Neumann problem (12), (15) from the Sobolev space $W_2^1(\Omega)$, and let $n > 2$.

Proposition. *If all the above conditions are satisfied, then $u_x \in L_p(\Omega)$ for some $p > 2$.*

Proof. Let $x^0 \in \overline{\Omega}$, $2R < \operatorname{diam}\Omega$, and let ζ be the standard cut-off function in $B_{2R}(x^0)$, $\zeta = 1$ in $B_R(x^0)$, $|\zeta_x| \leqslant 2/R$. The solution u belongs to $W_2^1(\Omega)$ and satisfies

$$\int_\Omega (A_{kl}^{\alpha\beta} u_{x_\beta}^l h_{x_\alpha}^k + a_0 u^k h^k + f_k h^k)\,dx = \int_{\partial\Omega} \varphi_k h^k\,d\Gamma$$

for all $h \in W_2^1(\Omega)$. Choose $h = (u - u_{2R})\zeta^2$, where $u_{2R} = u_{2R,x^0}$. Then the integral relation implies that

$$\begin{aligned}\int_{\Omega_R} (|u_x|^2 + |u|^2)\,dx \leqslant \frac{C}{R^2}\int_{\Omega_{2R}} |u - u_{2R}|^2\,dx + C\left(\int_{\Omega_{2R}} |f|^{\frac{2n}{n+2}}\,dx\right)^{\frac{n+2}{n}} \\ + C\left(\int_{\Gamma_{2R}} |\varphi|^{\frac{2(n-1)}{n}}\,d\Gamma\right)^{\frac{n}{n-1}} + C|u_{2R}|^2|\Omega_{2R}|,\end{aligned} \tag{16}$$

where $\Gamma_{2R} = \partial\Omega \cap B_{2R}(x^0)$, $\Omega_{2R} = \Omega \cap B_{2R}(x^0)$. Note that the integrals containing φ and f are estimated with the help of embedding theorems.

Moreover,

$$\int_{\Omega_{2R}} |u - u_{2R}|^2\, dx \leqslant C\Big(\int_{\Omega_{2R}} |u_x|^{\frac{2n}{n+2}}\, dx\Big)^{\frac{n+2}{n}}.$$

The Hölder inequality gives

$$|u_{2R}|^2|\Omega_{2R}| \leqslant \frac{C}{R^2}\Big(\int_{\Omega_{2R}} |u|^{\frac{2n}{n+2}}\, dx\Big)^{\frac{n+2}{n}}.$$

Hence from (16) we have

$$\int_{\Omega_R} (|u_x|^2 + |u|^2)\, dx \leqslant \frac{C}{R^2}\Big(\int_{\Omega_{2R}} (|u_x|^{\frac{2n}{n+2}} + |u|^{\frac{2n}{n+2}})\, dx\Big)^{\frac{n+2}{n}} + C\Big(\int_{\Omega_{2R}} |f|^{\frac{2n}{n+2}}\, dx\Big)^{\frac{n+2}{n}} + C\Big(\int_{\Gamma_{2R}} |\varphi|^{\frac{2(n-1)}{n}}\, d\Gamma\Big)^{\frac{n}{n-1}}.$$

Denote $g = |u_x|^{\frac{2n}{n+2}} + |u|^{\frac{2n}{n+2}}$, $q = \frac{n+2}{n}$. Then the previous inequality gives

$$\int_{\Omega_R} g^q\, dx \leqslant CR^{-2}\Big(\int_{\Omega_{2R}} g\, dx\Big)^q + \Big(\int_{\Omega_{2R}} F\, dx\Big)^q + \Big(\int_{\Gamma_{2R}} \Phi\, d\Gamma\Big)^{\frac{n}{n-1}}, \tag{17}$$

where $F = C^{\frac{n}{n+2}}|f|^{\frac{2n}{n+2}}$, $\Phi = C^{\frac{n-1}{n}}|\varphi|^{\frac{2(n-1)}{n}}$, $\Omega_R = \Omega_R(x^0)$, $\forall x^0 \in \overline{\Omega}$.

One can see that, as compared to (10°) for $\theta = 0$, $m = q$, $\bar{k} = n-1$, $l = n/(n-1)$, inequality (17) contains integrals over the domains Ω_R rather than over the balls B_R.

Fix $y^0 \in \partial\Omega$ and a cube Q with center y^0. Let g and F equal zero in $Q \setminus \Omega$. Then by (17) the prolonged functions $\widetilde{g}$ and $\widetilde{F}$ satisfy

$$\int_{B_R(x^0)} \widetilde{g}^q\, dx \leqslant BR^{-2}\Big(\int_{B_{5R}(x^0)} \widetilde{g}\, dx\Big)^q + C_0\Big(\int_{B_{5R}(x^0)} \widetilde{F}\, dx\Big)^q + C_0\Big(\int_{\Gamma_{5R}(x^0)} \Phi\, d\Gamma\Big)^{\frac{n}{n-1}}, \qquad \forall x^0 \in Q,\ R < \tfrac{1}{5}d(x^0, \partial Q),$$

where B and C_0 are positive constants (see **[A1]** or **[A5]**). By Theorem 2, these inequalities imply that $\widetilde{g} \in L_{\widehat{p}}(Q')$ for $Q' \Subset Q$ and some $\widehat{p} > q$. Therefore a constant $p > 2$ exists such that $u_x \in L_p(\Omega')$, $\Omega' = \Omega \cap Q'$. So the proposition is proved. □

COROLLARY. *The proposition remains true under the Legendre-Hadamard condition:*

$$A_{kl}^{\alpha\beta}(x)\xi_\alpha\xi_\beta\eta^k\eta^l \geqslant \nu|\xi|_n^2|\eta|_N^2, \qquad \forall \xi \in \mathbb{R}^n,\ \forall \eta \in \mathbb{R}^N,$$

where $\nu = \text{const} > 0$, $x \in \Omega$, $A_{kl}^{\alpha\beta} \in C(\overline{\Omega})$.

It is remarkable that if $A_{kl}^{\alpha\beta}$ depends also on u and is continuous in both arguments, then the strong ellipticity condition (13) is neccessary (**[GiSo]**).

We say a few words about the value of the exponent $p > 2$. An example of a linear system with $a_0 = f = 0$ exists (**[M]**) which shows that the difference $p - 2$ may be arbitrarily small. Nevertheless the fact of L_p-integrability is not trivial if we consider the example of linear system (12) ($a_0 = f = 0$, $n = 3$, $A_{kl}^{\alpha\beta} \in L_\infty(\mathbb{R}^3)$) that has a countable singular set dense in $\mathbb{R}^3$ **[So]**.

Let us pass to the author's results on L_p-integrability for Neumann type boundary value problems.

Let Ω be a bounded strongly Lipschitz domain in $\mathbb{R}^n$, $n > 2$, and suppose that the following three conditions hold:

I. The functions $A_{kl}^{\alpha\beta}(x, v)$ are measurable in Ω for all $v \in W_2^1(\Omega)$, and

$$\sum_{\substack{\beta,\alpha\leqslant n\\ l,k\leqslant N}} |A_{kl}^{\alpha\beta}(x,v)| \leqslant L, \qquad A_{kl}^{\alpha\beta}(x,v)\xi_\alpha^k\xi_\beta^l \geqslant \nu|\xi|^2, \quad \forall \xi \in \mathbb{R}^{nN},$$

where L and ν are positive constants.

II. The functions $a_k^\alpha(x, v)$ and $b_k(x, v, v_x)$ are measurable in Ω for all $v \in W_2^1(\Omega)$, and the estimates

$$\sum_{\substack{\alpha\leqslant n\\ k\leqslant N}} |a_k^\alpha(x,v)| \leqslant L|v|^r + f_1(x), \qquad r \leqslant \frac{n}{n-2}, \; f_1 \in L_s(\Omega), \; s > 2,$$

$$\sum_{k\leqslant N} |b_k(x,v,p)| \leqslant L(|p|^{r_1} + |v|^{r_0}) + f_0(x),$$

$$r_1 \leqslant \frac{n+2}{n}, \quad r_0 \leqslant \frac{n+2}{n-2}, \quad f_0 \in L_{s_0}(\Omega), \quad s_0 > \frac{2n}{n+2},$$

hold for $x \in \Omega$, $v \in \mathbb{R}^N$, $p \in \mathbb{R}^{Nn}$.

III. The functions $\varphi_k(x, v)$ are measurable on $\partial\Omega$ for all $v \in L_{\frac{2(n-1)}{n-2}}(\partial\Omega)$, and

$$\sum_{k\leqslant N} |\varphi_k(x,v)| \leqslant L|v|^{m_0} + \Phi(x),$$

$$m_0 \leqslant \frac{n}{n-2}, \quad \Phi \in L_m(\partial\Omega), \quad m > \frac{2(n-1)}{n}$$

for $x \in \Omega$, $v \in \mathbb{R}^N$.

THEOREM 3 ([**A5**, Theorem 2]). *Let $u \in W_2^1(\Omega)$ be a solution of* (12), (15) *and let conditions* I–III *hold. Then there exists a constant $p > 2$ such that $u \in W_p^1(\Omega)$.*

REMARK. Condition II implies controllable growth of the functions a_k^α and b_k in v and p.

Theorem 3 enables one to prove partial regularity of any solution of (12), (15). The exact result is stated in Theorem 5 (see §3).

For nonlinear elliptic systems under the uncontrollable growth conditions the L_p-estimate of the gradient of any solution of the Neumann problem was established in [**A2**].

Let $u \in W_m^1(\Omega) \cap L_\infty(\Omega)$, $m \in (1, n)$, be a weak solution of the boundary problem:

$$-\frac{d}{dx_\alpha} a_i^\alpha(x, u, u_x) + b_i(x, u, u_x) = 0, \qquad x \in \Omega, \tag{18}$$

$$a_i^\alpha(x, u, u_x)\cos(\mathbf{n}(x), x_\alpha) \restriction_{\partial\Omega} = g_i(x, u), \qquad i \leqslant N. \tag{19}$$

Assume that the following three conditions hold:

(a) The functions $a^\alpha = \{a_i^\alpha(x,u,p)\}_{i\leqslant N}$ and $b = \{b_i(x,u,p)\}_{i\leqslant N}$ satisfy the Carathéodory conditions for $x \in \Omega$, $u \in \mathbb{R}^N$, $p \in \mathbb{R}^{nN}$.

(b) For some positive δ the conditions

$$\begin{aligned} a^\alpha p_\alpha &\geqslant \nu(|u|)|p|^m - f_0(x), && f_0 \in L_{1+\delta}(\Omega),\\ \sum_{\alpha\leqslant n} |a^\alpha| &\leqslant \mu_1(|u|)|p|^{m-1} + f_1(x), && f_1 \in L_{\frac{m}{m-1}+\delta}(\Omega),\\ |b| &\leqslant \mu_2(|u|)|p|^m + f_2(x), && f_2 \in L_{1+\delta}(\Omega), \end{aligned}$$

hold for $x \in \Omega$, $u \in \mathbb{R}^n$, $p \in \mathbb{R}^{nN}$; the functions ν, μ_i are continuous, $\nu(t) > 0$ is nonincreasing, $\mu_i(t) \geqslant 0$, and μ_2 is nondecreasing for $t \in [0, +\infty)$.

(c) $g = \{g_i(x,u)\}_{i\leqslant N}$ is a Carathéodory function for $x \in \partial\Omega$ and $u \in \mathbb{R}^N$, and

$$|g| \leqslant \varphi_1(x)|u|^\alpha + \varphi_2(x),$$

where

$$\alpha \in [0, n(m-1)/(n-m)),$$
$$\varphi_1 \in L_{\frac{m(n-1)}{n(m-1)-\alpha(n-m)}+\delta}(\partial\Omega), \qquad \varphi_2 \in L_{\frac{m(n-1)}{n(m-1)}+\delta}(\partial\Omega).$$

REMARK. Condition (b) implies uncontrollable growth for system (18).

Let $u \in W_m^1(\Omega) \cap L_\infty(\Omega)$ be any solution of (18), (19). Then

$$\int_\Omega [a^\alpha(x,u,u_x)h_{x_\alpha} + b(x,u,u_x)h]\,dx = \int_{\partial\Omega} g(x,u)h\,d\Gamma, \quad \forall h \in W_m^1(\Omega) \cap L_\infty(\Omega). \tag{20}$$

THEOREM 4 ([**A2**, Theorem 2]). *Let conditions* a)–c) *hold, let* $u \in W_m^1(\Omega) \cap L_\infty(\Omega)$ *be a solution of* (20), *and let* $M = \operatorname{ess\,sup}_\Omega |u|$. *Then* $u_x \in L_p(\Omega)$ *for some* $p > m$ *if*

$$2\mu_2(M)M < \nu(M). \tag{21}$$

REMARK. If the boundary condition (19) holds on $\Gamma \Subset \partial\Omega$, then we have L_p-integrability in some neighborhood of Γ.

Note that the inequality $\mu_2 M < \nu$ is neccessary for $u_x \in L_{p,\mathrm{loc}}(\Omega)$, $p > m$ (see an example in [**HW**]). Assuming the growth of $b(x,u,p)$ in p not to exceed $|p|^{m-\varepsilon}$, $\varepsilon > 0$, Theorem 4 holds without condition (21).

Clearly, Theorem 2 implies L_p-estimates for higher-order derivatives. In particular, it is shown in [**A2**] that solutions of the Signorini problem with nonzero condition for the conormal derivative on $\Gamma \subset \partial\Omega$ belong to $W_{2+\delta}^2$, $\delta > 0$, in some neighborhood of Γ.

§3. Boundary regularity for solutions of elliptic Neumann-type problems

As mentioned in the Introduction, following the papers [**Mor, GM, G**], the partial regularity of solutions of elliptic systems was extensively studied. Most of the work deals with regularity inside the domain (see [**Gi3**] and [**JS**]).

Partial regularity of the solutions of the Dirichlet problem was studied in **[CGP, V]**. In particular, it was shown that any solution of system (14) with smooth boundary data $u \restriction_{\partial\Omega}$ has a singular set

$$\Sigma \subset \left\{ x \in \overline{\Omega} \mid \varliminf_{R\to 0} R^{2-n} \int_{\Omega_R(x)} |u_x|^2\,dx > \varepsilon \right\} \cup \left\{ x \in \Omega \mid \sup_R |u_{R,x}| = +\infty \right\},$$

for some positive ε; $\dim_{\mathscr{H}} \Sigma \leqslant n-p$, $p>2$. The example of M. Giaquinta **[Gi4]** shows that in general for nondiagonal quasilinear systems the singular set for the Dirichlet problem is not separated from the boundary. (The study of the regularity in the vicinity of the boundary for diagonal elliptic systems and related references can be found in **[F]**.)

The regularity of solutions of Neumann-type boundary problem (12), (15) with $\varphi_k \equiv 0$ (under conditions I, II of Theorem 3 in §2) was studied in **[GiMo2]**. The regularity theorems in Morrey and Campanato classes were proved for linear elliptic systems with zero conormal derivative on the boundary. In particular, solutions of Neumann-type problem for a system with constant coefficients were also analysed there. Concerning the quasilinear system, the interior regularity result was obtained in **[GiMo2]**, but the behavior of the solution in the neighborhood of the boundary was not described there.

The regularity theorem for solutions of (12), (15) in the case of nonregular $\varphi_k \neq 0$ was proved in **[A5]**. We assumed in **[A5]** that Ω is a C^1-domain and noted that the result was local. In case when $\Omega = B_1^+$ and the boundary condition is stated on $\Gamma_1 = B_1 \cap \{x_n = 0\} \subset \partial B_1^+$, the result can be formulated as follows.

THEOREM 5 (**[A5,** Theorem 3]). *Assume that the following conditions hold*:

(1) *The* $A_{kl}^{\alpha\beta}$ *are continuous functions satisfying condition* I *in* $B_1^+ \times \mathbb{R}^n$.

(2) *The* a_k^α *and* b_k *satisfy condition* II *in* $B_1^+ \times \mathbb{R}^n \times \mathbb{R}^{nN}$ *with* $s>n$, $s_0 > \frac{n}{2}$.

(3) *The* φ_k *satisfy condition* III *in* $\Gamma_1 \times \mathbb{R}^N$ *with* $m > n-1$ *for all* k.

(*See conditions* I–III *of Theorem* 3 *in* §2.) *Let* $u \in W_2^1(B_1^+)$ *be a solution of*

$$\int_{B_1^+} [(A_{kl}^{\alpha\beta} u_{x_\beta}^l + a_k^\alpha) h_{x_\alpha}^k + b_k h^k]\,dx + \int_{\Gamma_1} \varphi_k h^k\,d\Gamma = 0, \tag{22}$$
$$\forall h \in W_2^1(B_1^+), \qquad h \restriction_{\partial B_1^+ \setminus \Gamma_1} = 0.$$

Then there exists a set $\Omega_0 \subset \Omega$, *open in* $\widehat{B}_1 = B_1^+ \cup \Gamma_1$, *such that* $u \in C^\alpha(\Omega_0)$, $\alpha = \min\{2 - n/s_0,\ 1 - n/s,\ 1-(n-1)/m\}$, *the singular set*

$$\Sigma = \widehat{B}_1 \setminus \Omega_0$$
$$\subset \left\{ x \in \widehat{B}_1 \mid \varliminf_{R\to 0} R^{2-n} \int_{\Omega_R(x)} |u_x|^2\,dx > 0 \right\} \cup \left\{ x \in \widehat{B}_1 \mid \sup_R |u_{R,x}| = +\infty \right\},$$

and $\mathscr{H}_{n-p}(\Sigma) = 0$ *for some* $p > 2$.

It is easy to show that the derivatives of u are Hölder continuous in some neighborhood of a regular point $x^0 \in \Omega_0$. Indeed, fix $x^0 \in \Omega_0 \subset \widehat{B}_1$ and assume that $u \in C^\gamma(\overline{\Omega_R(x^0)})$, $\overline{\Omega_R(x^0)} \subset \widehat{B}_1$, $\gamma > 0$. Denote

$$B_{kl}^{\alpha\beta}(x) = A_{kl}^{\alpha\beta}(x, u(x)), \qquad p_k^\alpha(x) = a_k^\alpha(x, u(x)),$$
$$Q_k(x, u_x) = b_k(x, u(x), u_x), \qquad x \in \Omega_R(x^0);$$
$$\psi_k(x) = \varphi_k(x, u(x)), \qquad x \in \Gamma_R(x^0) = \Omega_R(x^0) \cap \Gamma_1.$$

Let u be a weak solution of the problem

$$
(23)\qquad \begin{aligned} -(B^{\alpha\beta}_{kl}(x)u^l_{x_\beta} + p^\alpha_k(x))_{x_\alpha} + Q_k(x, u_x) = 0, \qquad & x \in \Omega_R(x^0), \\ -(B^{n\beta}_{kl}(x)u^l_{x_\beta} + p^n_k(x)) \restriction_{\Gamma_R} = \psi_k(x), & \end{aligned}
$$

where

$$
(24)\qquad |Q_k(x, p)| \leqslant L(|p|^{\frac{n+2}{n}} + f_0(x)).
$$

The following theorem can be proved with the help of the regularity result obtained in **[GiMo2]** for the linear Neumann problem (see (23) with $Q_k \equiv 0$).

THEOREM 6 (**[A5, Theorem 4]**). *Let u be a weak solution of the quasilinear problem (22) and let $u \in C^\gamma(\overline{\Omega_R(x^0)})$, $\gamma > 0$. Denote*

$$
M = \max_{\overline{\Omega}_R} |u|, \qquad K_M = \{\, u \in \mathbb{R}^N \mid |u| \leqslant M \,\}
$$

and assume that

$$
A^{\alpha\beta}_{kl}, a^\alpha_k \in C^\theta(\overline{\Omega}_R \times K_M), \qquad \varphi_k \in C^\theta(\Gamma_R \times K_M), \qquad \theta > 0.
$$

Let $f_0 \in L_{s_0}(\Omega_R)$, $s_0 > n$ (see (24)). Then

$$
u_x \in C^{\theta_0}(\overline{\Omega_r(x^0)}), \qquad \theta_0 = \min\{\, \theta,\ 1 - n/s_0 \,\}, \qquad r < R.
$$

In conclusion, let us discuss problem (12), (15) for quasilinear systems under uncontrollable growth conditions. As far as we know, there are no results on partial regularity near the boundary in this case. Consider a simple example. Let u be a weak solution of the problem

$$
(25)\qquad \begin{aligned} -(A^{\alpha\beta}_{kl}(u)u^l_{x_\beta})_{x_\alpha} = a_k|u_x|^2, \qquad & x \in B^+_1, \\ A^{\alpha\beta}_{kl}(u)u^l_{x_\beta} \restriction_{\Gamma_1} = 0, \qquad & \Gamma_1 = B_1 \cap \{\, x_n = 0 \,\},\ k \leqslant N, \end{aligned}
$$

where $A^{\alpha\beta}_{kl}$ are continuous bounded functions in $\mathbb{R}^n$ satisfying the elipticity condition (13). The natural definition of solutions in the sense of distributions of (25) requires $u \in W^1_2(B^+_1) \cap L_\infty(B^+_1)$. If the condition $2aM < \nu$ holds ($M = \operatorname{ess\,sup}_{B^+_1} |u|$, ν is the ellipticity constant, $a = \sqrt{\sum_{k\leqslant N} a^2_k}$), then it is easy to prove that $u_x \in L_p(B^+_\sigma)$, $p > 2$, $\forall \sigma < 1$. Let us use the method of "frozen coefficients". Fix any $x^0 \in \Gamma_1$ and $B^+_R(x^0) \subset B^+_1$. For

$$
u_0 = \fint_{B^+_R(x^0)} u\, dx
$$

we have a model problem in $B^+_R(x^0)$

$$
(26)\qquad \begin{aligned} (A^{\alpha\beta}_{kl}(u_0)v^l_{x_\beta})_{x_\alpha} = 0 \qquad & \text{in } B^+_R(x^0), \\ A^{n\beta}_{kl}(u_0)v^l_{x_\beta} \restriction_{\Gamma_R} = 0, \qquad & v \restriction_{\partial B^+_R \setminus \Gamma_R} = u; \end{aligned}
$$

u is the solution under consideration. We need to be sure that the solution of (26) is bounded in B^+_R, but, as yet, we do not know any results of this type. An estimate of the "maximum principle" type for system (26) in a domain with smooth boundary in the case of the Dirichlet condition was obtained in **[Ca, Can]**.

§4. Parabolic reverse Hölder inequalities with additional anisotropic terms

Let $u(x,t)$, $(x,t) \in \Omega \times (0,T) = Q_T$, be a weak solution of a parabolic system. The higher integrability theorems for the gradient $u_x(x,t)$ can be proved for an arbitrary cylinder $\overline{Q}_0 \subset Q_T$ with the help of reverse Hölder inequalities in the parabolic metric. Here is the simplest statement about the reverse inequalities used by M. Giaquinta and M. Struwe [**GiS**].

THEOREM 7. *Let g be a nonnegative function from $L_q(Q_T)$, $q > 1$, and*

$$\fint_{Q_R(z^0)} g^q\,dz \leqslant \left(\fint_{Q_{4R}(z^0)} g\,dx\right)^q + \theta \fint_{Q_{4R}(z^0)} g^q\,dz,$$
$$\forall z^0 \in Q, \quad Q_{4R}(z^0) \subset Q, \quad B > 1.$$

Then there exists a constant $\theta_0 = \theta_0(q,n)$ such that $\theta < \theta_0$ implies $g \in L_{p,\mathrm{loc}}(Q_T)$ for $p \in [q, q+\varepsilon)$ and

$$\left(\fint_{Q_R} g^p\,dz\right)^{1/p} \leqslant C\left(\fint_{Q_{4R}} g^q\,dz\right)^{1/q}, \qquad \forall Q_{4R} \subset Q_T;$$

the constants C and $\varepsilon > 0$ depend only on b, q, θ, and n.

Even in the linear case it is natural to study parabolic systems with data from anisotropic functional spaces. Let $L_{m,r}(Q)$ be a space of functions supplied with the norm

$$\|u\|_{m,r,Q_T} = \left(\int_0^T \left(\int_\Omega |u|^m\,dx\right)^{r/m} dt\right)^{1/r}, \qquad m, r \geqslant 1,$$

(see [**LSU,** Chapters II and III]).

Our approach to the regularity problem for parabolic systems in terms of $L_{m,r}$-spaces resulted in the higher integrability theorem with additional anisotropic terms proved in [**A3**].

Let us introduce some special notation. We denote

$$P_R(0) = \{\, x \in \mathbb{R}^n \mid |x_i| < R, i = 1, \dots, n \,\}, \qquad n \geqslant 2;$$
$$Q = P_{3/2}(0) \times (-\tfrac{5}{4}, \tfrac{5}{4}), \qquad Q_0 = P_{1/2}(0) \times (-\tfrac{1}{4}, \tfrac{1}{4}),$$
$$\Lambda_R(t^0) = (t^0 - R^2, t^0 + R^2), \quad Q_R(z^0) = B_R(x^0) \times \Lambda_R(t^0), \quad z^0 = (x^0, t^0),$$
$$|Q_R| = \mathrm{meas}_n\, B_R \cdot 2R^2 = O(R^{n+2});$$
$$\fint_{Q_r(z^0)} h\,dz = \frac{1}{|Q_r|}\int_{Q_r(z^0)} h\,dz$$

and $\delta(z^1, z^2) = \max\{\, |x^1 - x^2|, |t^1 - t^2|^{1/2} \,\}$ is the parabolic metric.

Let $\gamma \subset P_{\frac{3}{2}}(0)$, $\mathscr{H}_{\bar{k}}(\gamma) < +\infty$, $\bar{k} < n$ (for the additional condition on $\bar{k}$ see (30)). Let γ satisfy (9), and denote $\gamma_r(x^0) = \gamma \cap B_r(x^0)$, $\Gamma = \gamma \times (-\frac{5}{4}, \frac{5}{4})$.

THEOREM 8 ([**A3,** Theorem 1]). *Let q, Q_0, and Γ be as determined above and let $g \in L_q(Q)$, $q > 1$, $f \in L_{m_1,m_2}(Q)$, and $\varphi \in L_{l_1,l_2}(\Gamma)$ be nonnegative functions. Suppose that for $R < \frac{1}{a}\delta(z^0, \partial Q)$ the estimate*

$$
\begin{aligned}
\fint_{Q_R(z^0)} g^q\,dz \leqslant \theta \fint_{Q_{aR}(z^0)} g^q\,dz + B\Big(\fint_{Q_{aR}(z^0)} g\,dQ\Big)^q \\
+ R^{-(n+2)}\Big(\int_{\Lambda_{aR}(t^0)}\Big(\int_{B_{aR}(x^0)} f^{m_1}\,dx\Big)^{m_2/m_1} dt\Big)^{r/m_2} \\
+ R^{-(n+2)}\Big(\int_{\Lambda_{aR}(t^0)}\Big(\int_{\gamma_{aR}(x^0)} \varphi^{l_1}\,d\mathscr{H}_{\bar{k}}\Big)^{l_2/l_1} dt\Big)^{s/l_2},
\end{aligned} \tag{27}
$$

holds for almost all $z^0 \in Q$ with some $\theta \in [0,1)$, $a > 1$, $B > 1$ and

$$
\frac{n}{2m_1} + \frac{1}{m_2} \geqslant \frac{n+2}{2r}, \qquad r \geqslant m_2 \geqslant m_1 \geqslant 1,
$$

$$
\frac{\bar{k}}{2l_1} + \frac{1}{l_2} \geqslant \frac{n+2}{2s}, \qquad s > l_2 \geqslant l_1 \geqslant 1.
$$

Then there exist constants $p_0 > q$ and $C_0 > 0$ such that $g \in L_p(Q_0)$ for $p \in [q, p_0)$ and

$$
\|g\|_{p,Q_0} \leqslant C_0\Big\{\|g\|_{q,Q} + \|f\|^{r/q}_{m_1+\varepsilon_1,m_2+\varepsilon_1,Q} + \|\varphi\|^{s/q}_{l_1+\varepsilon_2,l_2+\varepsilon_2,\Gamma}\Big\},
$$
$$
\varepsilon_1 = \frac{r}{q}(p-q), \quad \varepsilon_2 = \frac{s}{q}(p-q).
$$

The constants p_0 and C_0 depend on q, n, $\bar{k}$, B, θ, m_i, l_i, r, s, and a; Land ρ are from condition (9); *the constant C_0 depends also on $\delta(Q_0, \partial Q)$.*

REMARK 1. This L_p-estimate for g holds if $f \in L_{m_1+\varepsilon,m_2+\varepsilon}(Q)$, $\varphi \in L_{l_1+\varepsilon,l_2+\varepsilon}(\Gamma)$ for some $\varepsilon > 0$ and $\varepsilon_i \leqslant \varepsilon$, $i = 1, 2$.

REMARK 2. The assumptions of Theorem 8 on the structure of Γ arise from the simplest examples of parabolic boundary value problems. They may be modified. For instance, we can assume that $\Gamma = \{\gamma(t), t\}$, $t \in [-5/4, 5/4]$, and $\gamma(t)$ satisfies the (L, ρ)-condition uniformly in $[-5/4, 5/4]$. The manifolds $\Gamma \subset Q$ may be considered with $\mathscr{H}_m(\Gamma; \delta) < +\infty$ ($\mathscr{H}_m(\,\cdot\,; \delta)$ being the m-dimensional Hausdorff measure in the metric δ). Then the last term in (27) must be replaced by

$$
R^{-(n+2)}\Big(\int_{\Gamma_{aR}(z^0)} \varphi\, d\mathscr{H}_m(\delta)\Big)^s, \qquad \Gamma_R(z^0) = Q_R(z^0) \cap \Gamma,
$$

and the (L, ρ)-condition aquires the form

$$
\mathscr{H}_m(\Gamma_r(z^0); \delta) \leqslant L r^m, \qquad \forall z^0 \in \Gamma,\ r \leqslant \rho,
$$

where $ms \geqslant n + 2$.

REMARK 3. Suppose all the assumptions of Theorem 8 are valid but instead of (27) we have

$$\begin{aligned}\fint_{Q_R(z^0)} g^q\,dz + \psi_R(z^0) \leqslant \theta\Big\{\fint_{Q_{aR}(z^0)} g^q\,dz + \psi_{aR}(z^0)\Big\} + B\Big(\fint_{Q_{aR}} g\,dz\Big)^q \\ + R^{-(n+2)}\bigg(\int_{\Lambda_{aR}(t^0)}\Big(\int_{B_{aR}(x^0)} f^{m_1}\,dx\Big)^{\frac{m_2}{m_1}}dt\bigg)^{\frac{r}{m_2}} \\ + R^{-(n+2)}\bigg(\int_{\Lambda_{aR}(t^0)}\Big(\int_{\gamma_{aR}(x^0)} \varphi^{l_1}\,d\mathscr{H}_{\bar{k}}\Big)^{\frac{l_2}{l_1}}dt\bigg)^{\frac{s}{l_2}},\end{aligned} \tag{28}$$

where $\psi_\rho(z)$ is nonnegative function defined for almost all $z \in Q$, $\rho \in (0, d_0]$, $d_0 = \operatorname{diam} Q$, and

$$m_\psi(Q) = \sup_{z^0\in Q}\sup_{\rho\leqslant d_0}(\psi_\rho(z)\rho^{n+2}) < +\infty. \tag{29}$$

Then there exist constants $p_0 > g$ and $C_0 > 0$ such that $g \in L_p(Q_0)$ for $p \in [q, p_0)$ and

$$\|g\|_{p,Q_0} \leqslant C_0\{\|g\|_{q,Q} + \|f\|^{\frac{r}{q}}_{m_1+\varepsilon_1, m_2+\varepsilon_1, Q} + \|\varphi\|^{\frac{s}{q}}_{l_1+\varepsilon_2, l_2+\varepsilon_2, \Gamma} + m_\psi^{\frac{1}{q}}\}. \tag{30}$$

It appears that this modification of Theorem 8 is very efficient for the study of parabolic systems with controllable growth conditions. It can be proved easily by analysing the proof of [**A3,** Theorem 1].

Certainly we could add $\psi_\rho(z)$ in inequality (10) of Theorem 2 and derive the estimate for $\|g\|_{p,Q'}$ depending on

$$m_\psi = \sup_{x\in Q}\sup_{\rho\leqslant d_0}(\psi_\rho(x)\cdot\rho^n) < +\infty.$$

§5. L_p-estimates for the gradients of solutions of parabolic boundary-value problems

Using Theorem 8 we shall obtain the proof of higher integrability of gradients of weak solutions of parabolic boundary-value problems. As an example we have considered a problem with given boundary data for a parabolic system satisfying the controllable growth conditions [**A3**]. We assumed there that the functions a and b do not depend on u. The result can be formulated as follows.

Let u be a weak solution of the problem

$$\begin{aligned} u_t^k - \frac{d}{dx_\alpha}a_k^\alpha(z, u_x) + b_k(z, u_x) = 0, \qquad & z \in Q_T, \\ a_k^\alpha, (z, u_x)\cos(\mathbf{n}, x_\alpha) \restriction_{\Gamma_T} = \varphi_k(z), \qquad & \Gamma_T = \partial\Omega \times (0, T),\end{aligned} \tag{31}$$

where Ω is a strongly Lipschitz domain in $\mathbb{R}^n$, $n \geqslant 2$, and $\mathbf{n} = \mathbf{n}(x)$ is the outward normal at $x \in \partial\Omega$. Clearly, $\Gamma_T = \gamma \times (0, T)$, $\gamma \Subset \partial\Omega$, can be considered to prove the higher integrability of the gradient near Γ_T.

Assume that $a_k^\alpha(x,p)$ and $b_k(z,p)$ are Carathéodory functions on $Q_T \times \mathbb{R}^{nN}$ and

$$a_k^\alpha(z,p)p_\alpha^k \geqslant \nu|p|^2 - \Phi_1(z), \qquad |a_k^\alpha(z,p)| \leqslant \mu_1|p| + \Phi_2(z),$$
$$|b_k(z,p)| \leqslant \mu_2|p|^\beta + \Phi_3(z), \qquad \beta \leqslant (n+4)/(n+2). \tag{32}$$

Here $\nu > 0$, $\mu_i \geqslant 0$, $\Phi_1, \Phi_2^2 \in L_1(Q_T)$, $\Phi_3 \in L_{m,r}(Q_T)$ and

$$\frac{1}{r} + \frac{n}{2m} = 1 + \frac{n}{4}; \qquad \begin{array}{lll} m \in [\frac{2n}{n+2}, 2], & r \in [1,2] & \text{if } n \geqslant 3, \\ m \in (1,2], & r \in [1,2) & \text{if } n = 2. \end{array} \tag{33}$$

Assume that in the boundary condition (29) we have $\varphi_k \in L_{s,l}(\Gamma)$ and

$$\frac{1}{l} + \frac{n-1}{2s} = \frac{1}{2} + \frac{n}{4}; \qquad \begin{array}{lll} s \in \left[\dfrac{2(n-1)}{n}, \dfrac{2(n-1)}{n-2}\right], & l \in [1,2] & \text{if } n \geqslant 3, \\ s \in (1,\infty], & l \in [1,2) & \text{if } n = 2. \end{array} \tag{34}$$

Fix $(t_1,t_2) \subset (0,T)$ and denote $Q' = \Omega \times (t_1,t_2)$. Consider the space $V(Q')$ of functions continuous in t in the $L_2(\Omega)$-norm and supply it with the norm

$$|u|_{Q'} = \max_{[t_1,t_2]} \|u(\cdot,t)\|_{L_2(\Omega)} + \|u_x\|_{L_2(Q')}.$$

THEOREM 9 ([**A3**, Theorem 2]). *Let $u \in V(Q')$ be a weak solution of* (31) *and let a_k^α, b_k, φ_k satisfy conditions* (32)–(34). *Let*

$$\Phi_1, \Phi_2^2 \in L_{1+\varepsilon}(Q'), \qquad \Phi_3 \in L_{m+\varepsilon, r+\varepsilon}(Q'), \qquad \varphi_k \in L_{s+\varepsilon, l+\varepsilon}(\Gamma'),$$

for some $\varepsilon > 0$. Here $\Gamma' = \partial\Omega \times (t_1,t_2)$; $r \geqslant m$ and $l \geqslant s$ satisfy (33), (34). *Then there exists a constant $p_0 > 2$ such that $u_x \in L_p(Q'')$ for all $p \in [2,p_0)$, $\overline{Q}'' \subset Q' \cup \Gamma'$.*

REMARK. The linear scalar problem ($N = 1$) is an important particular case of (31):

$$u_t - (a_{ij}(z)u_{x_i})_{x_j} = f(z), \qquad z \in Q_T,$$
$$a_{ij}(z)u_{x_i}\cos(\mathbf{n}, x_j) \restriction_{\Gamma_T} = \varphi(z), \tag{31°}$$

where $a_{ij} \in L_\infty(Q_T)$, $a_{ij}(z)\xi_i\xi_j \geqslant \nu|\xi|^2$ for all $\xi \in \mathbb{R}^n$, $z \in Q_T$; $\nu = \text{const} > 0$. The assumptions $f \in L_{m,r}(Q')$ and $\varphi \in L_{s,l}(\Gamma')$ together with (33), (34) exactly define weak solutions of (31°) in the space $\widehat{V}(Q')$ with the norm

$$|u|_{Q'} = \operatorname{ess\,sup}_{(t_1,t_2)} \|u(\cdot,t)\|_{L_2(\Omega)} + \|u_x\|_{L_2(Q')}. \tag{35}$$

Under these conditions one can also prove that any solution from $\widehat{V}(Q')$ belongs to $V(Q')$ (see [**LSU**, Chapter 3, §4]).

The author considered an example of a quasilinear parabolic system under uncontrollable growth conditions and mixed boundary condition in [**A4**]. She chose mixed conditions in order to obtain the L_p-estimate for the gradient near the boundary both for the first and for the second boundary condition simultaneously.

Let Ω be a bounded srongly Lipschitz domain in $\mathbb{R}^n$, $n \geqslant 2$, and let $\partial\Omega$ consist of two parts γ_1 and γ_2 with Lipschitz' surface division, $\Gamma_i = \gamma_i \times (0, T)$. We study the initial-boundary-value problem:

$$(36)\qquad \begin{gathered} u_t^k - (a_{kl}^{\alpha\beta}(z) u_{x_\beta}^l)_{x_\alpha} = f_k(z, u, u_x), \qquad z \in Q_T, \\ a_{kl}^{\alpha\beta}(z) u_{x_\beta}^l \cos(\mathbf{n}, x_\alpha) \restriction_{\Gamma_1} = \psi_k(z), \\ u^k \restriction_{\Gamma_2} = 0, \qquad u^k \restriction_{t=0} = \varphi^k(x), \qquad x \in \Omega,\ k = 1, \dots, N. \end{gathered}$$

Let $a_{kl}^{\alpha\beta} \in L_\infty(Q_T)$ and

$$a_{kl}^{\alpha\beta} \xi_\alpha^k \xi_\beta^l \geqslant \nu |\xi|^2, \qquad \forall \xi \in \mathbb{R}^{nN},\ \nu = \text{const} > 0.$$

Fix any $M > 0$. Assume that $f(z, v, p)$ is a Carathéodory function for $z \in Q_T$, $v \in \mathbb{R}^n$, $|v| \leqslant M$, $p \in \mathbb{R}^{nN}$, and

$$(37)\qquad \begin{gathered} |f| \leqslant a_0(M)|p|^2 + f_0(z), \qquad f_0 \in L_{1+\varepsilon}(Q_T), \quad \varepsilon > 0, \\ 2a_0(M)M < \nu. \end{gathered}$$

Let

$$\begin{gathered} \psi_k \in L_{s+\varepsilon, l+\varepsilon}(\Gamma_1), \quad k \leqslant N, \qquad \frac{1}{l} + \frac{n-1}{2s} = \frac{n+2}{4}, \\ s \in \left[\frac{2(n-1)}{n}, \frac{2(n+1)}{n+2}\right], \quad l \in \left[\frac{2(n+1)}{n+2}, 2\right] \quad \text{if } n \geqslant 3; \\ 1 < s \leqslant l < 2 \quad \text{if } n = 2. \end{gathered}$$

The functions $\varphi_k \in W_{\frac{2n}{n+2}+\varepsilon}(\Omega)$, $\varphi_k \restriction_{\gamma_2} = 0$, $k = 1, \dots, N$, and $\varepsilon > 0$ may be taken to be the same everywhere.

Note that $\widehat{V}(Q_T)$ is a functional space with norm defined by (35). Now we are ready to state the following result.

THEOREM 10 ([**A4,** Theorem 3]). *Let $u \in \widehat{V}(Q_T) \cap L_\infty(Q_T)$ be a weak solution of (36). Let the data satisfy all the above conditions, and let $M = \operatorname{ess\,sup}_{Q_T} |u|$ in (37). Then there exist $p_0 > 2$ and $C_0 > 0$ such that $u_x \in L_p(Q_T)$, $p \in [2, p_0)$, and*

$$\begin{gathered} \|u_x\|_{p, Q_T} \leqslant C_0 \{ \|u_x\|_{2, Q_T} + \|f_0\|_{\frac{p}{2}}^{1/2} + \|\psi\|_{s+\gamma, l+\gamma, \Gamma_1} + \|\varphi_x\|_{\frac{2n}{n+2}+\gamma, \Omega} \}, \\ \gamma = p - 2 > 0. \end{gathered}$$

REMARK. Consider the linear problem (36): $f = f_0(z)$, $f_0 \in L_{m,r}(Q_T)$, and

$$\begin{gathered} \frac{1}{r} + \frac{n}{2m} = 1 + \frac{n}{4}, \\ m \in \left[\frac{2n}{n+2}, \frac{2(n+2)}{n+4}\right], \quad r \in \left[\frac{2(n+2)}{n+4}, 2\right] \quad \text{if } n \geqslant 3, \\ 1 < m \leqslant r < 2 \quad \text{if } n = 2. \end{gathered}$$

(These conditions are equivalent to (34) with the additional condition $r \geqslant m$; see Theorem 9.) If the remaining conditions of Theorem 10 hold, then for the solution

$u \in \widehat{V}(Q_T)$ of the linear problem the statement of Theorem 10 is valid with the following estimate:

$$\|u_x\|_{p,Q_T} \leqslant C_0\{\|u_x\|_{2,Q_T} + \|f_0\|_{m+\gamma,r+\gamma;Q_T} + \|\psi\|_{s+\gamma,l+\gamma,\Gamma_1} + \|\varphi_x\|_{\frac{2n}{n+2}+\gamma,\Omega}\}, \qquad \gamma = p-2>0.$$

The function u is proved to be continuous in t in the L_p-norm, and there is an estimate of $\max_{[0,T]}\|u\|_{p,\Omega}$ (see **[A4,** Theorem 4]).

Clearly, Theorem 8 provides L_p-estimates for much broader classes of nonlinear problems than those mentioned in the examples above. The L_p-estimates enable one to study the regularity of solutions of quasilinear parabolic systems. As pointed out in **[SJM]**, even for smooth data singularities may appear for $t>0$. That is why only partial regularity can be expected (about singularities see also **[JS]**).

Later we proved the higher integrability result for a general class of quasilinear and nonlinear parabolic systems under controllable growth conditions (including the limit degrees of growth):

$$\begin{aligned} &u_t^k - (A_{kl}^{\alpha\beta}(z,u)u_{x_\beta}^l + a_k^\alpha(z,u))_{x_\alpha} + b_k(z,u,u_x) = 0, \\ &\qquad\qquad z \in Q_T = \Omega\times(0,T), \\ &(A_{kl}^{\alpha\beta}(z,u)u_{x_\beta}^l + a_k^\alpha(z,u))\cos(\mathbf{n},x_\alpha) + \varphi_k(z,u)|_{\Gamma_T} = 0, \\ &\qquad\qquad \Gamma_T = \partial\Omega\times(0,T). \end{aligned} \tag{38}$$

For $Q=\Omega\times(t_1,t_2)$, $\Gamma=\partial\Omega\times(t_1,t_2)$, $(t_1,t_2)\subset(0,T)$, we make the following thre assumptions:

I. The functions $A_{kl}^{\alpha\beta}(z,u)$ are measurable in Q for all $u\in V(Q)$, and

$$\sum_{k,l\leqslant N}\sum_{\alpha,\beta\leqslant n}|A_{kl}^{\alpha\beta}(z,u)| \leqslant L, \qquad A_{kl}^{\alpha\beta}(z,u)\xi_\alpha^k\xi_\beta^l \geqslant \nu|\xi|^2,$$

$$\forall\xi\in\mathbb{R}^{nN}, \quad \nu=\mathrm{const}>0, \quad (z,u)\in Q\times\mathbb{R}^N.$$

II. The functions $a_k^\alpha(z,u)$ and $b_k(z,u,u_x)$ are measurable in Q for all $u\in V(Q)$, and for $(z,u,p)\in Q\times\mathbb{R}^N\times\mathbb{R}^{nN}$

$$\sum_{\substack{k\leqslant N\\ \alpha\leqslant n}}|a_k^\alpha(z,u)| \leqslant L|u|^r + f(z), \qquad r\leqslant\frac{n+2}{n}, \ f\in L_2(Q),$$

$$\sum_{k\leqslant N}|b_k(z,u,p)| \leqslant L(|u|^{r_0}+|p|^{r_1}) + f_0(z), \qquad r_0\leqslant\frac{n+4}{n}, \ r_1\leqslant\frac{n+4}{n+2},$$

$$s_1\in[2n/(n+2),2] \ \text{ and } \ s_2\in[1,2], \qquad \text{if } n\geqslant3,$$

$$s_1\in(1,2] \ \text{ and } \ s_2\in[1,2), \qquad \text{if } n=2.$$

III. The functions $\varphi_k(z,u)$ are measurable on Γ for all $u \in L_{2(n+1)/n}(\Gamma)$, and for $(z,u) \in \Gamma \times \mathbb{R}^N$

$$\sum_{k \leqslant N} |\varphi_k(z,u)| \leqslant L|u|^{m_0} + \phi(z),$$

$$m_0 \leqslant \frac{n+2}{n}, \quad \phi \in L_{S_3,S_4}(\Gamma), \quad \frac{1}{s_4} + \frac{n-1}{2s_3} = \frac{1}{2} + \frac{n}{4},$$

$$s_3 \in \left[\frac{2(n-1)}{n}, \frac{2(n-1)}{n-2}\right], \quad s_4 \in [1,2], \quad \text{if} \quad n \geqslant 3,$$

$$s_3 \in (1,\infty], \quad s_4 \in [1,2), \qquad\qquad \text{if} \quad n = 2.$$

(L is the same positive constant for all conditions.)

By using Remark 3 to Theorem 8 the following result was proved.

THEOREM 11 [**A6,** Theorem 2]. *Let Ω be a strongly Lipschitz domain in $\mathbb{R}^n$, $n \geqslant 2$, let the conditions* I–III *be fulfilled, and let $u \in V(Q)$ be a solution of* (38). *Moreover let $f \in L_{2+\varepsilon}(Q)$, $f_0 \in L_{S_1+\varepsilon,S_2+\varepsilon}(Q)$, $\phi \in L_{S_3+\varepsilon,S_4+\varepsilon}(\Gamma)$ for some $\varepsilon > 0$ and $s_2 \geqslant s_1$, $s_4 \geqslant s_3$. Then there exist constants $p_0 > 2$, $c_0 > 0$ such that*

$$u_x, |u|^{\frac{n+2}{n}} \in L_p(\widehat{Q}) \quad \textit{for all } p \in [2,p_0), \ \forall \widehat{Q} = \Omega \times \Lambda, \ \Lambda \subset (t_1,t_2)$$

and

$$\| |u_x| + |u|^{\frac{n+2}{n}} \|^2_{p,\widehat{Q}} \leqslant C_0 \{ \| |u_x| + |u|^{\frac{n+2}{n}} \|^2_{2,Q} + \|f\|^2_{2+\varepsilon,Q} + \|f_0\|^2_{s_1+\varepsilon,s_2+\varepsilon,Q}$$
$$+ \|\phi\|_{s_3+\varepsilon,s_4+\varepsilon,\Gamma} + \max_{[t_1,t_2]} \|u\|^2_{2,\Omega} \}, \qquad \varepsilon = p - 2 > 0.$$

We have already mentioned the papers by S. Campanato [**C4–C7**], where nonlinear parabolic systems under controllable growth conditions were studied (actually strictly controllable growth conditions were investigated). For weak solutions of such systems local L_p- and $\mathscr{L}_{p,\alpha}$-estimates and partial regularity were proved there.

The L_p-estimate of u_x permitted the author to prove the partial regularity of solutions (38) [**A6**].

Boundary regularity for Dirichlet condition was studied in [**C7**].

Partial regularity of the solution of the quasilinear system

$$u_t^k - (A_{kl}^{\alpha\beta}(z,u)u_{x_\beta}^l)_{x_\alpha} = f^k(z,u,u_x), \qquad k \leqslant N, \ z \in Q_T, \tag{39}$$

where the $A_{kl}^{\alpha\beta}$ are continuous bounded functions in $Q_T \times \mathbb{R}^N$ satisfying the parabolicity condition

$$A_{kl}^{\alpha\beta}(z,u)\xi_\alpha^k \xi_\beta^l \geqslant |\xi|^2, \qquad \forall \xi \in \mathbb{R}^{nN}, \quad \nu = \text{const} > 0;$$
$$|f(z,u,p)| \leqslant a|p|^2 + b, \qquad a,b = \text{const} > 0,$$
$$2a \operatorname{ess\,sup}_{Q_T} |u| \leqslant \nu.$$

was proved in [**GiS**]. The solution to problem (39) is shown to have a singular set

$$\Sigma \subset \left\{ z \in Q_T \;\middle|\; \varliminf_{R \to 0} R^{-n} \int_{Q_R(z)} |u_x|^2 dz > \varepsilon \right\}$$

with $\mathscr{H}_{n-\gamma}(\Sigma;\delta) = 0$ for some positive ε and γ.

Note that the regularity of solutions of linear parabolic systems with given conormal derivative was studied in detail in **[GiMo3]**. In particular, model parabolic Neumann-type problems in Campanato spaces were studied there.

Finally we would like to remark that in the case that we are interested in, the partial regularity of the Neumann problem for a parabolic system under uncontrollable growth conditions, the same difficulty arises as in the elliptic situation: the absence of a "maximum principle" type result for the model linear problem with mixed boundary conditions. (For a "maximum principle" result under the Dirichlet condition see **[Gis, Mau]**.)

§6. Proof of Theorem 2

From now on in the hypotheses of Theorem 2 we assume that

$$Q = Q_{\frac{3}{2}}(0), \qquad Q' = Q_{\frac{1}{2}}(0),$$

where

$$Q_R(0) = \{\, x \in \mathbb{R}^n \mid |x_i| < R, i = 1, \dots, n \,\}.$$

The proof essentially uses ideas of **[GiMo1, Gi3, Gi5]**. We need the following two lemmas.

Lemma 1 (Calderon–Zygmund lemma, see **[S,** Chapter I, §3, Theorem 4]). *Let Q_0 be a cube in $\mathbb{R}^n$,*

$$f \in L_1(Q_0), \qquad f \geqslant 0, \qquad \fint_{Q_0} f\,dx \leqslant \alpha.$$

Then there exists a sequence of n-cubes $\{Q_i\}$ with sides parallel to the axes and with disjoint interiors such that

(1) $f(x) \leqslant \alpha$ *for almost all* $x \in Q_0 \setminus \bigcup_j Q_j$,

(2) $\alpha < \fint_{Q_i} f\,dx \leqslant 2^n\alpha$.

Lemma 2 ((**[A1,** Lemma 2])). *Assume that $h(t)$, $H_i(t)\colon [t_0, \infty) \to [0, \infty)$, $t_0 > 0$, are nonincreasing functions with the following properties*:

(1) $\lim_{t\to+\infty} h(t) = \lim_{t\to+\infty} H_i(t) = 0$, $i = 1, \dots, m$,

(2)

$$-\int_t^\infty \tau^r\,dh(\tau) \leqslant b\{t^r h(t) + \sum_{i=1}^m t^{\alpha_i} H^{\beta_i}(t)\}, \qquad \forall t \geqslant t_0,$$

where $r > 0$, $\alpha_i \in [0, r]$, $\beta_i \geqslant 1$, and $b > 1$.

Then

$$-\int_{t_0}^\infty t^{\varkappa} dh(t) \leqslant \frac{r}{br - \varkappa(b-1)}\Big(-\int_{t_0}^\infty t^r\,dh(t)\Big) + \sum_{i=1}^m C_i H_i^{\beta_i - 1}(t_0)\Big(-\int_{t_0}^\infty t^{\varkappa - r + \alpha_i}\,dH_i(t)\Big),$$

where

$$C_i = \frac{b\varkappa(\varkappa - r)}{(\varkappa - r + \alpha_i)(br - \varkappa(b-1))}, \qquad \varkappa \in [r, br/(b-1)).$$

Lemma 2 is a modification of a singular lemma concerning the Stieltjes integral (see **[Ge]**). Denote

$$M = \|g\|_{q,Q} + \|F\|_{1,Q}^{m/q} + \|\Phi\|_{1,\Gamma}^{l/q},$$

$$G = M^{-1}\theta^{\frac{1}{2q}} g, \qquad F_0 = M^{-\frac{q}{m}}\theta^{\frac{1}{2m}} F, \qquad \Phi_0 = M^{-\frac{q}{l}}\theta^{\frac{1}{2l}}\Phi \quad \text{for } \theta > 0$$

$$G = M^{-1} g, \qquad F_0 = M^{-\frac{q}{m}} F, \qquad \Phi_0 = M^{-\frac{q}{l}}\Phi \quad \text{for } \theta = 0.$$

The reverse inequality (10) takes the form

$$\begin{aligned} \fint_{B_R(x)} G^q\,dx \leqslant B\biggl(\fint_{B_{aR}(x)} G\,dx\biggr)^q + \theta \fint_{B_{aR}(x)} G^q\,dx + R^{(m-1)n} \\ \times \biggl(\fint_{B_{aR}(x)} F_0\,dx\biggr)^m + R^{-n}\mathscr{H}^l_{\bar{k}}(\gamma_{aR}(x))\biggl(\fint_{\gamma_{aR}(x)} \Phi_0\,d\mathscr{H}_{\bar{k}}\biggr)^l, \end{aligned} \tag{40}$$

$R \leqslant \frac{1}{a}\min\{ d(x),\ R_0 \}$, where $d(x) = d(x, \partial Q)$. Besides, we have

$$\max\{ \|G\|^q_{q,Q},\ \|F_0\|^m_{1,Q},\ \|\Phi_0\|^l_{1,\Gamma} \} \leqslant \begin{cases} \sqrt{\theta}, & \theta \in (0,1), \\ 1, & \theta = 0. \end{cases} \tag{41}$$

Denote

$$C_0 = Q_{\frac{1}{2}}(0), \qquad C_k = \{ x \in Q \mid 2^{-k} \leqslant d(x) < 2^{-k+1} \}, \quad k \geqslant 1;$$

then $Q = \bigcup_{k\geqslant 0} C_k$. Let P_k be a maximal cube $P_k \subset C_k$ with side a_k; clearly, $a_k = 2^{-k} = d(C_k, \partial Q)$. For the functions $\zeta(x) = d(x)^n$ the following inequalities hold:

$$\begin{aligned} |P_k| \leqslant \zeta(x) \leqslant 2^n|P_k|, \quad x \in C_k, \\ 2^{-n}|P_k| \leqslant \zeta(x) \leqslant 4^n|P_k|, \quad x \in C_{k-1}\cup C_k \cup C_{k+1}. \end{aligned} \tag{42}$$

The estimates (42) guarantee that for all k the function $\zeta(x)$ is equivalent to the measure of P_k in C_k and in $C_{k-1}\cup C_k\cup C_{k+1}$. For some $m_0 \in N$ consider all the cubes $D_k \subset P_k$ with side $2^{-m_0}a_k$, $|D_k| = 2^{-m_0 n}|P_k|$. For each cube D_k there exists a ball $B_{r_k}(y^k) \supset D_k$, $r_k = 2^{-m_0}\sqrt{n}a_k$, $y^k \in C_k$. Now choose m_0 so that $2^{-m_0}a\sqrt{n} < \min\{1, R_0\}$. Then the inequalities (40) hold in B_{r_k} and B_{ar_k} and since m_0 is fixed, the function $\zeta(x)$ is equivalent to the measure of D_k in C_k and in $C_{k-1}\cup C_k\cup C_{k+1}$. From (40), (41) we derive

$$\begin{aligned} \biggl(\fint_{D_k} (G\zeta)^q\,dx\biggr)^{\frac{1}{q}} &\leqslant C|D_k|\biggl(\fint_{B_{r_k}} G^q\,dx\biggr)^{\frac{1}{q}} \\ &\leqslant C|D_k|\biggl\{B\biggl(\fint_{B_{ar_k}} G\,dx\biggr)^q + \theta\fint_{B_{ar_k}} G^q\,dx \\ &\qquad + r_k^{-n}\biggl(\int_{B_{ar_k}} F_0\,dx\biggr)^m + r_k^{-n}\biggl(\int_{\gamma_{ar_k}} \Phi_0\,d\mathscr{H}_{\bar{k}}\biggr)^l\biggr\}^{1/q} \\ &\leqslant C\{\|G\|_{q,Q} + \|F_0\|^{\frac{m}{q}}_{1,Q} + \|\Phi_0\|^{\frac{l}{q}}_{1,\Gamma}\} \leqslant t_0, \qquad t_0 = t_0(n, m_0, q, a, b) \geqslant 1. \end{aligned}$$

Hence for any cube D_k

$$\fint_{D_k} (G\zeta)^q\,dx \leqslant t_0^q.$$

Now fix $t > t_0$ and $s = \lambda t$, $\lambda > 1$, to be chosen later. Applying Lemma 1 to the function $f = (G\zeta)^q$, $Q_0 = D_k$, $\alpha = s^q$, we obtain a sequence of n-cubes $\{Q_k^j\}_{j\in\mathbb{N}}$ in D_k such that $G\zeta \leqslant s$ a.e. in $D_k \setminus \bigcup_j Q_k^j$ and

$$s^q < \fint_{Q_k^j} (G\zeta)^q\, dx \leqslant 2^n s^q. \tag{43}$$

Since C_k is a finite union of cubes D_k, we conclude that there exists a sequence of cubes (we keep the same notation $\{Q_k^j\}_{j\in\mathbb{N}}$ for it) such that $G\zeta \leqslant s$ a.e. in $C_k \setminus \bigcup_j Q_k^j$, and hence the inequalities (43) hold.

Hence for $E_s(G\zeta) = \{x \in Q | G\zeta > s\}$ we obtain

$$\int_{E_s(G\zeta)} (G\zeta)^q\, dx \leqslant \sum_{j,k} \int_{Q_k^j} (G\zeta)^q\, dx \leqslant \sum_{j,k} |Q_k^j| \fint_{Q_k^j} (G\zeta)^q\, dx \leqslant 2^n s^q \sum_{j,k} |Q_k^j|. \tag{44}$$

To estimate $\sum_{j,k} |Q_k^j|$, fix any $x \in Q_k^j$ and the ball $B_{r_k^j}(x) \supset Q_k^j$, $r_k^j = \sqrt{n} a_k^j$, where a_k^j is the side of Q_k^j. Then by using the left inequality (43) we derive

$$s^q < \fint_{Q_k^j} (G\zeta)^q\, dx \leqslant \widehat{C} |D_k|^q \fint_{B_{r_k^j}(x)} G^q\, dx, \qquad \widehat{C} = \widehat{C}(n, q), \tag{45}$$

and therefore

$$\frac{s^q \sqrt{\theta}}{\widehat{C} |D_k|^q} < \sqrt{\theta} \fint_{B_{r_k^j}(x)} G^q\, dx \leqslant \sqrt{\theta} \sup_R \fint_{B_R(x)} G^q\, dx < \sup_R \fint_{B_R(x)} G^q\, dx.$$

By the definition of the supremum, there exists a constant $\rho_0 > 0$ such that

$$\sqrt{\theta} \sup_R \fint_{B_R(x)} G^q\, dx < \fint_{B_{\rho_0}(x)} G^q\, dx. \tag{46}$$

Next we show that the inequality

$$a\rho_0 < \min\{\, 1/2^{k+1},\ R_0 \,\}$$

holds for suitable λ. Indeed,

$$\frac{s^q \sqrt{\theta}}{\widehat{C} |D_k|^q} < \fint_{B_{\rho_0}} G^q\, dx < \frac{1}{|B_{\rho_0}|} \int_Q G^q\, dx \leqslant \frac{\sqrt{\theta}}{|B_{\rho_0}|},$$

and hence $|B_{\rho_0}| < \widehat{C} |D_k|^q s^{-q}$ or $\rho_0 < (\frac{\widehat{C}}{\omega_n})^{\frac{1}{n}} 2^{-(m_0+k)q} \lambda^{-q/n}$ ($\widehat{C}$ is the constant from (45)). If λ is chosen from the inequality

$$a(\widehat{C}/\omega_n)^{\frac{1}{n}} 2^{-(m_0+k)q} \lambda^{-q/n} < \min\{\, 2^{-(k+1)},\ R_0 \,\},$$

then $a\rho_0 < \min\{2^{-(k+1)}, R_0\}$. From (46) we derive

$$\sqrt{\theta} \sup_R \fint_{B_R(x)} G^q\, dx < \sup_{R \leqslant \widehat{R}_a(x)} \fint_{B_R(x)} G^q\, dx, \tag{47}$$
$$\widehat{R}_a(x) = a^{-1} \min\{\, d(x),\ R_0 \,\}.$$

Taking the supremum over all $R > 0$ in the right-hand part of (40) and over $R \leqslant \widehat{R}_a(x)$ (for fixed x) in the left-hand part, we obtain

$$\sup_{R\leqslant \widehat{R}_a(x)} \fint_{B_R(x)} G^q\,dx \leqslant \theta \sup_R \fint_{B_R(x)} G^q\,dx + \sup_R\left\{ B\left(\fint_{B_R(x)} G\,dx\right)^q + R^{-n}\left(\int_{B_R(x)} F_0\,dx\right)^m + R^{-n}\left(\int_{\gamma_R(x)} \Phi_0\,d\mathscr{H}_{\bar{k}}\right)^l\right\} \tag{48}$$

From (47) and (48) we obtain

$$\sup_{R\leqslant \widehat{R}_a(x)} \fint_{B_R(x)} G^q\,dx \leqslant \frac{1}{1-\sqrt{\theta}} \sup_R\{\dots\} \tag{49}$$

where $\{\cdots\}$ is the same as in (48). From (45) for the chosen λ we obtain

$$ar_k^j < \min\{R_0,\ 2^{-(k+1)}\}$$

and hence

$$\fint_{B_{r_k^j}(x)} G^q\,dx \leqslant \sup_{R\leqslant \widehat{R}_a(x)} \fint_{B_R(x)} G^q\,dx.$$

From (45) and (49) we derive

$$s^q < \widehat{C}|D_k|^q \frac{1}{1-\sqrt{\theta}} \sup_R\{\cdots\}.$$

There exists an $R > 0$ such that

$$s^q < C_1|D_k|^q\left\{ B\left(\fint_{B_R(x)} G\,dx\right)^q + R^{-n}\left(\int_{B_R(x)} F_0\,dx\right)^m + R^{-n}\left(\int_{\gamma_R(x)} \Phi_0\,d\mathscr{H}_{\bar{k}}\right)^l\right\}, \tag{50}$$

where $C_1 = \widehat{C}/(1-\sqrt{\theta})$ (for $\theta = 0$ inequality (50) immediately follows from (40) and (45))

Next we use (50) to estimate $|B_R|$:

$$|B_R| < \frac{C_2|D_k|^q}{\lambda^q t_0^q}\{\|G\|_{q,Q}^q + \|F_0\|_{1,Q}^m + \|\Phi_0\|_{1,\Gamma}^l\},$$

and for sufficiently large λ this implies $R < \min\{1/2^{k+1}, \rho/2\}$ where ρ is the constant from the (L,ρ)-condition for Γ. Since $B_R(x)$ intersects at most with C_{k-1}, C_k, C_{k+1}, it follows from (50) that

$$s \leqslant C_3\left\{ \fint_{B_R(x)} G\zeta\,dx + R^{-n/q}\left(\int_{B_R(x)} F_0\zeta^{q/m}\,dx\right)^{m/q} + R^{-n/q}\left(\int_{\gamma_R(x)} \Phi_0\zeta^{q/l}\,d\mathscr{H}_{\bar{k}}\right)^{l/q}\right\},$$

and hence

$$\begin{aligned}\lambda t|B_R| \leqslant C_4\Bigg\{ t|B_R| &+ \int_{B_R\cap E_t(G\zeta)} G\zeta\,dx \\ &+ R^{n(1-\frac{1}{q})}\left(\int_{B_R\cap E_{t^{q/m}}(F_0\zeta^{q/m})} F_0\zeta^{q/m}\,dx\right)^{m/q} \\ &+ R^{n(1-1/q)}\left(\int_{\gamma_R\cap \mathbf{e}_{t^{q/l}}(\Phi_0\zeta^{q/l})} \Phi_0\zeta^{q/l}\,d\mathscr{H}_{\bar{k}}\right)^{l/q}\Bigg\},\end{aligned}$$

where $C_4 = C_4(n,q,a,\theta,L)$ and $\mathbf{e}_\tau(h) = \{\, x\in\Gamma \mid h(x) > \tau \,\}$. To estimate the surface integral we have used the (L,ρ)-condition: since $\gamma_R(x)\subset\gamma_{2R}(\widehat{x})$, $\widehat{x}$ being the nearest point to x on Γ, we have

$$\mathscr{H}_{\bar{k}}(\gamma_R(x)) \leqslant \mathscr{H}_{\bar{k}}(\gamma_{2R}(\widehat{x})) \leqslant L(2R)^{\bar{k}}, \quad 2R\leqslant\rho.$$

Increasing, if necessary, the value of λ, we get $\lambda > 2C_4$.

Denote $F_1 = F_0\zeta^{q/m}$, $\Phi_1 = \Phi_0\zeta^{q/l}$. Then the integral term with F_1 (see the previous inequality) can be estimated by the Cauchy inequality with parameter $\varepsilon > 0$:

$$R^{n\frac{q-1}{q}}\left(\int_{B_R\cap E_{t^{q/m}}(F_1)} F_1\,dx\right)^{m/q} \leqslant \frac{C_\varepsilon}{t^{q-1}}\left(\int_{B_R\cap E_{t^{q/m}}(F_1)} F_1\,dx\right)^m + \varepsilon t R^n.$$

A similar estimate holds for the term with Φ_1. Choose ε so that $2\varepsilon C_4/\omega_n = \lambda/4$. Then

$$\tag{51} \frac{\lambda}{4}t|B_R| \leqslant C_5\left\{ \int_{B_R\cap E_t(G\zeta)} G\zeta\,dx + \frac{1}{t^{q-1}}\left(\int_{B_R\cap E_{t^{q/m}}(F_1)} F_1\,dx\right)^m + \frac{1}{t^{q-1}}\left(\int_{\gamma_R\cap\mathbf{e}_{t^{q/l}}(\Phi_1)} \Phi_1\,d\mathscr{H}_{\bar{k}}\right)^l\right\}.$$

Inequality (50) gives an estimate of $|B_R(x)|$ for almost all $x\in Q_k^j$; the family of these balls covers the whole set $D = \bigcup_{j,k} Q_k^j$. By the well-known covering theorem ([**S**, Chapter I, §1]) there exists a disjoint sequence of balls $\{B_i\}$ such that

$$|D| \leqslant 5^n \sum_{i=1}^{\infty} |B_i|.$$

From (51) we get

$$|D| = \sum_{j,k} |Q_k^j| \leqslant C_6 \left\{ \frac{1}{t} \int_{E_t(G\zeta)} G\zeta \, dx + \frac{1}{t^q} \left(\int_{E_{t^{q/m}}(F_1)} F_1 \, dx \right)^m + \frac{1}{t^q} \left(\int_{\mathbf{e}_{t^{q/l}}(\Phi_1)} \Phi_1 \, d\mathscr{H}_{\bar{k}} \right)^l \right\}. \tag{52}$$

By (44) we estimate $\int_{E_t(G\zeta)} (G\zeta)^q \, dx$:

$$\begin{aligned}\int_{E_t(G\zeta)} (G\zeta)^q \, dx &\leqslant \int_{E_s(G\zeta)} (G\zeta)^q \, dx + (\lambda t)^{q-1} \int_{E_t \setminus E_s} (G\zeta)^q \, dx \\ &\leqslant 2^n \lambda^q t^q \sum_{j,k} |Q_k^j| + \lambda^{q-1} t^{q-1} \int_{E_t(G\zeta)} G\zeta \, dx.\end{aligned}$$

The last inequality combined with (52) implies

$$\int_{E_t(G\zeta)} (G\zeta)^q \, dx \leqslant C_7 \left\{ t^{q-1} \int_{E_t(G\zeta)} G\zeta \, dx + \left(\int_{E_{t^{q/m}}(F_1)} F_1 \, dx \right)^m + \left(\int_{\mathbf{e}_{t^{q/l}}(\Phi_1)} \Phi_1 \, d\mathscr{H}_{\bar{k}} \right)^l \right\}, \quad \forall t \geqslant t_0. \tag{53}$$

Denote

$$h(t) = \int_{E_t(G\zeta)} G\zeta \, dx, \quad H_1(t) = \int_{E_{t^{q/m}}(F_1)} F_1 \, dx, \quad H_2(t) = \int_{\mathbf{e}_{t^{q/l}}(\Phi_1)} \Phi_1 \, d\mathscr{H}_{\bar{k}}.$$

By a well-known property of Lebesgue integrals,

$$\int_{E_t(G\zeta)} (G\zeta)^q \, dx = -\int_t^\infty \tau^{q-1} dh(\tau).$$

So (53) takes the form

$$-\int_t^\infty \tau^{q-1} dh(\tau) \leqslant C_7 \{ t^{q-1} h(t) + H_1(t)^m + H_2(t)^l \}, \quad t \geqslant t_0.$$

It is clear that h, H_1 and H_2 satisfy the conditions of Lemma 2 with $r = q-1$, $m = 2$, $\alpha_1 = \alpha_2 = 0$, $\beta_1 = m$, $\beta_2 = l$, $b = C_7$. Hence for all $\varkappa \in [q-1, \frac{C_7(q-1)}{C_7-1})$ we have

$$-\int_{t_0}^\infty t^{\varkappa} dh(t) \leqslant C_8 \left[-\int_{t_0}^\infty t^{q-1} dh(t) + H_1^{m-1}(t_0) \left(-\int_{t_0}^\infty t^{\varkappa-q+1} dH_1(t) \right) + H_2^{l-1}(t_0) \left(-\int_{t_0}^\infty t^{\varkappa-q+1} dH_2(t) \right) \right]. \tag{54}$$

Denote $p = \varkappa + 1 > q$, $\widetilde{H}_1(\tau) = \int_{E_\tau(F_1)} F_1 \, dx = H_1(t)$, $\tau = t^{q/m}$. Then

$$\begin{aligned}-\int_{t_0}^\infty t^{p-q} dH_1(t) &= -\int_{\tau_0}^\infty \tau^{\frac{m}{q}(p-q)} d\widetilde{H}_1(\tau) \\ &= \int_{E_{\tau_0}(F_1)} F_1^{\frac{m}{q}(p-q)+1} \, dx, \qquad \tau_0 = t_0^{\frac{q}{m}}.\end{aligned}$$

Similar calculations show that

$$-\int_{t_0}^{\infty} t^{p-q}\,dH_2(t) = \int_{\mathbf{e}_{t_0^{q/l}}(\Phi_1)} \Phi_1^{\frac{l}{q}(p-q)+1}\,d\mathscr{H}_{\bar{k}}.$$

So (54) takes the form

$$\int_{E_{t_0}(G\zeta)} (G\zeta)^p\,dx \leqslant C_8\Biggl\{\int_{E_{t_0}(G\zeta)} (G\zeta)^q\,dx + \Biggl(\int_{E_{t_0^{q/m}}(F_1)} F_1\,dx\Biggr)^{m-1}\int_{E_{t_0^{q/m}}(F_1)} F_1^{\frac{m}{q}(p-q)+1}\,dx + \Biggl(\int_{\mathbf{e}_{t_0^{q/l}}(\Phi_1)} \Phi_1\,d\mathscr{H}_{\bar{k}}\Biggr)^{l-1}\int_{\mathbf{e}_{t_0^{q/l}}(\Phi_1)} \Phi_1^{\frac{l}{q}(p-q)+1}\,d\mathscr{H}_{\bar{k}}\Biggr\}.$$

Hence

$$\int_Q (G\zeta)^p\,dx \leqslant C_g\Biggl\{\int_Q (G\zeta)^q\,dx + \Biggl(\int_Q F_1^{\frac{m}{q}(p-q)+1}\,dx\Biggr)^{\frac{mp}{q[\frac{m}{q}(p-q)+1]}} + \Biggl(\int_\Gamma \Phi_1^{\frac{l}{q}(p-q)+1}\,d\mathscr{H}_{\bar{k}}\Biggr)^{\frac{lp}{q[\frac{l}{q}(p-q)+1]}}\Biggr\}.$$

Recall that $F_1 = F_0\zeta^{q/m}$ and $\Phi_1 = \Phi_0\zeta^{q/l}$. By the properties of ζ we derive

$$\|G\|_{p,Q'}^p \leqslant C_{10}\Bigl\{\|G\|_{q,Q}^q + \|F_0\|_{1+\frac{m}{q}(p-q),Q}^{\frac{mp}{q}} + \|\Phi_0\|_{1+\frac{l}{q}(p-q),\Gamma}^{\frac{lp}{q}}\Bigr\}. \tag{55}$$

Inequality (55) implies the validity of estimate (11) for g, F, and Φ, so Theorem 2 is proved.

References

[A1] A. A. Arkhipova, *Reverse Hölder inequalities with the surface integrals and L_p-estimates in Neumann-type problems*, Embedding Theorems and Their Applications to Problems of Mathematical Physics (S. K. Godunov, editor), Inst. Mat. Sibirsk. Otdel. Akad. Nauk SSSR, Novosibirsk, 1989, pp. 3–17. (Russian)

[A2] ______, *Some applications of reverse Hölder inequalities with the surface integrals*, Problemy Mat. Anal., vyp. 12, St. Peterburg. Univ., St. Petersburg, 1992, pp. 13–28; English transl., J. Math. Sci. (to appear).

[A3] ______, *Reverse Hölder inequalities in parabolic problems with anisotropic space data*, Some Applications of Functional Analysis to Problems of Mathematical Physics (S. K. Godunov, editor), Inst. Mat. Sibirsk. Otdel. Akad. Nauk SSSR, Novosibirsk, 1992, pp. 3–22. (Russian)

[A4] ______, *L_p-estimates for the gradients of solutions in boundary-value problems for parabolic quasilinear systems*, Problemy Mat. Anal., vyp. 13, St. Peterburg. Univ., St. Petersburg, 1992, pp. 5–18; English transl., J. Math. Sci. (to appear).

[A5] ______, *Partial regularity of the solutions of quasilinear elliptic systems with nonsmooth conormal derivative.*, Mat. Sb. **184** (1993), no. 2, 87–104; English transl. in Math. USSR-Sb. **78** (1994).

[A6] ______, *On the Neumann problem for the quasilinear parabolic systems under controllable growth conditions.* Part I: *L_p-regularity results.* Part II: *Partial Hölder continuity of the solution*, Preprints.

[BI] B. V. Bojarski and T. Iwaniec, *Analytical foundations of the theory of quasi-conformal mappings in $\mathbb{R}^n$*, Ann. Acad. Sci. Fenn. Ser A I Math. **8** (1983), 257–324.

[C] A. Canale, *A higher integrability theorem from reverse Hölder inequalities with different*, Boll. Un. Mat. Ital. A (7) **5** (1991), no. 2, 137–146.

[C1] S. Campanato, *Hölder continuity of the solutions of some nonlinear elliptic systems*, Adv. Math. **48** (1983), 16–43.

[C2] ———, *A maximum principle for nonlinear elliptic systems: boundary fundamental estimates*, Adv. Math. **66** (1987), 291–317.

[C3] ———, *$\mathscr{L}^{2,\lambda}$-theory for nonlinear nonvariational differential systems*, Rend. Mat. Appl. (7) **10** (1990), 531–550.

[C4] ———, *On the nonlinear parabolic systems in divergence form. Hölder continuity and partial Hölder continuity of the solutions*, Ann. Mat. Pura Appl. (4) **137** (1984), 83–122.

[C5] ———, *L^p-regularity and partial Hölder continuity for solutions of second order parabolic systems with strictly controlled growth*, Ann. Mat. Pura Appl. (4) **128** (1981), 287–316.

[C6] ———, *L^p-regularity for weak solutions of parabolic systems*, Ann. Scuola Norm Sup. Pisa Cl. Sci. (4) **7** (1980), 65–85.

[C7] ———, *Partial Hölder continuity of solutions of quasilinear parabolic systems of second order with linear growth*, Rend. Sem. Mat. Univ. Padova **64** (1981), 59–75.

[CGP] F. Colombini, E. De Giorgi, and F. Piccinini, *Frontiere orientate di misura minima e questioni collegate*, Scuola Norm. Sup., Pisa, 1972.

[Ca] A. Canfora, *Teorema del massimo modulo e teorema di esistenza per il problema di Dirichlet relativo ai sistemi fortemente ellittici*, Recerche Mat. **15** (1966), 249–294.

[Can] P. Cannarsa, *On a maximum principle for elliptic systems with constant coefficients*, Rend. Sem. Mat. Univ. Padova **64** (1981), 77–92.

[F] M. Fuchs, *Some remarks on the boundary regularity for minima of variational problems with obstacles*, Manuscripta Math. **54** (1985), 107–119.

[G] E. Giusti, *Regolarità parziale delle soluzione di sistemi ellittici quasi-lineari di ordine abitrario*, Ann. Scuola Norm. Sup. Pisa (3) **23** (1969), 115–141.

[Ge] F. W. Gehring, *L_p-integrability of the partial derivatives of a quasi conformal mapping*, Acta Math. **130** (1973), 265–277.

[Gi1] M. Giaquinta, *The problem of the regularity of minimizers*, Proc. Internat. Congr. Math. (Berkeley, CA, 1986), Vol. 2, Amer. Math. Soc., Providence, RI, 1987, pp. 1072–1083.

[Gi2] ———, *Quasiconvexity, growth conditions and partial regularity,*, Partial Differential Equations and Calculus of Variations, Lecture Notes in Math., vol. 1357, Springer-Verlag, Berlin, 1988, pp. 211–237.

[Gi3] ———, *Multiple integrals in the calculus of variations and nonlinear elliptic systems*, Princeton Univ. Press, Princeton, NJ, 1983.

[Gi4] ———, *A counter-example to the boundary regularity of solutions to elliptic quasilinear systems*, Manuscripta Math. **26** (1978), 217–220.

[Gi5] ———, *The regularity problem of extremals of variational integrals*, Systems of Nonlinear Partial Differential Equations (J. M. Ball, editor), NATO Adv. Sci. Inst. Ser. C: Math. Phys. Sci., vol. 111, Reidel, Dordrecht, 1983, pp. 115–145.

[GiMo1] M. Giaquinta and G. Modica, *Regularity results for some classes of higher order nonlinear elliptic systems*, J. Reine Angew. Math. **311/312** (1979), 145–169.

[GiMo2] ———, *Nonlinear systems of the type of the stationary Navier–Stokes systems*, J. Reine Angew. Math. **330** (1982), 173–214.

[GiMo3] ———, *Local existence for quasilinear parabolic systems under nonlinear boundary conditions*, Ann. Mat. Pura Appl. (4) **149** (1987), 41–59.

[GiSo] M. Giaquinta and Y. Souček, *Caccioppoli's inequality and Legendre–Hadamard condition*, Math Ann. **270** (1985), 105–107.

[GiS] M. Giaquinta and M. Struwe, *On the partial regularity of weak solutions of nonlinear parabolic systems*, Math. Z. **179** (1982), 437–451.

[GM] E. Giusti and M. Miranda, *Sulla regolarità delle soluzioni deboli di una classe di sistemi ellittici quasi-lineari*, Arch. Rational Mech. Anal. **31** (1968), 173–184.

[HW] S. Hildebrandt and K.-O. Widman, *Some regularity results for quasilinear elliptic systems of second order*, Math. Z. **142** (1975), 67–86.

[I] T. Iwaniec, *On L_p-integrability in PDE's and quasiregular mappings for large exponents*, Ann. Acad. Sci. Fenn. Ser A I Math. **7** (1982), 301–322.

[JS] O. John and Y. Stara, *On the regularity and non-regularity of elliptic and parabolic systems*, Equadiff-7 (Proc. Conf., Prague, 1989; J. Kurzweil, editor), Teubner-Texte Math., vol. 118, Teubner, Leipzig, 1990, pp. 28–36.

[LSU] O. A. Ladyzhenskaya, V. A. Solonnikov, and N. N. Ural′tseva, *Linear and quasilinear equations of parabolic type*, "Nauka", Moscow, 1967; English transl., Amer. Math. Soc., Providence, RI, 1968.

[M] N. Meyers, *An L_p-estimate for the gradients of solutions of second-order elliptic divergence equations*, Ann. Scuola Norm. Sup. Pisa (3) **17** (1963), 189–206.

[Ma] O. Martio, *On the integrability of the derivative of a quasiregular mapping*, Math. Scand. **35** (1974), 43–48.

[Mau] A. Maugeri, *Regularity results for parabolic systems*, Equadiff-8 (Proc. Conf., Prague, 1991) (to appear).

[ME] N. Meyers and A. Elcrat, *Some results on regularity for solutions of nonlinear elliptic systems and quasi-regular functions*, Duke Math. J. **42** (1975), 121–136.

[Mo] C. Modica, *Qusiminimi di alcuni funzionali degeneri*, Ann. Mat. Pura Appl. (4) **142** (1985), 121–143.

[Mor] Ch. Morrey, *Partial regularity results for non-linear elliptic systems*, J. Math. Mech. **17** (1967/68), 649–670.

[Mu] B. Muckenhoupt, *Weighted norm inequalities for the Hardy maximal function*, Trans. Amer. Math. Soc. **165** (1972), 207–226.

[S] E. Stein, *Singular integrals and differentiability properties of functions*, Princeton Univ. Press, Princeton, NJ, 1970.

[Sb] C. Sbordone, *On some integral inequalities and their applications to the calculus of variations*, Boll. Un. Mat. Ital. A (7) **1** (1986), 73–94.

[SbF] C. Sbordone and N. Fusco, *Higher integrability from reverse Jensen inequalities with different supports*, Partial Differential Equations and the Calculus of Variations, Vol. II, Birkhäuser, Basel, 1989, pp. 541–562.

[SJM] J. Stará, O. John, and J. Malý, *Counterexample to the regularity of weak solution of the quasilinear parabolic system*, Comment. Math. Univ. Carolin. **27** (1986), 123–136.

[So] Y. Souček, *Singular solution to linear elliptic systems*, Comment Math. Univ. Carolin. **25** (1984), 273–281.

[St] E. W. Stredulinsky, *Higher integrability from reverse Hölder inequalities*, Indiana Univ. Math. J. **29** (1980), 408–417.

[V] E. Viszus, *On the regularity up to the boundary for higher order quasilinear elliptic systems*, Comment. Math. Univ. Carolin. **31** (1990), 295–306.

Department of Mathematics and Mechanics, St. Petersburg State University

Translated by M. PANKRATOV

Amer. Math. Soc. Transl.
(2) Vol. **164**, 1995

Quasilinear Parabolic Equations with Small Parameter in a Hilbert Space

Ya. Belopol′skaya

The main purpose of this paper is to study solutions of the Cauchy problem for the quasilinear parabolic equation

$$\frac{\partial u_\varepsilon}{\partial t} = \frac{1}{2}\operatorname{Tr} u''_\varepsilon(A_x, A_x) + (u'_\varepsilon, a_x) + \frac{1}{\varepsilon}c_x(u)u, \qquad u_\varepsilon(0, x) = f(x), \tag{0.1}$$

in an infinite-dimensional Hilbert space, and especially their behavior as $\varepsilon \longrightarrow 0$. In the finite-dimensional case the corresponding results have been obtained by M. Freidlin [**F1**]. In this paper we extend Freidlin's results to the infinite-dimensional case under more general assumptions as compared with [**F1**].

To investigate various problems concerning infinite-dimensional parabolic equations both in the whole space and in a bounded domain, the methods of the theory of stochastic differential equations turn out to be very effective. The theory of stochastic differential equations in infinite-dimensional spaces was developed by various authors. The pioneering results here are due to Yu. Daletskiĭ and his students [**D, BD1**]. In those papers it was demonstrated that infinite-dimensional linear parabolic equations and systems may be solved provided the corresponding diffusion processes could be constructed.

Starting from the papers by M. Freidlin [**F2**] and H. McKean [**McK**], it was revealed that a similar connection does exist between a class of diffusion processes and quasilinear parabolic equations. Later, Ya. Belopol′skaya and Yu. Daletskiĭ extended the results of [**F2**] to the finite-dimensional case and constructed diffusion processes associated with nonlinear parabolic equations and systems [**BD2, BD3, BD4**].

To deal with small parameter problems for quasilinear parabolic equations, a special approach was developed by M. Freidlin [**F3, GF**]. This approach is based on the large deviation theory for diffusion processes. Large deviation theory for this class of stochastic processes has been developed in its turn by A. Wentzell and M. Freidlin [**FW**], D. Stroock [**S**], S. Varadhan [**V**], and others. In this paper we extend the results of this theory to the infinite-dimensional case and apply them to investigate the solution of problem (0.1).

The first section of the paper deals with the large deviation principle for Wiener measure in Hilbert space. In the second section, the distribution of the solution of an infinite-dimensional stochastic differential equation is proved to obey the large

1991 *Mathematics Subject Classification*. Primary 35K55, 35R15.

0065-9290/95/$1.00 + $.25 per page

deviation principle. In the last two sections we deal with a probabilistic representation of the solution of a quasilinear parabolic equation and use it to investigate the behavior of $u_\varepsilon(t,x)$ as $\varepsilon \to 0$, applying the results of previous sections.

In particular, in the last section we prove that, in dealing with the solution of an infinite-dimensional quasilinear equation, one reveals a phenomenon that may be interpreted as a wave front spreading. This phenomenon may be easily illustrated in the case of a one-dimensional parabolic equation with constant coefficients. Consider the Cauchy problem

$$\begin{aligned} \frac{\partial v_\varepsilon}{\partial t} &= \frac{\varepsilon}{2}\sigma^2 \frac{\partial^2 v_\varepsilon}{\partial x^2} + a\frac{\partial v_\varepsilon}{\partial x} + \frac{1}{\varepsilon} c v_\varepsilon, \\ v_\varepsilon(0,x) &= \chi(x<0) = \begin{cases} 1, & x<0, \\ 0, & x>0. \end{cases} \end{aligned} \tag{0.2}$$

It is well known that the solution of (0.2) may be represented in the form

$$v_\varepsilon(t,\beta t) = e^{ct\varepsilon^{-1}} \int_{\beta t\varepsilon^{-1}}^{\infty} \frac{\varepsilon^{-1/2}}{\sqrt{2\pi\sigma^2 t}} e^{-\frac{(z-a)^2\varepsilon}{2\sigma^2 t}}\, dz,$$

which results in

$$\lim_{\varepsilon\to 0} v_\varepsilon(t,\beta t) = \begin{cases} 0 & \text{if } \beta > \sigma\sqrt{2c}, \\ \infty & \text{if } \beta < \sigma\sqrt{2c}. \end{cases}$$

This means that for small ε the velocity of propagation of the domain of large values of v_ε is asymptotically equal to $\sigma\sqrt{2c}$.

It will be shown that under certain assumptions on the coefficients and the Cauchy data, the function $u_\varepsilon(t,x)$ solving (0.1) behaves in a similar way. Namely, for ε small enough, u_ε is close to a step function $\varkappa(t,x)$ having two values 0 and 1. The set $F_t = \{x \in H : \varkappa(t,x) = 1\}$, being the support of the function $\varkappa$, may be used in the preliminary study to describe the evolution of the function $u_\varepsilon(t,x)$, while the boundary of this set may be interpreted as a wave front.

§1. Large deviation principle for the Wiener measure in Hilbert space

To start our investigation of the solutions of infinite-dimensional stochastic differential equations, let us recall some basic notions concerning the topics to be discussed.

Let H be a real separable Hilbert space, $\{e_k\}$ an orthonormal basis in H, $\|\cdot\|_H$ the norm generated by the inner product $(\cdot,\cdot)_H$, and $\mathfrak{B}_H$ a Borel σ-algebra in H. For a pair of Hilbert spaces H, B denote by $L_{12}(H,B)$ the Hilbert space of Hilbert-Schmidt operators mapping H into B with norm $\sigma^2(A) = \{\sum_{k=1}^\infty \|Ae_k\|_B^2\}^{1/2}$. Denote by $L_{12}(H,H)$ by $L_{12}(H)$. Consider a triple of Hilbert spaces $H_+ \subset H \subset H_-$ densely imbedded by means of a Hilbert-Schmidt imbedding operator J. Let $Z_+ \subset Z \subset Z_-$ be another triple of Hilbert spaces possessing a similar imbedding operator. Below we shall take for the Z-triple a triple of Sobolev spaces

$$Z_+ = W_2([0,T],H_+), \quad Z = L_2([0,T],H), \qquad Z_- = Z_+^*.$$

Put $\mathscr{H}_+ = Z_+ \otimes H_+$, $\mathscr{H} = Z \otimes H$, $\mathscr{H}_- = \mathscr{H}_+^*$, which means that

$$\mathscr{H}_+ = W_2([0,T],H_+), \qquad \mathscr{H} = L_2([0,T],H).$$

Given a canonical Gaussian measure ν on the space $(\mathscr{H}_-, \mathfrak{B}_{\mathscr{H}_-})$, let us define some random variables on the probability space $(\mathscr{H}_-, \mathfrak{B}_{\mathscr{H}_-}, \nu)$.

Let $\mathrm{id}_{\mathscr{H}_-}$ denote the identity operator in $\mathscr{H}_-$ and $\varkappa$ the random variable generated by the mapping $\mathrm{id}_{\mathscr{H}_-}$. Consider the mapping $S_\alpha\colon \mathscr{H} \longrightarrow H$, $\alpha \in Z$, defined by the relation $(\varphi, S_\alpha h)_H = (\alpha \otimes \varphi, h)_{\mathscr{H}}$, $\varphi \in H$. The obvious estimate $\|S_\alpha\| \leqslant \|\alpha\|_Z$ shows that $S_\alpha \varkappa$ is an H_--valued random variable for any $\alpha \in Z$. Indeed the composition of S_α with the imbedding operator $J\colon H_0 \longrightarrow H_-$ defines an operator $JS_\alpha \in L_{12}(\mathscr{H}, H_-)$, which may be extended to the whole $\mathscr{H}_-$ as a measurable linear mapping. The parameters of the resulting random variable distribution are defined by the relations

$$\mathbf{E}(\varphi, S_\alpha \varkappa) = \int_{\mathscr{H}_-} (\varphi, S_\alpha \varkappa) \nu(d\varkappa) = 0,$$

$$\mathbf{E}(\varphi, S_\alpha \varkappa)(\psi, S_\beta \varkappa) = \int_{\mathscr{H}_-} (\varphi, S_\alpha \varkappa)(\psi, S_\beta \varkappa) \nu(d\varkappa) = (\varphi, \psi)_H (\alpha, \beta)_Z.$$

Thus, the linear continuous map $\alpha \mapsto S_\alpha$ from Z into a set of Gaussian random variables, canonical with respect to $H_+ \subset H \subset H_-$ and having parameters $(0, \|\alpha\|_Z^2)$, corresponds to the rigged space $\mathscr{H}_+ \subset \mathscr{H} \subset \mathscr{H}_-$.

The random variable $\varkappa$ is called a *white noise* corresponding to the Hilbert space H, while $(\mathscr{H}_-, \mathfrak{B}_{\mathscr{H}_-}, \nu)$ is called a *Wiener space*. Notice that if $w(t)$ is a Wiener process valued in H, then $(h, \varkappa) = \int_0^T ((h(t), dw)_H$ for any $h \in \mathscr{H}$. Moreover, if $F\colon \mathscr{H}_- \longrightarrow L_{12}(\mathscr{H}_-, H)$, then $\big(F(\varkappa)\varkappa, \varphi\big) = \int_0^T \big(\varphi(t), F(t, \varkappa)\big)\, dt$.

Consider $Q \in L_{12}(\mathscr{H})$ and represent it in the form

$$(Q\varphi)(t) = \int_0^t q(t, \tau) \varphi(\tau)\, d\tau,$$

where $q(t, \tau)$ is an L_{12}-valued kernel. Notice that

$$\sigma_2^2(Q) = \int_0^T \int_0^T \sigma_2^2\big(q(t, \tau)\big)\, d\tau\, dt.$$

A measurable mapping $h\colon \mathscr{H}_- \longrightarrow H$ is called *nonanticipated* if for any $y \in \mathscr{H}_-$ with a support $\operatorname{supp} y \in [t, T]$

$$H(\varkappa + y, t) = h(\varkappa, t) \quad \nu\text{-a.e.}$$

In other words, h is nonanticipated if $h(t, x)$ does not depend on the values $\varkappa(\tau)$ for $\tau > t$. For the mapping $\Phi(\varkappa)$ defined by

$$\Phi(\varkappa) h = \int_0^T A(t, \varkappa) h(t)\, dt,$$

where $A\colon [0, T] \times \mathscr{H}_- \longrightarrow L_{12}(\mathscr{H})$ is a nonanticipated mapping, we shall write

$$\Phi(\varkappa)\varkappa = \int_0^T A(t, \varkappa)\, dw(t)$$

and call it a *stochastic integral.*

A function $I_T\colon \mathscr{H} \longrightarrow [0, \infty]$ is called a *rate function* if it possesses the following properties:

1. $I_T \not\equiv \infty$;
2. I_T is lower semicontinuous;
3. for any $K > 0$ the set $\{\, h \in \mathscr{H} : I_T(h) < K \,\}$ is compact.

A family μ_ε of probability measures defined on $(\mathscr{H}, \mathfrak{B}_{\mathscr{H}})$ is said to satisfy the *large deviation principle* with rate I_T if

$$\overline{\lim_{\varepsilon\to 0}}\, \varepsilon \ln \mu_\varepsilon(Q) \leqslant -\inf_{\varphi\in Q} I_T(\varphi)$$

for all closed sets in $\mathscr{H}$ and

$$\underline{\lim_{\varepsilon\to 0}}\, \varepsilon \ln \mu_\varepsilon(G) \geqslant -\inf_{\varphi\in G} I_T(\varphi)$$

for all nonempty open G in $\mathscr{H}$.

Let us prove that the function

$$I_T(\varphi) = \begin{cases} \frac{1}{2}\int_o^T \|A^{-1}\dot{\varphi}(t)\|^2\,dt, & \varphi(0)=0, \quad \varphi\in\mathscr{H}_A = A\mathscr{H}, \\ \infty, & \text{otherwise}, \end{cases} \tag{1.1}$$

may serve for a rate function in the large deviation principle for distributions of the stochastic process $\xi_\varepsilon(t)$ of the form

$$\xi_\varepsilon(t) = \varepsilon^{1/2}\int_0^t A(s)\,dw(s), \tag{1.2}$$

where $A \in L_{12}(H)$.

THEOREM 1.1. *The family* $\mu_\varepsilon = \mathbf{P}\circ\xi_\varepsilon^{-1}$ *satisfies the large deviation principle with rate function* I_T *of the form* (1.1).

The proof will be presented as a series of lemmas.

LEMMA 1.1. *The function* I_T *of the form* (1.1) *is a rate function.*

PROOF. Let us check that the function I_T is lower semicontinuous with respect to the norm $\|\psi\| = \sup_{0\leqslant t\leqslant T}\|\psi(t)\|$ and that for any $K>0$ the set $\{\,\psi\in\mathscr{H} : I_T(\psi)<K\,\}$ is compact.

First, note that the lower semicontinuity of $I_T(\psi)$ easily follows from the relation

$$2I_T(\psi) = \sup_n n\sum_{k=1}^n \|A^{-1}[\psi(\frac{k}{n}) - \psi(\frac{k-1}{n})]\|^2, \tag{1.3}$$

if $\psi(0)=0$, because $\{\,\psi\in\mathscr{H} : \psi(0)=0\,\}$ is closed and the variable

$$n\sum_{k=1}^n \|A^{-1}[\psi(\frac{k}{n}) - \psi(\frac{k-1}{n})]\|^2$$

is continuous for each n with respect to ψ.

To prove (1.3), consider the set $M = \{\,\psi\in\mathscr{H} : I_T(\psi)<\infty\,\}$ and put

$$\alpha_n(\psi) = n\sum \|A^{-1}[\psi(\frac{k}{n}) - \psi(\frac{k-1}{n})]\|^2.$$

Notice that the conditions $\psi(0)=0$ and $\alpha(\psi) = \sup_n \alpha_n(\psi) < \infty$ imply $\psi\in M$.

Indeed, choose $\psi^{(n)}$ to be the polygonal interpolation between points k/n for the curve ψ. Then

$$\alpha_n(\psi) = \int_0^T \|A^{-1}\dot{\psi}^{(n)}(t)\|^2\,dt.$$

Hence, $\alpha(\psi) < \infty$ implies

$$\sup_n \int_0^T \|A^{-1}\dot{\psi}(t)\|^2\,dt = Q < \infty.$$

Thus, for any $\varphi \in \mathscr{H}_A \cap \mathscr{H}_+$,

$$\begin{aligned}\int_0^T \big(A^{-1}\dot{\psi}(t), A^{-1}\varphi(t)\big)\,dt &= \lim_{n\to\infty}\int_0^T \big(A^{-1}\psi^{(n)}(t), A^{-1}\dot{\varphi}(t)\big)\,dt\\ &\leqslant \lim_{n\to\infty}\int_0^T \big(A^{-1}\dot{\psi}^{(n)}(t), A^{-1}\varphi(t)\big)\,dt\\ &\leqslant Q^{1/2}\left[\int_0^T \|A^{-1}\varphi(t)\|^2\,dt\right]^{1/2}.\end{aligned}$$

Finally, $\psi \in \mathscr{H}_A$ since ψ has a distributional derivative. It follows from the Schwarz inequality that $\alpha_n(\psi) \leqslant 2I_T(\psi)$ for $\psi \in M$ and arbitrary n. By the triangle inequality,

$$|\alpha_n^{1/2}(\psi) - \alpha_n^{1/2}(\varphi)| \leqslant \big[\alpha_n(\psi - \varphi)\big]^{1/2}$$

for any $\varphi, \psi \in M$ and $n > 0$. In particular, $\alpha(\psi_n) \to \alpha(\psi)$ if $\psi_n, \psi \in M$ and $I_T(\psi_n - \varphi) \to 0$ as $n \to \infty$. Thus, we must prove (1.3) only for a set of functions which is dense in $\mathscr{H}_A$ in the topology associated with I_T. As a matter of fact, one may take $\psi \in C^\infty([0,T], H_+)$ with $\psi(0) = 0$. For such a function

$$2I_T(\psi) \geqslant \alpha(\psi) \geqslant \lim_{n\to\infty} n\sum_{k=1}^{n}\Big\|A^{-1}\Big[\psi\Big(\frac{k}{n}\Big) - \psi\Big(\frac{k-1}{n}\Big)\Big]\Big\|^2 = 2I_T(\psi). \tag{1.4}$$

It follows from Lemma 1.1 that, given a closed nonempty set $Q \in \mathfrak{B}_{\mathscr{H}}$, one has

$$\inf_{\psi\in Q^\delta} I_T(\psi) \to \inf_{\psi\in Q} I_T(\psi) \tag{1.5}$$

as $\delta \to 0$. Here $Q^\delta = \{\psi : (\exists\psi' \in Q) : \|\psi - \psi'\| < \delta\}$. Indeed, to prove (1.5), put $i_\delta = \inf_{\psi\in Q^\delta} I_T(\psi)$, $i = \inf_{\psi\in Q} I_T(\psi)$. It is obvious that i_δ increases as $\delta \to 0$. Assume that $i_\delta \leqslant i' \leqslant i$ for any $\delta \geqslant 0$. One may choose $\psi_n \in Q^{1/n}$ in such a way that $I_T(\psi_n) \leqslant i_{1/n} + \frac{1}{n} \leqslant i' + \frac{1}{n}$. It has been proved in Lemma 1.1 that one may choose a convergent consequence $\psi_{n'}$ of ψ_n as well. Finally choose ψ such that $\|\psi_{n'} - \psi\| \to 0$. Obviously, $\psi \in Q$ while $I_T(\psi) \leqslant \lim I_T(\psi_{n'}) \leqslant i' \leqslant i$. Thus, if (1.5) is not valid, we obtain a contradiction.

The second assertion easily results from the Arzelà-Ascoli theorem and the fact that $I_T(\varphi)$ is lower semicontinuous. One may apply the Arzelà-Ascoli theorem, since $\{\varphi \in \mathscr{H}_A : \varphi = A\psi,\ \psi \in \mathscr{H}\}$ is compact and

$$|\varphi(t_2) - \varphi(t_1)| \leqslant \{2I_T(\varphi)\}^{1/2}(t_2 - t_1)^{1/2}.$$

□

Consider the stochastic process $\xi_\varepsilon(t)$ and let μ_ε be its distribution.

LEMMA 1.2. *Let* $s_0, \delta, \gamma > 0$. *Then*

$$\mathbf{P}\{\|\xi_\varepsilon(\cdot) - \varphi\| < \delta\} \geqslant \exp\{-\varepsilon^{-1}(I_T(\varphi) + \gamma)\}$$

for $\varepsilon > 0$ *small enough uniformly with respect to* φ, *satisfying the estimate* $I_T(\varphi) \leqslant s < \infty$. *If* $\Phi(s) = \{\varphi \in \mathscr{H} : I_T(\varphi) < s\}$, *then*

$$\mathbf{P}\{\|\xi_\varepsilon(\cdot) - \Phi(s)\| \geqslant \delta\} \leqslant \exp\{-\varepsilon^{-1}(s - \gamma)\}$$

for $s < s_0$.

PROOF. Let $I_T(\varphi) \leqslant K < \infty$; that is, $\varphi \in M$. Consider a process $\eta_\varepsilon(t)$ of the form $\eta_\varepsilon(t) = \xi_\varepsilon(t) - \varphi(t)$. This process has a distribution in $\mathscr{H}$ absolutely continuous with respect to the distribution of $\xi_\varepsilon(t)$, and the density may be represented in the form

$$\begin{aligned}(1.6)\qquad &\frac{d\mu_\eta}{d\mu_\xi}(\varepsilon^{1/2}Aw)\\ &\quad= \exp\left\{-\varepsilon^{-1/2}\int_0^T (A^{-1}\dot\varphi(s), dw) - \frac{\varepsilon^{-1}}{2}\int_0^T \|A^{-1}\dot\varphi(s)\|^2\, ds\right\}\end{aligned}$$

due to the Girsanov formula [**BD1**]. It follows that

$$\begin{aligned}\mathbf{P}\{\|\xi_\varepsilon(\cdot) - \varphi(\cdot)\| < \delta\} &= \mathbf{P}\{\|\eta_\varepsilon(\cdot)\| < \delta\}\\ &= \exp\left\{-\frac{\varepsilon^{-1}}{2}\int_0^T \|A^{-1}\dot\varphi(s)\|^2\, ds\right\}\\ &\quad\times \int_{\|\xi_\varepsilon(\cdot)\|<\delta} \exp\left\{\varepsilon^{-1/2}\int_0^T (A^{-1}\dot\varphi(s), dw)\right\}\mathbf{P}(d\omega).\end{aligned}$$

It is easy to check that $\mathbf{P}\{\|\xi_\varepsilon(\cdot)\| < \delta\} \to 1$ uniformly with respect to $T \leqslant \theta$. In particular, by the Chebyshev inequality,

$$\begin{aligned}(1.7)\qquad \mathbf{P}\{\|\xi_\varepsilon(\cdot)\| < \delta\} &= \mathbf{P}\left\{\|\xi(\cdot)\| < \frac{\delta^2}{\varepsilon}\right\}\\ &\geqslant 1 - \mathbf{E}\|\xi(T)\|^2\frac{\varepsilon}{\delta^2} = 1 - \varepsilon\delta^{-2}\operatorname{Tr} A^*AT \geqslant \frac{3}{4}\end{aligned}$$

if $\varepsilon < \delta^2/4T\sigma_2^2(A)$. Moreover, using the Chebyshev inequality once again, one has

$$\begin{aligned}&\mathbf{P}\left\{-\varepsilon^{-1/2}\int_0^T (A^{-1}\dot\varphi(s), dw) \leqslant -2\sqrt{2}\varepsilon^{-1/2}\sqrt{I_T(\varphi)}\right\}\\ &\quad\leqslant \mathbf{P}\left\{\left|\varepsilon^{-1/2}\int_0^T (A^{-1}\dot\varphi(s), dw)\right| \geqslant 2\sqrt{2}\varepsilon^{-1/2}\sqrt{I_T(\varphi)}\right\}\\ &\quad\leqslant \frac{\varepsilon^{-1}\mathbf{E}\left(\int_0^T (A^{-1}\dot\varphi(s), dw)\right)^2}{8\varepsilon^{-1}I_T(\varphi)} = \frac{1}{4},\end{aligned}$$

which leads to the estimate

$$(1.8)\qquad \mathbf{P}\left\{\exp\left\{-\varepsilon^{-1/2}\int_0^T (A^{-1}\dot\varphi(s), dw)\right\} \geqslant \exp\{-2\sqrt{2}\varepsilon^{-1/2}\sqrt{I_T(\varphi)}\}\right\} \geqslant \frac{3}{4}.$$

It follows from (1.7), (1.8) that

$$\int_{\|\xi_\varepsilon\|<\delta} \exp\Big\{-\varepsilon^{-1/2}\int_0^T (A^{-1}\dot\varphi(s), dw)\Big\}\mathbf{P}(d\omega) > \frac{1}{2}\exp\Big\{-2\sqrt{2}\varepsilon^{-1/2}\sqrt{I_T(\varphi)}\Big\}$$

and

$$\mathbf{P}\{\|\eta_\varepsilon\| < \delta\} > \frac{1}{2}\exp\{-\varepsilon^{-1}I_T(\varphi) - 2\sqrt{2}\varepsilon^{-1/2}\sqrt{I_T(\varphi)}\}.$$

To prove the second assertion of the lemma, we must estimate the probability of the following event: {trajectories of the process $\xi_\varepsilon(t)$ are far enough from the set $\Phi(s)$}.

Consider an orthonormal basis $\{e_k\}$ in H consisting of eigenfunctions of the operator $B = A^*A$, and denote by $\{\lambda_k\}$ the corresponding eigenvalues. Let ξ_k denote the coordinates of ξ in the basis $\{e_k\}$, $\xi_k = (\xi, e_k)$.

Consider a finite-dimensional vector $\widetilde{\xi^n} = (\xi_1, \xi_2, \dots, \xi_n, 0, \dots)$. It is easy to check that

$$\mathbf{P}\{\|\xi_\varepsilon - \Phi(s)\| \geqslant \delta\} \leqslant \mathbf{P}\{\widetilde{\xi^n_\varepsilon} \notin \Phi(s)\} + \mathbf{P}\{\|\xi_\varepsilon - \widetilde{\xi^n_\varepsilon}\| \geqslant \delta\}. \tag{1.9}$$

For the first term in the right hand side of (1.9) one has

$$\begin{aligned}\mathbf{P}\{\widetilde{\xi^n_\varepsilon} \notin \Phi(s)\} &= \mathbf{P}\{I_T(\widetilde{\xi^n_\varepsilon}) > s\} = \mathbf{P}\{I_T(\widetilde{\xi^n}) > s\varepsilon^{-1}\}\\ &= \mathbf{P}\Big\{\sum_{i=1}^n \lambda_i^{-1}(\xi_i)^2 > 2s\varepsilon^{-1}\Big\}.\end{aligned}$$

Note that the ξ_i are independent random variables having normal distributions with parameters $(0,1)$. Applying the exponential Chebyshev inequality, we get

$$\begin{aligned}\mathbf{P}\{\widetilde{\xi^n_\varepsilon} \notin \Phi(s)\} &= \mathbf{P}\Big\{\sum_{i=1}^n \lambda_i^{-1}(\xi_i)^2 > 2s\varepsilon^{-1}\Big\} \leqslant \frac{\mathbf{E}\exp\frac{1-\alpha}{2}\sum\lambda_i^{-1}(\xi_i)^2}{\exp\{\varepsilon^{-1}s(1-\alpha)\}}\\ &= \frac{\prod_{i=1}^n \mathbf{E}\exp\frac{1-\alpha}{2}\lambda_i^{-1}(\xi_i)^2}{\exp\{\varepsilon^{-1}s(1-\alpha)\}} = \prod_{i=1}^n (1-(1-\alpha))\exp\{-\varepsilon^{-1}s(1-\alpha)\}\end{aligned}$$

since the ξ_i are independent. The last equality is valid for $0 < \alpha < 1$. Hence one can choose N, α_0 so that, for $n < N$ and $\alpha > \alpha_0$,

$$\mathbf{P}\{\widetilde{\xi^n_\varepsilon} \notin \Phi(s)\} \leqslant C\exp\{-\varepsilon^{-1}s(1-\alpha)\}.$$

To estimate the second term in the right-hand side of (1.9), denote by $a(s,t)$ the kernel of the operator A and represent A in the form

$$(A\varphi)(t) = \int_0^T a(s,t)\varphi(s)\,ds.$$

Let $a_n(s,t)$ be a symmetric positive definite kernel continuously differentiable in the square $[0,T]\times[0,T]$ and satisfying the relation

$$\int_0^T\int_0^T \sigma_2^2(a(s,t) - a_n(s,t))\,ds\,dt < \frac{1}{n}.$$

Denote by A_n the corresponding operator. For a symmetric square-integrable kernel $b_n(s,t) = a(s,t) - a_n(s,t)$, denote by $e_k(s)$ the eigenfunctions and by μ_k the corresponding eigenvalues of B_n. In this case one has the following easily verified relations:

$$b_n(s,t) = \sum_{k=1}^{\infty} \mu_k e_k(s) \otimes e_k(t),$$

$$\sum \mu_k^2 = \int_0^T \int_0^T \sigma_2^2(b_n(s,t))\, ds\, dt,$$

$$\int_0^T b_n(s,t) dw(s) = \sum \mu_k e_k(t) \int_0^T (e_k(s), dw(s)),$$

$$\int_0^T \left\| \int_0^T b_n(t,s)\, dw \right\|^2 dt = \sum \mu_k^2 \left| \int_0^T (e_k(s), dw(s)) \right|^2.$$

The random variables $\xi_k = \int_0^T (e_k(s), dw(s))$ possess normal distributions with parameters $(0, 1)$, and they are independent for different k. Keeping the above relations in mind, let us show that

$$\begin{aligned} \gamma &= \mathbf{E}\exp(\alpha\|\xi - \widetilde{\xi}^n\|^2) = \mathbf{E}\exp\left(\alpha \int_0^T \left\| \int_0^T b_n(s,t)\, dw \right\|^2 dt \right) \\ &= \mathbf{E}\exp(\alpha \sum \mu_k^2 \xi_k^2) = \prod_{k=1}^{\infty} \mathbf{E}\exp(\alpha\mu_k^2\xi_k^2) = \prod_{k=1}^{\infty} (1 - 2\alpha\mu_k^2)^{-1/2}. \end{aligned} \tag{1.10}$$

The last inequality is valid if $2\alpha\mu_k^2 < 1$ for each k. Since

$$\sum \mu_k^2 = \int_0^T \int_0^T \sigma_2^2(b_n(s,t))\, ds\, dt \to 0$$

as $n \to \infty$, (1.10) is valid for any $n > N = N(\alpha)$. The convergence of the series $\sum \mu_k^2$ results from the convergence of the infinite product in (1.10). Thus, if $n > N$, the value in the right-hand side of (1.10) is finite and so one gets $\gamma < \infty$. Now to derive the estimate for $\mathbf{P}\{\|\xi_\varepsilon - \widetilde{\xi}_\varepsilon^n\| \geqslant \delta\}$, one may use the exponential Chebyshev inequality. It implies the estimate

$$\begin{aligned} \mathbf{P}\{\|\xi_\varepsilon - \widetilde{\xi}_\varepsilon^n\| \geqslant \delta\} &= \mathbf{P}\{\|\xi - \widetilde{\xi}^n\|^2 \geqslant \delta^2/\varepsilon\} = \mathbf{P}\{\exp\alpha\|\xi - \widetilde{\xi}^n\|^2 \geqslant e^{\alpha\delta^2\varepsilon^{-1}}\} \\ &\leqslant e^{-\alpha\delta^2\varepsilon^{-1}} \mathbf{E}\exp(\alpha\|\xi - \widetilde{\xi}^n\|^2\} \leqslant \operatorname{const}\exp\{-\alpha\delta^2\varepsilon^{-1}\}, \end{aligned}$$

which proves the required inequality for $\delta\alpha = s$ and n large enough. □

Below we shall need some more auxiliary results.

Let $\xi(t) = Aw(t)$, $K = \sigma_2^2(A)$ and

$$\xi_\theta(t) = \exp\{(\theta, \xi(t)) - \tfrac{1}{2}\|A^*\theta\|^2 t\}.$$

Using Itô's formula, one may check that $\xi_\theta(t)$ is a martingale and, thus, $\mathbf{E}\xi_\theta(t) = 1$.

Apply Doob's inequality to the martingale $\xi_{\lambda\theta}(t)$ for $\lambda > 0$. Then, given θ such that $\|\theta\|_H = 1$,

$$\mathbf{P}\{\sup_{0\leqslant t\leqslant T}(\theta, \xi(t)) \geqslant \delta\} \leqslant \mathbf{P}\{\sup_{0\leqslant t\leqslant T}\xi_{\lambda\theta}(t) \geqslant \exp\{\lambda\delta - \frac{\lambda^2 T}{2}B\}\}$$
$$\leqslant \exp\{-\lambda\delta + \frac{\lambda^2 T}{2}B\}.$$

Taking $\lambda = \delta/TB$, we find that

$$\mathbf{P}\{\sup_{0\leqslant t\leqslant T}(\theta, \xi(t)) \geqslant \delta\} \leqslant \exp\{-\frac{\delta^2}{2TB}\}.$$

Finally, consider the process

$$\xi_\varepsilon(t) = \varepsilon^{1/2}Aw(t) + \int_0^t a_{\xi_\varepsilon(\tau)}\,d\tau \tag{1.11}$$

and the functional

$$I_T(\psi) = \begin{cases} \frac{1}{2}\int_0^T \|A^{-1}[\dot\psi(t) - a_{\psi(t)}]\|^2\,dt, & \psi(0) = 0,\ \psi, a_\psi \in \mathscr{H}_A \\ \infty, & \text{otherwise.}\end{cases} \tag{1.12}$$

LEMMA 1.3. *The distribution μ_ε of the process* (1.11) *satisfies the large deviation principle with rate function* (1.12).

PROOF. The assertion of the lemma is an easy consequence of Theorem 1.1, since the Girsanov formula allows us to replace (1.6) by

$$\frac{d\mu_\eta}{d\mu_\xi}(\varepsilon^{1/2}Adw) = \exp\Big\{\varepsilon^{-1/2}\int_0^T (A^{-1}[\dot\varphi(s) - a_{\varphi(s)}], dw)$$
$$- \frac{\varepsilon^{-1}}{2}\int_0^T \|A^{-1}[\dot\varphi(s) - a_{\varphi(s)}]\|^2\,ds\Big\}$$

for $\eta_\varepsilon(t) = \xi_\varepsilon(t) - \varphi(t)$. □

§2. Large deviation principle for solutions of stochastic equations in Hilbert space

Consider the stochastic equation

$$\xi_\varepsilon(t) = x + \int_0^t a_{\xi_\varepsilon(\tau)}\,d\tau + \varepsilon^{1/2}\int_0^t A_{\xi_\varepsilon(\tau)}\,dw \tag{2.1}$$

and assume that its coefficients $a_x = A_x b$, $b \in H$, $A_x \in L_{12}(H)$, satisfy the following conditions:

C1. $\|a_x - a_y\|^2 + \sigma_2^2(A_x - A_y) \leqslant L\|x-y\|^2$, $\|a_x\| \leqslant A$, $\sigma_2^2(A_x) \leqslant B$;

C2. $\|a_x - a_y\|^2 + \sigma_2^2(A_x - A_y) \leqslant L\|x-y\|^2$, $(a_x, x) \leqslant [\rho_0 + \rho_1\|x\|]\|x\|$, $\sigma_2^2(A_x) \leqslant \rho[1 + \|x\|^2]$.

It is known **[BD1]** that under the above conditions there exists a solution $\xi_\varepsilon(t)$ of equation (2.1) possessing the Markov properties.

Let $\mu_\varepsilon = \mathbf{P}\circ\xi_\varepsilon^{-1}$ be a distribution of the process $\xi_\varepsilon(t)$.

We shall prove in this section that the measure μ_ε satisfies the large deviation principle with rate function

(2.2)
$$I_T(\varphi) = \begin{cases} \frac{1}{2}\int_0^T \|A^{-1}_{\varphi(t)}[\dot\varphi(t) - a_{\varphi(t9)}]\|^2\,dt, & \text{if } \varphi(0) = x,\ \varphi \in \mathscr{H}_A,\ a_\varphi = A_\varphi b, \\ \infty, & \text{otherwise,} \end{cases}$$

both under conditions **C1** and **C2** if, in addition, $x \in H_+$.

Let us start with condition **C1** and consider the auxiliary process

$$\xi^n_\varepsilon(T) = x + \int_0^T a_{\xi^n_\varepsilon(t)}\,dt + \varepsilon^{1/2}\int_0^T A_{\xi^n_\varepsilon\left([nt]/n\right)}\,dw, \tag{2.3}$$

where $[\theta]$ denotes the integer part of the number θ.

Notice that $\xi^n_\varepsilon(\cdot)$ may be represented in the form

$$\xi^n_\varepsilon(\cdot) = F_n\left(\varepsilon^{1/2} w(\cdot)\right) = F_n(\varepsilon^{1/2}\varkappa), \tag{2.4}$$

where $F_n\colon \mathscr{H}_- \to \mathscr{H}$ is defined by the relation

$$F_n(\varphi)(t) = F_n(\varphi)\Big(\frac{k}{n}\Big) + \int_{k/n}^t a_{F_n(\varphi)(s)}\,ds + A_{F_n(\varphi)\left(\frac{k}{n}\right)}\Big(\varphi(t) - \varphi\Big(\frac{k}{n}\Big)\Big), \quad \frac{k}{n} \leqslant t \leqslant \frac{k+1}{n}. \tag{2.5}$$

Let $I^n_T(\varphi)$ have the form

(2.6)
$$I^n_T(\varphi) = \begin{cases} \frac{1}{2}\int_0^T \|A^{-1}_{\varphi\left(\frac{[nt]}{n}\right)}\left[\dot\varphi(t) - a_{\varphi(t)}\right]dt, & \text{if } \varphi(0) = x,\ \varphi \in \mathscr{H}_A,\ a_\varphi = A_\varphi b, \\ \infty, & \text{otherwise.} \end{cases}$$

Lemma 2.1. *Both functions $I_T(\varphi)$ of the form* (2.2) *and $I^n_T(\varphi)$ of the form* (2.6) *are rate functions, and for each closed set $G \subseteq \mathscr{H}_A$ we have*

$$\inf_{\varphi\in G} I^n_T(\varphi) \to \inf_{\varphi\in G} I_T(\varphi) \quad \textit{as } n \to \infty.$$

Proof. The first assertion of the lemma is proved similarly to Lemma 1.1. To prove the second claim, notice that for any $\varphi \in \mathscr{H}$, $I_T(\varphi) = \infty$ if and only if $I^n_T(\varphi) = \infty$ for all n. Thus, we assume that $\inf I_T(\varphi) < \infty$. In this case, there exists an element $\varphi_0 \in G$ such that $I_T(\varphi_0) = \inf I_T(\varphi)$ and, evidently,

$$\varlimsup_{n\to\infty} \inf_{\varphi\in G} I^n_T(\varphi) \leqslant \varlimsup_{n\to\infty} I^n_T(\varphi_0) = I_T(\varphi_0).$$

To complete the proof, note that

$$\sup |I^n_T(\varphi) - I_T(\varphi)| \to 0$$

whenever

$$K_L = \left\{\varphi \in \mathscr{H} : \varphi(0) = x,\ \int_0^T \|A^{-1}\dot\varphi(t)\|^2\,dt \leqslant L\right\},$$

and

$$\inf_{\varphi\in G} I_T(\varphi) = \inf_{\varphi\in G\cap K_L} I_T(\varphi).$$

One can choose $L < \infty$ so that

$$\inf_{\varphi \in G} I_T^n(\varphi) = \inf_{\varphi \in G \cap K_L} I_T^n(\varphi), \qquad n \geqslant 1,$$

and

$$\inf_{\varphi \in G} I_T(\varphi) = \inf_{\varphi \in G \cap K_L} I_T(\varphi),$$

since

$$\varlimsup_{n \to \infty} \inf_{\varphi \in G} \left(I_T^n(\varphi)\right) \leqslant \inf_{\varphi \in G} I_T(\varphi) < \infty.$$

Thus, choosing $\varphi_n \in G \cap K_L$ so that $I_T^n(\varphi_n) = \inf_{\varphi \in G} I_T^n(\varphi)$ and taking a subsequence $\{\varphi_{n'}\}$ of $\{\varphi_n\}$ such that

$$I_T^{n'}(\varphi_{n'}) \longrightarrow \varliminf_{n \to \infty} I_T^n(\varphi_n)$$

and $\varphi_{n'} \to \varphi$, we conclude that

$$\inf_{\varphi \in G} I_T(\varphi) \leqslant I_T(\varphi) \leqslant \varliminf_{n \to \infty} I_T(\varphi_{n'}) = \varliminf_{n' \to \infty} I_T^{n'}(\varphi_{n'}) = \lim_{n \to \infty} \inf_{\varphi \in G} I_T^n(\varphi).$$

To achieve our purpose, we must check that $\xi_\varepsilon^n(\cdot) \to \xi_\varepsilon(\cdot)$ at a speed sufficient to make it possible to extend large deviation results valid for ξ_ε^n to the limit process ξ_ε. □

LEMMA 2.2. *For $x \in H$, $T > 0$ and $\delta > 0$*

$$\lim_{n \to \infty} \varlimsup_{\varepsilon \to 0} \varepsilon \ln \mathbf{P}\Big\{ \sup_{0 \leqslant t \leqslant T} \|\xi_\varepsilon(t) - \xi_\varepsilon^n(t)\| > \delta \Big\} = -\infty.$$

PROOF. Denote $\eta_\varepsilon^n(t) = \xi_\varepsilon(t) - \xi_\varepsilon^n(t)$. Then

$$\eta_\varepsilon^n(t) = \int_0^t \left[a\left(\xi_\varepsilon(\tau)\right) - a\left(\xi_\varepsilon^n(\tau)\right)\right] d\tau + \int_0^t \left[A\left(\xi_\varepsilon(\tau)\right) - A\left(\xi_\varepsilon^n\left([n\tau]/n\right)\right)\right] dw(\tau).$$

For $\rho > 0$, define

$$\tau_{n,\rho}^\varepsilon = \inf\left\{ t \geqslant 0 : \|\xi_\varepsilon^n(t) - \xi_\varepsilon^n\left([nt]/n\right)\| \geqslant \rho \right\}, \qquad \eta_{\varepsilon,\rho}^n(t) = \eta_\varepsilon^n\left(t \wedge \tau_{n,\rho}^\varepsilon\right)$$

and

$$\theta_{n,\rho}^\varepsilon = \inf\left\{ t \geqslant 0 : \|\eta_{\varepsilon,\rho}^n(t)\| \geqslant \delta \right\}.$$

Then

$$\mathbf{P}\Big\{ \sup_{0 \leqslant t \leqslant T} \|\eta_\varepsilon^n(t)\| \geqslant \delta \Big\} \leqslant \mathbf{P}\left\{\tau_{n,\delta}^\varepsilon \leqslant T\right\} + \mathbf{P}\left\{\theta_{n,\rho}^\varepsilon \leqslant T\right\}. \tag{2.7}$$

For the first summand we have

$$\mathbf{P}\left\{\tau_{n,\delta}^\varepsilon < T\right\} \leqslant \sum_{k=0}^{[nT]} \mathbf{P}\left\{ \sup_{\frac{k}{n} \leqslant t \leqslant \frac{k+1}{n}} \|\xi_\varepsilon^n(t) - \xi_\varepsilon^n\left(\frac{k}{n}\right)\| \geqslant \delta \right\}. \tag{2.8}$$

Notice next that for a process $\gamma(t)$ of the form

$$\gamma(t) = \int_0^t \beta(s)\, dw, \qquad \beta(s) \in L_{12}(H),$$

the process $m(t) = \exp\{(\theta, \gamma(t)) - \frac{1}{2}\int_0^t \|\beta^*(s)\theta\|^2\,ds\}$ is a martingale and for fixed $\theta \in H$ such that $\|\theta\| = 1$, and all $\lambda > 0$, we have

$$\begin{aligned}
\mathbf{P}\{\sup_{0\leqslant t\leqslant T}(\lambda\theta, \bar{\gamma}(t)) \geqslant \lambda R\} &\leqslant \mathbf{P}\{\sup(\lambda\theta, \gamma(t)) \geqslant (R - AT)\lambda\} \\
&\leqslant \mathbf{P}\Big\{\sup_{0\leqslant t\leqslant T}\exp\Big\{(\lambda\theta, \gamma(t)) - \frac{1}{2}\lambda^2\int_0^t \|\beta^*(s)\theta\|^2\,ds\Big\} \\
&\qquad\qquad \geqslant \exp\Big\{\lambda(R - AT) - \frac{\lambda^2 B^2}{2}T\Big\}\Big\} \\
&\leqslant \exp\Big\{-\lambda(R - AT) + \frac{\lambda^2 B^2}{2}T\Big\}
\end{aligned}$$

for $\bar{\gamma}(t) = \int_0^t \alpha(s)\,ds + \int_0^t \beta(s)\,dw$, $\alpha(s) \in H$, if $\|\alpha(s)\| \leqslant A$, $\sigma_2(\beta(s)) \leqslant B$. Taking $\lambda = \frac{R-AT}{B^2T}$, one may see that

$$\mathbf{P}\{\sup_{0\leqslant t\leqslant T}(\theta, \gamma(t)) \geqslant R\} \leqslant \exp\Big\{-\frac{(R - AT)^2}{2B^2T}\Big\}. \tag{2.9}$$

Since the estimate is uniform with respect to θ, it is valid for θ_0 such that $(\theta_0, \gamma(t)) = \|\gamma(t)\|$. Thus, we obtain the estimate

$$\mathbf{P}\{\sup_{0\leqslant t\leqslant T}\|\gamma(t)\| \geqslant R\} \leqslant \exp\Big\{-\frac{(R - AT)^2}{2B^2T}\Big\}. \tag{2.10}$$

Coming back to (2.8), we see that

$$\mathbf{P}\{\tau^\varepsilon_{n,\delta} \leqslant T\} \leqslant n\exp\Big\{-\frac{\left(\delta - \frac{A}{n}\right)^2}{2B^2\varepsilon}n\Big\}.$$

Hence,

$$\lim_{n\to\infty}\varlimsup_{\varepsilon\to 0}\varepsilon\ln\mathbf{P}\{\tau^\varepsilon_{n,\delta} \leqslant T\} = -\infty, \qquad \delta > 0.$$

Given $0 < \rho < 1$ and $\lambda > 1$, set $z_\lambda(y) = (\rho^2 + \|y\|^2)^\lambda$. It follows from Itô's formula that

$$z_\lambda(\eta^n_\varepsilon(t)) - \int_0^t v^\varepsilon_\lambda(s)\,ds$$

is a martingale for

$$\begin{aligned}
v^\varepsilon_\lambda(s) &= 2\lambda(\rho^2 + \|\eta^n_\varepsilon(s)\|^2)^{\lambda-1}(\eta^n_\varepsilon(s), a(\xi_\varepsilon(s)) - a(\xi^n_\varepsilon(s))) \\
&\quad + 2\lambda(\lambda - 1)\varepsilon\|[A(\xi_\varepsilon(s)) - A(\xi_\varepsilon([ns]/n))]^*\eta^n_\varepsilon(s)\|^2[\rho^2 + \|\eta^n_\varepsilon(s)\|^2]^{\lambda-2} \\
&\quad + 2\varepsilon\sigma_2^2\Big(A(\xi_\varepsilon(s)) - A(\xi_\varepsilon([ns]/n))\Big)[\rho^2 + \|\eta^n_\varepsilon(s)\|^2]^{\lambda-1}.
\end{aligned}$$

For $0 \leqslant t \leqslant \tau^\varepsilon_{n,\rho}$ one has $\|v^\varepsilon_\lambda(s)\| \leqslant k(\lambda(1+\varepsilon) + \lambda^2\varepsilon)z_\lambda$ where $k < \infty$ is independent of ε, n, and λ. Putting $\lambda = \varepsilon^{-1}$ and $Z^\varepsilon_{n,\rho}(t) = \mathbf{E}(\rho^2 + \|\eta^n_{\varepsilon,\rho}(t \wedge \theta^\varepsilon_{n,\rho})\|^2)^{1/\varepsilon}$, one has

$$Z^\varepsilon_{n,\rho}(t) \leqslant \rho^{2/\varepsilon} + 3\frac{k}{\varepsilon}\int_0^t Z^\varepsilon_{n,\rho}(s)\,ds$$

as long as $0 < \varepsilon \leqslant 1$. Hence, $Z^{\varepsilon}_{n,\rho}(t) \leqslant \rho^{2/\varepsilon} e^{3kt/\varepsilon}$. Note that

$$(\rho^2 + \delta^2)^{1/\varepsilon} \mathbf{P}\{\theta^{\varepsilon}_{n,\rho} \leqslant T\} \leqslant Z^{\varepsilon}_{n,\rho}(T)$$

and, as a result,

$$\mathbf{P}\{\theta^{\varepsilon}_{n,\rho} \leqslant T\} \leqslant \Big(\frac{\rho^2}{\rho^2 + \delta^2}\Big)^{1/\varepsilon} \exp \frac{3kT}{\varepsilon}$$

for $0 < \varepsilon \leqslant 1$. Thus,

$$\overline{\lim_{\varepsilon \to 0}}\, \varepsilon \ln \mathbf{P}\{\theta^{\varepsilon}_{n,\rho} \leqslant T\} \leqslant \ln \frac{\rho^2}{\rho^2 + \delta^2} + 3kT.$$

Finally, given $M > 0$, choose $0 < \rho < 1$ to satisfy

$$\ln \frac{\rho^2}{\rho^2 + \delta^2} + 3kT \leqslant -2M,$$

and choose N so that $\lim \varepsilon \ln \mathbf{P}\{\tau^{\varepsilon}_{n,\rho} \leqslant T\} \leqslant -2M$ for $n > N$. Then for $n > N$ there is an $0 < \varepsilon_n < 1$ such that $\mathbf{P}\{\tau^{\varepsilon}_{n,\rho} \leqslant T\} \leqslant -N/\delta$ and $\mathbf{P}\{\theta^{\varepsilon}_{n,\rho} < T\} \leqslant \exp\{-M\varepsilon^{-1}\}$ for $0 < \varepsilon \leqslant \varepsilon_n$. Thus, by (2.7) for $0 < \varepsilon < \varepsilon_n$,

$$\mathbf{P}\{\sup_{0 \leqslant t \leqslant T} \|\eta^n_\varepsilon(t)\| \geqslant \delta\} \leqslant 2e^{-M\varepsilon^{-1}},$$

and, for $n > N$,

$$\overline{\lim_{\varepsilon \to 0}}\, \varepsilon \ln \mathbf{P}\{\sup_{0 \leqslant t \leqslant T} \|\eta^n_\varepsilon(t)\| \geqslant \delta\} \leqslant -M. \qquad \square$$

THEOREM 2.1. *For each $T > 0$ and $x \in H_+$, the distribution $\mu_\varepsilon(\cdot) = \mathbf{P} \circ \xi_\varepsilon^{-1}(\cdot)$ on $\mathscr{H}$ satisfies the large deviation principle with rate function I_T of the form* (2.2).

PROOF. Let G be a closed subset in $\mathscr{H}$ and assume that $\varnothing \neq G \subset \{\varphi : \varphi(0) = x\}$. Given $\delta > 0$, one has

$$\mathbf{P}\{\xi_\varepsilon(\cdot) \in G\} \leqslant \mathbf{P}\{\xi^n_\varepsilon \in \overline{G}^\delta\} + \mathbf{P}\{\sup_{0 \leqslant t \leqslant T} \|\xi_\varepsilon(t) - \xi^n_\varepsilon(t)\| > \delta\}$$

and therefore

$$\begin{aligned}
&\overline{\lim_{\varepsilon \to 0}}\, \varepsilon \ln \mathbf{P}\{\xi_\varepsilon \in G\} \\
&\quad \leqslant \overline{\lim_{\varepsilon \to 0}}\, \varepsilon \ln \mathbf{P}\{\xi^n_\varepsilon \in \overline{G}^\delta\} \vee \overline{\lim_{\varepsilon \to 0}}\, \varepsilon \ln \mathbf{P}\{\sup \|\xi_\varepsilon(t) - \xi^n_\varepsilon(t)\| \geqslant \delta\} \\
&\quad \leqslant -\inf_{\varphi \in G^\delta} I^n_T(\varphi) \vee \overline{\lim_{\varepsilon \to 0}}\, \varepsilon \ln \mathbf{P}\{\sup_{0 \leqslant t \leqslant T} \|\xi_\varepsilon(t) - \xi^n_\varepsilon(t)\| > \delta\}.
\end{aligned}$$

Letting $n \to \infty$, we get for all $\delta > 0$

$$\overline{\lim_{\varepsilon \to 0}}\, \varepsilon \ln \mathbf{P}\{\xi_\varepsilon(\cdot) \in G\} \leqslant -\inf_{\varphi \in G^\delta} I_T(\varphi).$$

By arguments similar to those used in the proof of Lemma 1.1, we have

$$\inf_{\varphi \in \overline{G}^\delta} I_T(\varphi) \to \inf_{\varphi \in G} I_T(\varphi),$$

and the proof is completed.

Next, let Q be an open set in $\mathscr{H}$. Given $\varphi_0 \in Q$ such that $I_T(\varphi_0) < \infty$, put $B_\rho = \{\varphi : \sup_{0\leqslant t\leqslant T} \|\varphi(t) - \varphi_0(t)\| < \rho\}$, $\rho > 0$. Choose $\delta > 0$ so that $B_{2\delta} \subseteq Q$. Then

$$\mathbf{P}\{\xi_\varepsilon^n(\cdot) \in B_\delta\} \leqslant \mathbf{P}\{\xi_\varepsilon(\cdot) \in Q\} + \mathbf{P}\{\sup_{0\leqslant t\leqslant T} \|\xi_\varepsilon(t) - \xi_\varepsilon^n(t)\| \geqslant \delta\}.$$

As a result, we obtain

$$\begin{aligned} -I_T^n(\varphi_0) &\leqslant \varliminf_{\varepsilon\to 0} \varepsilon \ln \mathbf{P}\{\xi_\varepsilon^n(\cdot) \in B_\delta\} \\ &\leqslant \varliminf_{\varepsilon\to 0} \varepsilon \ln \mathbf{P}\{\xi_\varepsilon(\cdot) \in Q\} \vee \varlimsup_{\varepsilon\to 0} \varepsilon \ln \mathbf{P}\{\sup \|\xi_\varepsilon(t) - \xi_\varepsilon^n(t)\| \geqslant \delta\}. \end{aligned}$$

Letting $n \to \infty$, we get

$$\varliminf_{\varepsilon\to 0} \varepsilon \ln \mathbf{P}\{\xi_\varepsilon(\cdot) \in Q\} \geqslant -I_T(\varphi_0).$$

Thus we have proved the theorem under the assumption that $\sigma_2^2(A_x) \leqslant B^2$ and $\|a_x\| \leqslant A$. To get rid of this assumption, we must prove (2.10) assuming that conditions **C2** are valid. To this end, put $m_\lambda(y) = (1 + \|y\|^2)^\lambda$ and assume $x = 0$ without loss of generality. It follows from Itô's formula that

$$m_\lambda(\xi_\varepsilon(t)) - \int_0^t \beta_\lambda^e p(s)\,ds$$

is a martingale if

$$\begin{aligned} \beta_\lambda^\varepsilon(s) &= 2\lambda(1 + \|\xi_\varepsilon(s)\|^2)^{\lambda-1}(a(\xi_\varepsilon(s)), \xi_\varepsilon(s)) \\ &\quad + 2\lambda(\lambda-1)\varepsilon(1 + \|\xi_\varepsilon(s)\|^2)^{\lambda-2}\|A^*(\xi_\varepsilon(s))\xi_\varepsilon(s)\|^2 \\ &\quad + \varepsilon\lambda(1 + \|\xi_\varepsilon(s)\|^2)^{\lambda-1}\sigma_2^2(A(\xi_\varepsilon(s))) \\ &\leqslant c(\lambda + (\lambda+1)\varepsilon)m_\lambda(\xi_\varepsilon(s)). \end{aligned}$$

Set $\theta_R^\varepsilon = \inf\{t \geqslant 0 : \|\sigma_\varepsilon(t)\| \geqslant R\}$; then, for $0 \leqslant \varepsilon \leqslant 1$,

$$\mathbf{E}(1 + \|\xi_\varepsilon(t\wedge\theta_R^\varepsilon)\|^2)^{1/\varepsilon} \leqslant 1 + \frac{3c}{\varepsilon}\int_0^t \mathbf{E}[1 + \|\xi_\varepsilon(s\wedge\theta_R^\varepsilon)\|^2]^{1/\varepsilon}\,ds,$$

and Gronwall's lemma implies

$$\mathbf{E}[1 + \|\xi_\varepsilon(t\wedge\theta_R^\varepsilon)\|^2]^{1/\varepsilon} \leqslant e^{3ct/\varepsilon}.$$

In particular,

$$\mathbf{P}\{\theta_R^\varepsilon \leqslant T\} \leqslant (1 + R^2)^{-1/\varepsilon}e^{3cT/\varepsilon},$$

and so

$$\varlimsup_{\varepsilon\to 0} \varepsilon \ln \mathbf{P}\{\sup_{0\leqslant t\leqslant T} \|\xi^\varepsilon(t)\| \geqslant R\} \leqslant -\ln(1 + R^2) + 3cT. \qquad \square$$

THEOREM 2.2. *Assume that* **C2** *is valid. Then $I_T(\varphi)$ of the form* (2.2) *is a rate function. Moreover if $\xi_\varepsilon(t)$ is a solution of* (2.1) *and μ is the distribution of $\xi_\varepsilon(t)$,* *then for each* $x \in H_+$ *and* $T > 0$, μ_ε *satisfies the large deviation principle with rate function* $I_T(\varphi)$.

PROOF. Consider a nonnegative C^∞-function λ on H such that

$$\lambda(y) = \begin{cases} 1 & \text{if } \|y\| \leqslant 1, \\ 0 & \text{if } \|y\| \geqslant 2. \end{cases}$$

Given $R > 0$, put $\lambda_R(y) = \lambda(\frac{y}{R})$, $a_R(y) = \lambda_R(y)a(y)$, $A_R(y) = \lambda_R(y)A(y)$ and notice that $a_R(y)$ and $A_R(y)$ do satisfy **C1**. Denote by $\mu_\varepsilon^R(\cdot)$ the distribution of the solution $\xi_\varepsilon^R(t)$ of the equation

$$\xi_\varepsilon^R(t) = x + \int_0^t a_R\big(\xi_\varepsilon^R(s)\big)\,ds + \int_0^t A_R\big(\xi_\varepsilon^R(s)\big)\,dw.$$

For any $Q \in \mathfrak{B}_{\mathscr{H}}$, we have

$$\mu_\varepsilon^R\big(Q \cap \{\theta_R \geqslant T\}\big) = \mu_\varepsilon\big(Q \cap \{\theta_R \geqslant T\}\big),$$

where $\theta_R = \inf\{\, t \geqslant 0 : \|\xi_\varepsilon(t)\| \geqslant R \,\}$. Finally, $I_T^R(\varphi) = I_T(\varphi)$ for any $\varphi \in \mathscr{H}_A$ satisfying $\sup_{0 \leqslant t \leqslant T} \|\varphi(t)\| \leqslant R$.

Let us show that for any open $Q \in \mathscr{F}_{\mathscr{H}}$,

$$\underline{\lim}_{\varepsilon \to 0} \varepsilon \ln \mu_\varepsilon(Q) \geqslant -\inf_{\varphi \in Q} I_T(\varphi). \tag{2.12}$$

Evidently, (2.12) will be proved if for all $\varphi \in Q$ such that $\varphi(0) = x$ we have

$$\underline{\lim}\, \varepsilon \ln \mu_\varepsilon(Q) \geqslant -I_T(\varphi).$$

Take such a φ_0, choose $R > 0$ so that $\|\varphi_0\| < R$, and set $B_R = \{\, \varphi : \|\varphi\| < R \,\}$. Then, by Theorem 2.1,

$$\begin{aligned} \underline{\lim}_{\varepsilon \to 0} \varepsilon \ln \mu_\varepsilon(Q) &\geqslant \underline{\lim}_{\varepsilon \to 0} \varepsilon \ln \mu_\varepsilon(Q \cap B_R) = \lim \varepsilon \ln \mu_\varepsilon^R(Q \cap B_R) \\ &\geqslant -\inf_{\varphi \in Q \cap B_R} I_T^R(\varphi) \geqslant -I_T(\varphi_0). \end{aligned}$$

For a closed set G in $\mathscr{H}_A$ and $L < \infty$, we can use (2.11) to find R so that

$$\begin{aligned} \overline{\lim}_{\varepsilon \to 0} \varepsilon \ln \mu_\varepsilon(Q) &\leqslant \underline{\lim}_{\varepsilon \to 0} \varepsilon \ln\big[\mu_\varepsilon^R(G \cap \overline{B}_R) + \mu_\varepsilon^R(B_R^c)\big] \\ &\leqslant \big(-\inf_{\varphi \in G \cap \overline{B}_R} I_T^R(\varphi)\big) \vee (-L) \leqslant -\big\{\inf_{\varphi \in G} I_T(\varphi) \wedge L\big\}. \end{aligned}$$

Letting $L \to \infty$, we obtain

$$\overline{\lim}_{\varepsilon \to 0} \varepsilon \ln \mu_\varepsilon(G) \leqslant -\inf_{\varphi \in G} I_T(\varphi).$$

Finally we must show that $I_T(\varphi)$ of the form (2.2) is a rate function. To this end, choose $L < \infty$ and find $R > 0$ so that

$$\overline{\lim}_{\varepsilon \to 0} \varepsilon \ln \mu_\varepsilon(\overline{B}_R)^c \leqslant -L;$$

this may be done by (2.11). It follows from (2.12) that

$$\inf_{\varphi \in (\overline{B}_R)^c} I_T(\varphi) > L$$

and, as a result, $\{\varphi : I_T(\varphi) \leqslant L\} \subseteq \overline{B}_R$. But this means

$$\{\varphi : I_T(\varphi) \leqslant L\} = \{\varphi : I_T(\varphi) \leqslant L\} \cap \overline{B}_R = \{\varphi : I_T^R(\varphi) \leqslant L\} \cap \overline{B}_R$$

and the latter set is compact since $I_T^R(\varphi)$ is a rate function. □

§3. Diffusion processes associated with quasilinear parabolic equations

Consider the Cauchy problem for the quasilinear parabolic equation

$$\frac{\partial u}{\partial t} = \frac{1}{2} \operatorname{Tr} u''(A_x(u), A_x(u)) + (u', a_x(u)) + c_x(u)u = 0, \qquad u(0,x) = f(x) \tag{3.1}$$

in a Hilbert space H and assume that its coefficients satisfy the following conditions:

(a) $\|a_x(u) - a_y(v)\|^2 + \sigma_2^2(A_x(u) - A_y(v)) \leqslant C[\|x-y\|^2 + K_{u,v}\|u-v\|^2]$, $\|c_x(u) - c_y(v)\| \leqslant C\{\|x-y\| + K_{u,v}\|u-v\|\}$;

(b) $\sigma_2^2(A_x(u)) \leqslant \rho_2[1 + \|x\|^2 + \|u\|^{2p}]$, $(a_x(u), x) \leqslant [\rho_0 + \rho_1\|x\| + \rho_2\|u\|^p]\|x\|$, $c_x(u) \leqslant \rho_2\|u\|^p + \rho_3$;

(c) all the coefficients and the initial function are C_2-smooth functions with respect to its arguments, $|f(x)| \leqslant K_f < \infty$.

Along with (3.1), consider the system of equations

$$\xi(t) = x + \int_0^t a_{\xi(\tau)}(u(t-\tau, \xi(\tau)))\, d\tau + \int_0^t A_{\xi(\tau)}(u(t-\tau, \xi(\tau)))\, dw, \tag{3.2}$$

$$u(t,x) = \mathbf{E} f(\xi(t)) \exp \int_0^t c_{\xi(\tau)}(u(t-\tau, \xi(\tau)))\, d\tau; \tag{3.3}$$

let us prove that the function $u(t,x)$ satisfying (3.3) is a solution in the sense of distributions of (3.1), provided some assumptions are fulfilled.

To get the result we need some machinery developed in **[BD3]**. Namely, given a smooth positive bounded function $v(s,x)$ such that $|v(s,x)| \leqslant K_v(s) < \infty$ and $|v(s,x) - v(s,y)| \leqslant L_v(s)\|x-y\|$, we consider the stochastic equation

$$\xi(t) = x + \int_s^t a_{\xi(t)}(v(t-\tau, \xi(\tau)))\, d\tau + \int_s^t A_{\xi(\tau)}(v(t-\tau, \xi(\tau)))\, dw \tag{3.4}$$

and derive some estimates for its solution.

LEMMA 3.1. *Assume that* **C3** *is valid and* $v(s,x)$ *satisfies the above assumptions. Then there is a unique solution* $\xi_{s,x}^v(t)$ *of* (3.4), *and the following estimates hold*:

$$\mathbf{E}\|\xi_{s,x}^v(t)\|^2 \leqslant \left\{\|x\|^2 + 3\rho_2 \int_s^t K_v^2(t-\tau)\, d\tau\right\} e^{K(t-s)},$$

$$\mathbf{E}\|\xi_{s,x}^v(t) - \xi_{s,y}^v(t)\|^2 \leqslant \|x-y\|^2 \exp \int_s^t K_1[1 + L_v^2(t-\tau)]\, d\tau (1 + K^2 f),$$

$$\mathbf{E}\|\xi_{s,x}^v(t) - \xi_{s,y}^{v_1}(t)\|^2 \leqslant 3p \int_s^t \|v(t-\tau) - v_1(t-\tau)\|^2\, d\tau (1 + K_f^2) \times \exp \int_s^t K_1[1 + L_v^2(t-\tau)]\, d\tau.$$

The lemma is proved in **[BD3, BD4]**.

Consider a function

$$u(t-s,x)=\mathbf{E}f\left(\xi^v_{s,x}(t)\right)\exp\int_s^t C_{\xi^v_{s,x}(\tau)}\left(v(t-\tau,\xi^v_{s,x}(\tau))\right)d\tau \tag{3.5}$$

and denote by $\mathcal{M}$ the subset in $C(H,\mathbb{R}^1)$ consisting of all Lipschitz functions.

LEMMA 3.2. *Assume that $f, v\in\mathcal{M}$ and* **C3** *is valid. Then there exists an interval $\Delta=[s_1,T]$ such that any function u_τ of the form* (3.5) *belongs to $\mathcal{M}$ for $\tau\in\Delta$.*

PROOF. Let $\alpha(s)$ be a positive scalar function such that $K_v(s)\leqslant\alpha(s)$ for any $s\in[0,T]$. Then there exists an interval Δ such that $K_u(s)\leqslant\alpha(s)$ for any $s\in\Delta$. To construct the function $\alpha(s)$ possessing the desired properties, note that the function u of the form (3.5) obeys the estimate

$$|u(t-s,x)|^2\leqslant K_f^2\exp\int_s^t\left[\rho_3+\rho_2K_v^2(t-\tau)\right]d\tau \tag{3.6}$$

due to assumptions on f and Lemma 3.1. Let $\gamma(s)$ be the minimal γ such that $|u(t-s,x)|\leqslant\gamma(t-s)$ for any $s\in[0,t]$. Then (3.6) implies

$$\gamma(t-s)\leqslant K_f^2\exp\int_s^t\left[\rho_3+\rho_2K_v^2(t-\tau)\right]d\tau.$$

Let β be a solution of the equation

$$\beta(t-s)=K_f^2\exp\int_s^t\left[\rho_3+\rho_2\beta(t-\tau)\right]d\tau.$$

Note that β satisfies the equation

$$\frac{d\beta(t-s)}{ds}=\beta(t-s)\left[\rho_3+\rho_2\beta(t-s)\right]$$

as well, and may be represented in the form

$$\beta(t-s)=\frac{\rho_3\exp\rho_3(t-s)}{\rho_3+\rho_2K_f-\rho_2K_f\exp\rho_3(t-s)}.$$

Evidently, $K_v(s)\leqslant\beta(s)$ implies $K_u(s)\leqslant\beta(s)$.

The function β is bounded over the interval $\Delta=[s_1,t]$ whose length $|\Delta|=|t-s_1|$ satisfies the estimate

$$|t-s_1|<\frac{1}{\rho_3}\ln\left(\frac{\rho_3}{\rho_2K_f}\right). \tag{3.7}$$

To construct a function $l(s)$ such that $L_v(s)\leqslant l(s)$ implies $L_u(s)\leqslant l(s)$, we proceed in a similar way. Namely, note that the function (3.5) obeys the estimate

$$\begin{aligned}
&|u(t-s,x)-u(t-s,y)|^2\\
&\quad\leqslant L_f^2\mathbf{E}\|\xi^v_{s,x}(t)-\xi^v_{s,y}(t)\|^2\exp K(t-s)\\
&\qquad+K_f^2K_1\int_s^t\mathbf{E}\|\xi^v_{s,x}(\tau)-\xi^v_{s,y}(\tau)\|^2\left[1+L_v^2(t-\tau)\right]d\tau\\
&\quad\leqslant\|x-y\|^2\{L_f^2+K_1K_f^2\}\exp K\int_s^t\left[1+L_v^2(t-\tau)\right]d\tau
\end{aligned} \tag{3.8}$$

by Lemma 3.1. Using this estimate, one can show that

$$\gamma(t-s) \leqslant \left[L_f^2 + K_1 K_f^2\right] \exp K \int_s^t \left[1 + L_v^2(t-\tau)\right] d\tau,$$

where $\gamma(s)$ is the minimal γ such that

$$|u(t-s,x) - u(t-s,y)| \leqslant \gamma(t-s)\|x-y\|.$$

Denote by $\delta(s)$ the solution of the equation

$$\delta(t-s) = \left[L_f^2 + K_1 K_f^2\right] \exp \int_s^t K\left[1 + \delta(t-\tau)\right] d\tau.$$

Arguments similar to the above tell us that the function

$$\delta(t-s) = \frac{e^{K(t-s)}\left[L_f^2 + K_1 K_f^2\right]}{\left[1 + K_f - \exp\{K(t-s)\}\right]}$$

possesses the required property. The function δ is bounded over the interval $\Delta_1 = [s_2, t]$ such that

$$|t - s_2| \leqslant \frac{1}{K}\ln(1+K_f). \tag{3.9}$$

Thus we have proved that for any $s \in \Delta \cap \Delta_1$ the function $u(t-s,x)$ belongs to $\mathcal{M}$, provided $v(t-s,x) \in \mathcal{M}$.

Coming back to system (3.2), (3.3), note that (3.3) may be treated as a functional equation

$$u(s,x) = \Phi(s,x,u). \tag{3.10}$$

Let u_n be a system of successive approximations to the solution of (3.9).

THEOREM 3.1. *Assume that* **C3** *is valid. Then there exists a unique solution of* (3.9) *for any* $s \in \Delta_2$, *where* Δ_2 *is an interval of length* $|\Delta_2| \leqslant \min(|\Delta_1|, |\Delta|)$.

PROOF. Lemma 3.2 implies that the mapping Φ acts in $\mathcal{M}$. To estimate

$$\varkappa_n(s,x) = |u_{n+1}(t-s,x) - u_n(t-s,x)|,$$

let us use Lemma 3.1. Then

$$\varkappa_n(s,x) \leqslant C\left\{\int_s^t \|u_n(t-\tau) - u_{n-1}(t-\tau)\|^2\, d\tau\right\}^{1/2}.$$

Hence, $\varkappa_n(s) = \sup_x \varkappa_n(s,x)$ satisfies the inequality

$$\varkappa_n(s) \leqslant C\int_s^t \varkappa_{n-1}(s)\, ds \leqslant \cdots \leqslant C^n \int_s^t \cdots \int_{\tau_1}^t \|u_1(t-\tau) - f\|\, d\tau \dots d\tau_n$$

with $C = \left[K_f^2 K + L_2\right]^{1/2} \exp K(t-s)$. Since all the functions u_n are uniformly bounded, we obtain $\|u_1 - f\| \leqslant \text{const}$ and, finally,

$$\|u_{n+1}(t-s) - u_n(t-s)\| \leqslant \frac{C^n}{n!}\text{const}.$$

As a result, the sequence u_n converges in the sup-norm to a limit function

$$u(t-s,x) = \lim_{n\to\infty} u_n(t-s,x)$$

for any $s \in \Delta_2$. By Lemma 3.2 the limit function belongs to $\mathscr{M}$ as well.

To prove that the solution u is unique in $\mathscr{M}$, assume that there are two solutions u and v such that $u(0,x) = v(0,x) = f(x)$. Applying the estimates of Lemma 3.1, one can verify that

$$\|u(t-s) - v(t-s)\| \leqslant \text{const} \int_s^t \|u(t-\tau) - v(t-\tau)\|\, d\tau,$$

and so

$$\|u(t-s) - v(t-s)\| \equiv 0.$$

Given a bounded Lipschitz function u, Lemma 3.1 guarantees that there exists a unique random process satisfying (3.2). The uniqueness of the solution of (3.3) implies that

$$v(t-s,x) = \mathbf{E} f\big(\xi^u_{s,x}(t)\big) \exp \int_s^t C_{\xi^u_{s,x}(\tau)}\big(u(t-\tau, \xi^u_{s,x}(\tau))\big)\, d\tau = u(t-s,x). \qquad \square$$

REMARK. Given a bounded Lipschitz function $u(\tau, x)$, one may treat (3.2) as an equation with coefficients satisfying standard assumptions of the theory of stochastic differential equations. Thus all the results of this theory are valid for this equation. In particular, the solution $\xi(t)$ of (3.2) is a Markov process and the function u is a solution in the sense of distributions of the Cauchy problem (3.1). Moreover, if u is C_2-smooth, then u is a unique C_2-class solution of (3.1). Notice that, by results of **[BD3]**, u is a function of class C_2 if the coefficients and Cauchy data are smooth enough.

Below, we shall also need some results concerning the solution of a mixed problem for the parabolic equation (3.1) in a bounded domain G with smooth boundary ∂G. Consider the problem

$$(3.11)\qquad \begin{aligned} \frac{\partial u}{\partial t} &= \frac{1}{2}\operatorname{Tr} u''\big(A_x(u), A_x(u)\big) + \big(u', a_x(u)\big) + c_x(u) \\ &= \mathfrak{A}_x(u)u + c_x(u)u, \qquad x \in G \\ u(0,x) &= f(x), \quad x \in G, \qquad u(t,x) = \psi(t,x), \quad x \in \partial G, \end{aligned}$$

and assume that $\psi(0,x) = f(x)$ if $x \in \partial G$.

Define a function

$$\Pi(t,x) = \begin{cases} f(x) & \text{if } t = 0,\ x \in G, \\ \psi(t,x) & \text{if } t \geqslant 0,\ x \in \partial G. \end{cases}$$

To solve (3.11) let us construct a diffusion process as the solution of the system

$$(3.12)\qquad \xi(t) = x + \int_0^t a_{\xi(s)}\big(u(t-s, \xi(s))\big)\, ds + \int_0^t A_{\xi(s)}\big(u(t-s, \xi(s))\big)\, dw,$$

$$(3.13)\qquad u(t,x) = \mathbf{E}\Pi\big(t-\tau_t, \xi(\tau_t)\big) \exp \int_0^t c_{\xi(s)}\big(u(t-s, \xi(s))\big)\, ds,$$

where $\tau_t = t \wedge \tau$ and $\tau = \inf\{\, s \geqslant 0 : \xi(s) \notin D \,\}$.

To prove the existence and uniqueness theorem for the solution of (3.12)–(3.13) we proceed in a way similar to the above. First, note that a process $\xi_x^v(t)$ satisfying (3.6), in addition to the inequalities in Lemma 3.1, satisfies

$$\mathbf{E}\|\xi_x^v(\tau_t)\| \leqslant \left\{\|x\|^2 + 3\rho_2 \int_0^t K_v^2(t-s)\, ds\right\} e^{Kt},$$

$$\mathbf{E}\|\xi_x^v(\tau) - \xi_y^v(\tau)\|^2 \leqslant \|x-y\|^2 \exp K_1 \int_0^t \left[1 + L_v^2(t-s)\right] ds,$$

for a Markov moment $\tau \in [0,t]$.

Both estimates are easy consequences of the estimates in Lemma 3.1 and known martingale estimates provided $\tau \in [0,t]$, $t < \infty$. Let $v(t,x)$ be a bounded Lipschitz continuous function. To prove that the function

$$u(t,x) = \mathbf{E}\Pi\big(t-\tau_t, \xi(\tau_t)\big) \exp \int_0^t c_{\xi(s)}\big(v(t-s,\xi(s))\big)\, ds \tag{3.14}$$

with $v \in \mathcal{M}$ and $\Pi \in \mathcal{M}$ belongs to the set $\mathcal{M}$ defined above, under the conditions $\sigma_2\big(A_x(v)\big) > 0$ and $\|\Pi\| < \infty$, one needs estimates similar to (3.6) and (3.8). To prove (3.8), one can use the strong Markov property of the process $\xi_x^v(t)$ and write (3.14) in the form

$$u(t,x) = \mathbf{E}u\big(t-\tau_t, \xi_x^v(\tau_t)\big) \exp \int_0^t c_{\xi_x^v(s)}\big(v(t-s,\xi_x^v(s))\big)\, ds.$$

Denote $\tau = \tau_x \wedge \tau_y$ and $\tau_x = \inf\{s \geqslant 0, \xi_x(s) \notin G\}$. Then

$$\begin{aligned}
&|u(t,x) - u(t,y)| \\
&\quad \leqslant \mathbf{E}\Big\{ |u(t-\tau_t, \xi_x(\tau_t)) - \Pi(t-\tau_t, \xi_y(\tau_t))| \exp \int_{\tau_t}^t c_{\xi_x(s)}\big(v(t-s,\xi_x(s))\big)\, ds \\
&\qquad\quad + |\Pi(t-\tau_t, \xi_y(\tau_t))| \Big| \exp \int_{\tau_t}^t c_{\xi_x(s)}\big(v(t-s,\xi_x(s))\big)\, ds \\
&\qquad\qquad - \exp \int_{\tau_t}^t c_{\xi_y(s)}\big(v(t-s,\xi_y(s))\big)\, ds \Big| \Big\} \chi(\tau_y < \tau_x) \\
&\quad + \mathbf{E}\Big\{ |\Pi(t-\tau_t, \xi_x(\tau_t)) - u(t-\tau_t, \xi_y(\tau_t)| \exp \int_{\tau_t}^t c_{\xi_y(s)}\big(v(t-s,\xi_y(s))\big)\, ds \\
&\qquad\quad + |\Pi(t-\tau_t, \xi_x(\tau_t))| \Big| \exp \int_{\tau_t}^t c_{\xi_x(s)}\big(v(t-s,\xi_x(s))\big)\, ds \\
&\qquad\qquad - \exp \int_{\tau_t}^t c_{\xi_y(s)}\big(v(t-s,\xi_y(s))\big)\, ds \Big| \Big\} \chi(\tau_x < \tau_y).
\end{aligned}$$

Consider the function

$$m(t,x) = \mathbf{E}\|\xi_x^v(\tau_t) - y\|^2,$$

where $x \in G$, $y \in \partial G$ and $\tau = \inf\{s : \xi_x(s) \notin G\}$. Recall that a point $y \in \partial G$ is called a *regular boundary point* if there exists a function $l(t,x)$ (a *barrier*) such that $l(t,y) = 0$, $l(t,x) > 0$ and, given a neighborhood U of the point y for any

$x \in U \setminus y$, the inequality $\partial l/\partial t - \mathfrak{A}_x(v)l < 0$ is valid. The point $y \in \partial G$ is called *(α, β, h)-regular* if, in addition, $l(t,x)$ satisfies the estimates

$$C_1\|x-y\|^\alpha \leqslant l(t,x) \leqslant C_2\|x-y\|^\beta, \qquad x \in \{z \in U : \|z-y\| \leqslant h\}.$$

In what follows, we assume that $\sigma_2(A_x(v)) > 0$. This condition implies that all the points $y \in \partial G$ are $(2,h)$-regular. Applying Itô's formula to the function $m(t,x)$, one can prove that $m(t,x)$ solves the problem

$$\partial m/\partial t = \mathfrak{A}_x(v)m, \quad x \in G,$$
$$m(0,x) = \|x-y\|^2, \quad x \in G, \qquad m(t,x)\big|_{x\in\partial G} = \|x-y\|^2.$$

Next, due to the above assumption, there exists a neighborhood $U_h(y)$ for which a barrier $l(t,x)$ is defined. Choose $K>0$ large enough to make the inequality $Kl(t,x) > m(t,x)$ valid for $x \in G \cap U_h(y)$. Since solutions of the equation $\partial m/\partial t - \mathfrak{A}_x(v)m = 0$ obey the maximum principle, we conclude that

$$m(t,x) = \mathbf{E}\|\xi(\tau_t) - y\|^2 < Kl(t,x) \leqslant C_2K\|x-y\|^2. \tag{3.17}$$

Coming back to (3.15), note that (3.17) implies

$$\begin{aligned}
&\mathbf{E}|u(\tau, \xi_x(\tau_y)) - \Pi(t-\tau, \xi_y(\tau_y))| \\
&\quad = \mathbf{E}|\mathbf{E}_{\xi_x(\tau_y)}\Pi(t-\tau, \xi_x(\tau_y)) - \Pi(t-\tau, \xi_y(\tau_y))| \\
&\quad \leqslant C\{\mathbf{E}\|\xi_x(\tau_y) - \xi_y(\tau_y)\|^2\}^{1/2},
\end{aligned}$$

and (3.15) yields

$$|u(t,x) - u(t,y)| \leqslant K_2\|x-y\|^2 \exp K \int_0^t \left[1 + L_v^2(t-s)\right] ds.$$

Finally, (3.6) follows immediately from (3.14), and so a function $u(t,x)$ of the form (3.14) satisfies the assertion of Lemma 3.2. This implies the existence and uniqueness of the solution of (3.13) and hence of (3.12). □

REMARK. Consider the problem

$$\begin{aligned}
&\frac{\partial v}{\partial t} = \frac{1}{2}\operatorname{Tr} v''(A_x(v), A_x(v)) + (v', a_x(v)) + c_x(v)v + b_x(t), \quad x \in G, \\
&\qquad v(0,x) = f(x), \qquad v(t,x) = \psi(t,x) \quad \text{if } x \in \partial G.
\end{aligned} \tag{3.18}$$

Under above assumptions, provided that in addition $b_x(t)$ is a smooth function, one can prove that there exists an interval Δ such that given $t \in \Delta$, there exists a unique solution of problem (3.18) that may be constructed as the solution of the system

$$\xi(t) = x + \int_0^t a_{\xi(s)}(v(t-s, \xi(s)))\, ds + \int_0^t A_{\xi(s)}(v(t-s, \xi(s)))\, dw,$$
$$\begin{aligned}
v(t,x) = {}& \mathbf{E}\Pi(t-\tau_t, \xi(\tau_t)) \exp \int_0^t c_{\xi(s)}(v(t-s, \xi(s)))\, ds \\
&+ \mathbf{E}\int_0^{\tau_t} b_{\xi(s)}(t-s) \exp\left\{\int_0^s c_{\xi(s_1)}(v(t-s_1, \xi(s_1)))\, ds_1\right\} ds.
\end{aligned}$$

To prove this one proceeds as in the proof of the existence and uniqueness of the solutions of problems (3.1) and (3.11).

§4. Wave front propagation for solutions of a quasilinear equation with a small parameter

Consider the Cauchy problem

$$
\tag{4.1}
\begin{gathered}
\frac{\partial u}{\partial t} = \frac{\varepsilon}{2} \operatorname{Tr} u''(A_x, A_x) + \frac{1}{\varepsilon} c_x(u)u, \\
u(0,x) = f(x).
\end{gathered}
$$

It follows from the results of the previous section that the function u_ε satisfying the system

$$
\tag{4.2}
\xi(t) = x + \int_0^t a_{\xi_\varepsilon(\tau)}\, d\tau + \varepsilon^{1/2} \int_0^t A_{\xi_\varepsilon(\tau)}\, dw,
$$

$$
\tag{4.3}
u_\varepsilon(t,x) = \mathbf{E} f(\xi_\varepsilon(t)) \exp \int_s^t c_{\xi_\varepsilon(\tau)}\big(u_\varepsilon \varepsilon^{-1}(t-\tau, \xi(\tau))\big)\, d\tau
$$

gives the solution of (4.1). Due to the results of §2, one may take the function

$$
\tag{4.4}
I_T(\psi) = \begin{cases} \frac{1}{2}\int_0^T \|A^{-1}_{\psi(\tau)}\left[\dot\psi(\tau) - a_{\psi(\tau)}\right] d\tau, & \text{if } \psi(0) = x,\ a_\psi, \psi \in \mathscr{H}_A, \\ \infty, & \text{otherwise,} \end{cases}
$$

for the rate function and assert that for any $\psi \in \mathscr{H}_A$, $\psi(0) = x$, and arbitrary $\alpha, \beta > 0$, there exists an $\varepsilon_0 > 0$ such that, for $\varepsilon > \varepsilon_0 > 0$,

$$
\tag{4.5}
\mathbf{P}\{\rho_{0T}(\xi_\varepsilon, \psi) < \beta\} \geqslant \exp\{-\varepsilon^{-1}(I_T(\psi) + \alpha)\},
$$

where $\rho_{0T}(\xi, \psi) = \sup_{0 \leqslant t \leqslant T} \|\xi_\varepsilon(t) - \psi(t)\|$. Moreover, for any $s < \infty$, the set $\Phi(s) = \{\varphi \in \mathscr{H}_A : \varphi(0) = x, I_T(\varphi) \leqslant s\}$ is compact in $C([0,T], H)$ and, given α, β, one may choose $\varepsilon_0 > 0$ so that for any $\varepsilon > \varepsilon_0$

$$
\tag{4.6}
\mathbf{P}\{\rho_{0T}(\xi_\varepsilon, \Phi(s)) \geqslant \beta\} \leqslant \exp\{-\varepsilon^{-1}(s - \alpha)\}.
$$

Recall, in addition, that $I_T(\psi)$ is semicontinuous from below in $C([0,T], H)$, which means that

$$
I_T(\psi) \leqslant \varliminf_{n \to \infty} I_T(\psi_n)
$$

if the $\psi_n \in C([0,T], H)$ uniformly converge to ψ over $[0,T]$ for $n \to \infty$.

Consider one more function, namely

$$
\tag{4.7}
J_T(\varphi) = \int_0^T c(\varphi(s))\, ds - I_T(\varphi)
$$

and assume that $c(x) = c(x,0) = \max_{0 \leqslant u \leqslant 1} c(x,u) > 0$ and $c(x,u) < 0$ for $u > 1$.

LEMMA 4.1. *Let $f(x)$, $x \in H$, be a nonnegative bounded Lipschitz function, let $F_0 = \{x \in H : f(x) > 0\}$, and let f be continuous on the set (F_0) of inner points of F_0. Then*

$$
\tag{4.8}
\begin{aligned}
&\lim_{\varepsilon \to 0} \varepsilon \ln \mathbf{E} f(\xi_\varepsilon(t)) \exp\left\{\varepsilon^{-1} \int_0^t c(\xi_\varepsilon(s))\, ds\right\} \\
&\qquad = \sup\{J_T(\varphi) : \varphi(0) = x,\ \varphi(t) \in F_0\}.
\end{aligned}
$$

PROOF. Denote by M the upper bound of the right-hand side of (4.8). Since $c(x)$ is bounded, it is easy to check that $M < \infty$. Using the properties of the function

f and the semicontinuity from above of the functional $J_T(\varphi)$, one sees that for any $\delta > 0$ there exists $\widetilde{\varphi} \in C([0,T],H)$ such that $\widetilde{\varphi}(0) = x$, $\rho_{0T}(\widetilde{\varphi}, H \setminus F_0) = \alpha > 0$ and $J_T(\widetilde{\varphi}) > M - \delta$. Let θ, δ be positive numbers such that

$$\int_0^t |[c(\varphi(s)) - c(\psi(s))]|\, ds < \frac{\delta}{2}$$

if $\rho_{0T}(\varphi,\psi) < \theta$, and denote $\alpha_1 = \theta \wedge \alpha/2$.

Using (4.3) and the fact that $f(x)$ is continuous on (F_0), one can conclude that

$$\begin{aligned}
&\mathbf{E} f(\xi_\varepsilon(t)) \exp\Big\{-\varepsilon^{-1}\int_0^t c(\xi_\varepsilon(\tau))\, d\tau\Big\} \\
&\quad \geqslant \mathbf{E}\chi\big(\rho_{0T}(xi_\varepsilon, \widetilde{\varphi}) < \alpha_1\big) f(\xi_\varepsilon(t)) \exp\Big\{\varepsilon^{-1}\int_0^t c(\xi_\varepsilon(\tau))\, d\tau\Big\} \\
&\quad \geqslant \min |f(x)| \exp\Big\{\varepsilon^{-1}\int_0^t c(\widetilde{\varphi}(s))\, ds - \frac{de}{2\varepsilon}\Big\}\mathbf{P}\{\rho_{0T}(\xi_\varepsilon, \widetilde{\varphi}) < \alpha_1\} \\
&\quad \geqslant \exp\Big\{-\frac{2\delta}{\varepsilon} + \varepsilon^{-1}\int_0^t c(\widetilde{\varphi}(s))\, ds - \varepsilon^{-1} I_t(\widetilde{\varphi})\Big\} \geqslant e^{\varepsilon^{-1}[M-3\delta]}
\end{aligned}$$

for ε small enough.

To determine the upper bound, put $s = |M| + t\sup_{x\in H}|c(x)| + 1$ and

$$\mathbf{E} f(\xi_\varepsilon(t)) \exp\Big\{\varepsilon^{-1}\int_0^t c(\xi_\varepsilon(\tau))\, d\tau\Big\} = \gamma_1 + \gamma_2, \tag{4.9}$$

where

$$\begin{aligned}
\gamma_1 &= \mathbf{E} f(\xi_\varepsilon(t))\chi\big(\rho_{0T}(\xi_\varepsilon, \Phi(s)) < \tfrac{\theta}{2}\big) \exp\Big\{\varepsilon^{-1}\int_0^t c(\varphi(\tau))\, d\tau\Big\}, \\
\gamma_2 &= \mathbf{E} f(\xi_\varepsilon(t))\chi\big(\rho_{0T}(\xi_\varepsilon, \Phi(s)) \geqslant \tfrac{\theta}{2}\big) \exp\Big\{\varepsilon^{-1}\int_0^t c(\varphi(\tau))\, d\tau\Big\}, \\
\Phi(s) &= \big\{\varphi \in C([0,T],H), \varphi(0) = x, I_T(\varphi) \leqslant s\big\},
\end{aligned}$$

θ being defined as above. To get the estimate for γ_1 in (4.9), note that, since $\Phi(s)$ is compact, one may choose a finite $(\theta/2)$-net $\varphi_1, \dots, \varphi_n$ such that

$$\begin{aligned}
&\sup|f(x)| \sum_{i=1}^n \mathbf{E}\chi\big(\rho_{0t}(\xi_\varepsilon, \varphi_i) < \theta\big) \exp\Big\{\varepsilon^{-1}\int_0^t c(\varphi(s))\, ds\Big\} \\
&\quad \leqslant \sup|f(x)| \sum_{i=1}^n \exp\Big\{\varepsilon^{-1}\int_0^t c(\varphi_i(s))\, ds + \frac{\delta}{2\varepsilon}\Big\}\mathbf{P}\{\rho_{0t}(\xi_\varepsilon, \varphi_i) < \theta\}.
\end{aligned} \tag{4.10}$$

Denote $m_i = \inf\{I_t(\varphi) : \rho_{0t}(\varphi, \varphi_i) < \theta\} - \delta/4$, $i = 1, \dots, n$. Since $I_t(\varphi)$ is semicontinuous, one may choose $\varkappa > 0$ so that $\rho_{0t}(\varphi_i, \Phi(m_i)) \geqslant \varkappa + \theta$. Hence, it follows from (4.5) that

$$\mathbf{P}\{\rho_{0t}(\varphi_i, \xi_\varepsilon) < \theta\} \leqslant \mathbf{P}\{\rho_{0t}(\xi_\varepsilon, \Phi(m_i)) \geqslant \varkappa\} \leqslant \exp\{-\varepsilon^{-1}(m_i - \delta/4)\} \tag{4.11}$$

for ε small enough.

To get the bound for γ_2 in (4.9), apply (4.6). Indeed, given an arbitrary $\delta > 0$, we have

$$\gamma_2 < \exp\{\varepsilon^{-1} t \sup_{x\in H} |c(x)| - \varepsilon^{-1}(s-\delta)\} \leqslant \exp\{\varepsilon^{-1}(|M|+1-\delta)\}$$

for sufficiently small ε. Finally, by (4.10), (4.11), it follows that

$$\gamma_1 < \sum_{i=1}^{n} \exp\{\varepsilon^{-1} \sup\{ J_t(\varphi) : \rho_{0t}(\varphi, \varphi_i) < \theta \} + 2\delta\}.$$

Since δ is arbitrarily small, the lemma is proved. □

Now put

$$\begin{aligned} S(t,x,y) &= \sup\{ J_t(\varphi) : \varphi \in C([0,1], H),\ \varphi(0) = x,\ \varphi(t) = y \}, \\ S(t,x) &= \sup_{y\in F_0} S(t,x,y), \\ Q &= \{ (t,x) : t > 0,\ x \in H,\ S(t,x) < 0 \}, \\ \mathscr{K}_Q(x) &= \begin{cases} 1, & (t,x) \notin Q, \\ 0, & (t,x) \in Q, \end{cases} \end{aligned}$$

and note that $S(t,x,y)$ and $S(t,x)$ are continuous functions, increasing in the argument t, and $S(t,x,y) = S(t,y,x)$.

THEOREM 4.1. *Assume that the coefficients a_x, A_x, $c_x(u)$ obey* **C3** *and $c_x = c_x(0) = \max_u c(x,u)$ is bounded. Let $f(x)$ be a nonnegative bounded function with support F_0 belonging to the closure of the set (F_0) of its interior points. In addition, let $f(x)$ be continuous for $x \in (F_0)$. Suppose that, for $(t,x) \in Q$,*

$$\begin{aligned} S(t,x) = \sup\{ J_t(\varphi) : \varphi \in C([0,t], H),\ \varphi(0) = x,\ \varphi(t) \in F_0,& \\ S(t-s, \varphi(s)) < 0,\ s \in (0,t)\}.& \end{aligned} \tag{4.12}$$

Then, for any $T, \delta > 0$ and (t,x) such that $S(t,x) \neq 0$, the solution $u_\varepsilon(t,x)$ of (4.1) tends to $\mathscr{K}_Q$ as $\varepsilon \to 0$ uniformly in

$$(t,x) \in Q_\delta = \{ (t,x) : t \in [0,T],\ \|x\| < T,\ |S(t,x)| > \delta \}.$$

PROOF. It follows from §3 that the function u_ε satisfying (4.1) may be represented in the form

$$u_\varepsilon(t,x) = \mathbf{E} f(\xi_\varepsilon(t)) \exp \varepsilon^{-1} \int_0^t c_{\xi_\varepsilon(\tau)}(u(t-\tau, \xi_\varepsilon(\tau)))\, d\tau. \tag{4.13}$$

Therefore, since $c(x) = c(x,0) \geqslant c(x,u)$, one can deduce that

$$0 \leqslant u_\varepsilon(t,x) \leqslant \mathbf{E} f(\xi_\varepsilon(t)) \exp\left\{\varepsilon^{-1} \int_0^t c_{\xi_\varepsilon(\tau)}\, d\tau\right\}. \tag{4.14}$$

Relations (4.13) and (4.14) imply

$$\lim_{\varepsilon\to 0} \varepsilon \ln u_\varepsilon(t,x) \leqslant S(t,x),$$

and thus $\lim_{\varepsilon\to 0} u_\varepsilon(t,x) = 0$ provided $(t,x) \in Q$, where $S(t,x) < 0$. This convergence is uniform on the set $Q_\delta \cap Q$.

Now assume that $N = \{ (t,x) : S(t,x) > 0 \}$; our present goal is to prove that $\lim_{\varepsilon\to 0} u_\varepsilon(t,x) = 1$ for $(t,x) \in N$.

Let τ_1, τ_2 be Markov times defined by

$$\tau_1 = \inf\{ s : u_\varepsilon(t-s, \xi_\varepsilon(s)) > 1-\lambda \}$$

$$\tau_2 = \inf\{ s : S(t-s, \xi_\varepsilon(s)) = 0 \}, \qquad \tau = \tau_1 \wedge \tau_2$$

for a small positive ε. Since the process $\xi_\varepsilon(s)$ possesses the strong Markov property [**BD3**], one can easily check that

$$\begin{aligned}
u_\varepsilon(t,x) &= \mathbf{E} u_\varepsilon\big(t-\tau, \xi_\varepsilon(\tau)\big) \exp\Big\{\varepsilon^{-1}\int_0^\tau c\big(\xi_\varepsilon(s), u(t-s,\xi_\varepsilon(s))\big)\,ds\Big\}\\
&= \mathbf{E}\chi(\tau=\tau_1) u_\varepsilon\big(t-\tau_1, \xi_\varepsilon(\tau_1)\big) \exp\Big\{\varepsilon^{-1}\int_0^{\tau_1} c\big(\xi_\varepsilon(s), u(t-s,\xi_\varepsilon(s))\big)\,ds\Big\}\\
&\quad + \mathbf{E}\chi(\tau=\tau_2) u_\varepsilon\big(t-\tau_2, \xi_\varepsilon(\tau_2)\big) \exp\Big\{\varepsilon^{-1}\int_0^{\tau_2} c\big(\xi_\varepsilon(s), u_\varepsilon(t-s,\xi_\varepsilon(s))\big)\,ds\Big\}\\
&= \gamma_1 + \gamma_2.
\end{aligned}$$

Since $c(x,u)$ is nonnegative in the interval $0 \leqslant u \leqslant 1-\lambda$, the function γ_1 is bounded from below:

$$\gamma_1 > (1-\lambda)\mathbf{E}\xi(\tau=\tau_1) = (1-\lambda)\mathbf{P}\{\tau=\tau_1\}. \tag{4.15}$$

To get the bound for γ_2, note that for those points (t_0, x_0) for which $S(t_0,x_0) = 0$ we have

$$u_\varepsilon(t_0,x_0) > \exp\{\delta\varepsilon^{-1}\} \tag{4.16}$$

for any $\delta > 0$ if ε is small enough. To verify this, choose $\widetilde{\varphi} \in C([0,T],H)$ so that $\widetilde{\varphi}(0) = x_0$, $\widetilde{\varphi}(t_0) \in (F_0)$, and suppose that the distance ρ between the point $\big(t-s, \widetilde{\varphi}(s)\big)$ and the complement to the set Q is positive, while $J_t(\widetilde{\varphi}) > -de/2$, $s \in (\widetilde{\tau}, t_0 - \widetilde{\tau})$. Such a function may be chosen by assumption (4.12). Note that the number $\widetilde{\tau}$ may be taken arbitrary small. Certainly in this case the distance ρ will be small as well, but still positive. The set Q contains the subset $\big(t_0 - s, \widetilde{\varphi}(s)\big)$; moreover the distance between this subset and the complement of Q is positive for all $c \in (0, t_0)$ except for small neighborhoods of the points $s = 0$ and $s = t_0$. This implies that $u_\varepsilon\big(t_0 - s, \widetilde{\varphi}(s)\big)$ is close to 0 for $s \in (\widetilde{\tau}, t_0 - \widetilde{\tau})$ and small ε. Therefore, given ε small enough, we have

$$\sup_{\widetilde{\tau}<s<t_0-\widetilde{\tau}} \big(c(\widetilde{\varphi}(s)) - c(\widetilde{\varphi}(s), u_\varepsilon(t-s, \widetilde{\varphi}(s)))\big) < \frac{\delta}{4}.$$

Hence one can choose $\widetilde{\tau}$, γ so small that

$$\begin{aligned} u_\varepsilon(t_0,x_0) &= \mathbf{E}f(\xi_\varepsilon(t_0))\exp\Big\{\varepsilon^{-1}\int_0^{t_0} c(\xi_\varepsilon(s),u_\varepsilon(t-s,\xi_\varepsilon(s)))\,ds\Big\} \\ &\geqslant \mathbf{E}f(\xi_\varepsilon(t_0))\chi(\rho_{0,t_0}(\xi_\varepsilon,\widetilde{\varphi})<\gamma) \\ &\quad\times\exp\Big\{\varepsilon^{-1}\int_0^{t_0} c(\xi_\varepsilon(s),u_\varepsilon(t-s,\xi_\varepsilon(s)))\,ds\Big\} \\ &\geqslant \mathbf{P}\{\rho_{0,t_0}(\xi_\varepsilon,\widetilde{\varphi})<\gamma\}\exp\Big\{\varepsilon^{-1}\int_0^{t_0} c(\widetilde{\varphi}(s))\,ds-\frac{de}{4\varepsilon}\Big\} \\ &\geqslant \exp\Big\{\varepsilon^{-1}\{J_{t_0}(\varphi)-\frac{3\delta}{4}\}\Big\}\geqslant\exp\Big(-\frac{\delta}{\varepsilon}\Big) \end{aligned} \tag{4.17}$$

for sufficiently small ε, having in mind (4.15).

Denote $S_0=S(t,x)>0$ and choose $h>0$ so that

$$\inf\{S(\tau,y):|\tau-t|<h,\|x-y\|<h\}>\tfrac{1}{2}S_0.$$

Due to (4.16), for ε small enough, $u_\varepsilon(t-\tau_2,\xi_\varepsilon(\tau_2))>e^{-\delta\varepsilon^{-1}}$ for $\delta\in(0,\frac{m}{2})$, where

$$m=\min_{\substack{\|x-y\|\leqslant h\\ 0\leqslant u\leqslant 1-\lambda}} c(y,u).$$

Hence,

$$\gamma_2\geqslant\mathbf{E}\chi(\tau=\tau_2)\exp(-\frac{\delta+m}{\varepsilon})\geqslant\exp\frac{m}{2\delta}\cdot\mathbf{P}\{\tau=\tau_2\}, \tag{4.18}$$

since $S(t,x)$ and the process $\xi_\varepsilon(t)$ are continuous, and, therefore, $\mathbf{P}\{\tau_2>0\}=\gamma>0$. Finally, it follows from (4.15) and (4.18) that $u_\varepsilon(t,x)>1-\lambda$ for ε small enough.

It now remains to prove that $\overline{\lim}\,u_\varepsilon(t,x)=1$. For that purpose consider the set $\Lambda_\varepsilon=\{(t,x):t\geqslant 0,\ u_\varepsilon(t,x)\geqslant 1+\lambda\}$ and denote $\tau_3=\inf\{s:(t-s,\xi_\varepsilon(s))\notin\Lambda_\varepsilon\}$. Due to the strong Markov property of the process $\xi_\varepsilon(t)$ it follows from (4.13) that

$$\begin{aligned} u_\varepsilon(t,x) &= \mathbf{E}u_\varepsilon(t-\tau_3,\xi_\varepsilon(\tau_3))\exp\Big\{\varepsilon^{-1}\int_0^{\tau_3} c(\xi_\varepsilon(s),u_\varepsilon(\tau_3-s,\xi_\varepsilon(s)))\,ds\Big\} \\ &= \alpha_1+\alpha_2, \end{aligned}$$

where

$$\alpha_1=\mathbf{E}u_\varepsilon(t-\tau_3,\xi_\varepsilon(\tau_3))\chi(\tau_3<t)\exp\Big\{\varepsilon^{-1}\int_0^{\tau_3} c\,ds\Big\},$$

$$\alpha_2=\mathbf{E}f(\xi_\varepsilon(t))\chi(\tau_3=t)\exp\Big\{\varepsilon^{-1}\int_0^{t} c\,ds\Big\}.$$

Taking into account the properties of the functions c, u, one gets

$$\alpha_1\leqslant(1+\lambda)\mathbf{P}\{\tau_3<t\},$$

while

$$\alpha_2\leqslant\|f\|\exp\Big\{-\frac{t}{\varepsilon}\min_{\substack{2+\|f\|>u>1+\lambda\\ \|y-x\|<t}}|c(y,u)|\Big\}\mathbf{P}\{\tau_3=\tau\}.$$

This implies $\overline{\lim}_{\varepsilon\to 0}u_\varepsilon(t,x)\leqslant 1$, and so $\lim_{\varepsilon\to 0}u_\varepsilon(t,x)=1$ uniformly in Q_δ. $\square$

To make the phenomena of wave front propagation more evident, consider a case when $c(x) = c(x,0) = c = \text{const} > 0$; let us prove that, given $f(x)$ and $F_0 = \{x \in H,\ f(x) > 0\}$, we have $\lim_{\varepsilon\to 0} u_\varepsilon(t,x) = 1$ if $\rho(x,F_0) < t\sqrt{2c}$, and $\lim_{\varepsilon\to 0} u_\varepsilon(t,x) = 0$ if $\rho(x,F_0) > t\sqrt{2c}$. To prove this we need some auxiliary assertions.

LEMMA 4.2. *Let* $\mathscr{H}(t,x,y) = \{\varphi \in C([0,T],H),\ \varphi(0) = x,\ \varphi(t) = y\}$. *Then*

$$\inf\left\{\int_0^t \|A^{-1}(\varphi(s))\dot\varphi(s)\|^2\,ds, \varphi \in \mathscr{H}(t,x,y)\right\} = \frac{1}{t}d^2(x,y),$$

where

$$d(x,y) = \inf\left\{\int_0^1 \|A^{-1}(\varphi(s))\dot\varphi(s)\|\,ds, \varphi \in \mathscr{H}(1,x,y)\right\}$$

is the Riemannian metric in H. The bound is attained on the set of minimal geodesics that connect the points x and y under the assumption that the parameter along these geodesics is chosen to be proportional to arc length.

PROOF. Denote by $l_{0t}(\varphi)$ the length of the curve φ between the points $\varphi(0)$ and $\varphi(t)$, and put

$$l_{0t}(\varphi) = \int_0^t \|A^{-1}(\varphi(s))\dot\varphi(s)\|\,ds.$$

Assume that $\gamma(s)$, $0 \leqslant s \leqslant t$, is the minimal geodesic connecting the points x and y in H. Let the parameter s along the curve be proportional to arc length. Then

$$(B(\gamma(s))\dot\gamma(s),\dot\gamma(s)) = \frac{d^2(x,y)}{t^2} = \frac{1}{t}l_{0t}^2(\gamma), \qquad s \in [0,t],$$

$$\int_0^t (B(\gamma(s))\dot\gamma(s),\dot\gamma(s))\,ds = \frac{1}{t}d^2(x,y)$$

for $B = (A^{-1})^*A^{-1}$. Suppose $\varphi \in C([0,T],H)$ is an absolutely continuous function such that $\varphi(0) = x$, $\varphi(t) = y$. It follows from the Schwarz inequality that

$$\int_0^t \|A^{-1}(\varphi(s))\dot\varphi(s)\|^2\,ds \geqslant \frac{1}{t}l_{0t}^2(\varphi) \geqslant \frac{1}{t}d^2(x,y). \tag{4.19}$$

If the parameter along the curve is arc length, then

$$\|A^{-1}(\varphi(s))\dot\varphi(s)\|^2 = \text{const}$$

and the first inequality in (4.19) becomes equality. The second inequality in (4.19) becomes an equality when φ becomes a minimal geodesic connecting x and y. □

THEOREM 4.2. *Assume that conditions* **C3** *are valid and that* $c(x) = c = \text{const} > 0$. *Let* $f(x)$ *satisfy the assumptions of Theorem* 4.1. *Then*

$$\lim_{\varepsilon \to 0} u_\varepsilon(t,x) = 1 \quad \textit{if } d(x, F_0) < t\sqrt{2c},$$
$$\lim_{\varepsilon \to 0} u_\varepsilon(t,x) = 0 \quad \textit{if } f(x, F_0) > t\sqrt{2c}.$$

PROOF. It follows from Lemma 4.2 that

(4.20)
$$\begin{aligned} S(t,x,y) &= \sup\{\, J_t(\varphi) : \varphi \in C([0,T],H),\ \varphi(0) = x,\ \varphi(t) = y \,\} \\ &= ct - \inf\Big\{ \frac{1}{2}\int_0^t \|A^{-1}(\varphi(s))\dot{\varphi}(s)\|^2\,ds : \\ &\qquad\qquad \varphi \in C([0,T],H),\ \varphi(0) = x,\ \varphi(t) = y,\ \varphi \in \mathscr{H}_A \Big\} \\ &= ct - \frac{1}{2t} d^2(x,y). \end{aligned}$$

Hence

$$S(t,x) = ct - \frac{1}{2t} d^2(x, F_0),$$
$$Q = \{\, (t,x) : S(t,x) < 0 \,\} = \{\, (t,x) : d(x,F_0) > t\sqrt{2c} \,\}.$$

Since the upper bound in (4.20) is attained on the minimal geodesics connecting the points x and y, provided they are equipped with a parameter proportional to arc length, this implies condition (4.12). Thus it follows from Theorem 4.1 that

$$\lim_{\varepsilon \to 0} u_\varepsilon(t,x) = 1 \quad \text{if } d(x, F_0) < t\sqrt{2c},$$
$$\lim_{\varepsilon \to 0} u_\varepsilon(t,x) = 0 \quad \text{if } d(x, F_0) > t\sqrt{2c}. \qquad \square$$

Assume now that condition (4.12) is not valid. To investigate this situation, we shall consider the mixed problem in a domain $G \subset H$ with smooth boundary ∂G:

$$\begin{aligned} &\frac{\partial u_k}{\partial t} = \frac{\varepsilon}{2} \operatorname{Tr} u_k''(A_x, A_x) + (u_k', a_x) + \varepsilon^{-1} c_x^k(u) u_k, \qquad t > 0,\ x \in G, \\ &u_k(0,x) = f_k(x), \qquad u_k(t,x)\big|_{x \in \partial G} = \psi_k(t,x), \qquad k = 1,2, \end{aligned} \tag{4.21}$$

under the assumption that the functions $c_x^k(u)$ are bounded with respect to x and Lipschitz continuous with respect to x and u.

LEMMA 4.3. *Let* $c_x^1(u) < c_x^2(u)$, $x \in G$, $0 \leqslant f_1(x) \leqslant f_2(x) \leqslant K < \infty$ *and* $0 \leqslant \psi_1(t,x) \leqslant \psi_2(t,x) \leqslant M < \infty$. *Then* $u_1(t,x) < u_2(t,x)$ *for* $t \geqslant 0$, $x \in G$.

PROOF. Consider the difference $v(t,x) = u_2(t,x) - u_1(t,x)$. It may be easily proved that $v(t,x)$ satisfies the problem

$$\begin{aligned} &\frac{\partial v}{\partial t} = \frac{\varepsilon}{2} \operatorname{Tr} v''(A_x, A_x) + \varepsilon^{-1} b_x(t) v + \varepsilon^{-1} B(t,x), \\ &v(0,x) = \widetilde{f}(x), \qquad v(t,x)\big|_{x \in \partial G} = \widetilde{\psi}(t,x), \end{aligned} \tag{4.22}$$

where $b_x(t) = (u_2 - u_1)^{-1}\big(c_x^2(u_2)u_2 - c_x^2(u_1)u_1\big)$, $B(t,x) = c_x^2(u_1)u_1 - c_x^1(u_1)u_1$, $\widetilde{f}(x) = f_2(x) - f_1(x)$, $\widetilde{\psi}(s) = \psi_2(x) - \psi_1(x)$.

Let $\xi_\varepsilon(t)$ be a diffusion process with generator

$$\mathfrak{A}_\varepsilon f(x) = \frac{\varepsilon}{2} \operatorname{Tr} f''(A_x, A_x) + (f', a_x).$$

Let $\tau_\varepsilon = \inf\{ s : \xi_\varepsilon(s) \notin G \}$, and let $\Pi(t,x)$ be defined on $G \cup \{\partial G \times [0,\infty)\}$ by

$$\Pi(t,x) = \begin{cases} \widetilde{f}(x) & \text{if } t = 0, \\ \widetilde{\psi}(t,x) & \text{if } t > 0,\ x \in \partial G. \end{cases}$$

By results of §3 one may represent the solution of (4.22) in the form

$$\begin{aligned} v(t,x) &= \mathbf{E}\Pi\big(t - (t \wedge \tau_\varepsilon), \xi_\varepsilon(t \wedge \tau_\varepsilon)\big) \exp\Big\{ \varepsilon^{-1} \int_0^{t\wedge\tau_\varepsilon} b\big(t-s, \xi_\varepsilon(s)\big)\, ds \Big\} \\ &\quad + \varepsilon^{-1}\mathbf{E} \int_0^{t\wedge\tau_\varepsilon} B\big(t-s, \xi_\varepsilon(s)\big) \exp\Big\{ \varepsilon^{-1} \int_0^s b\big(t-s_1, \xi_\varepsilon(s_1)\big)\, ds_1 \Big\}\, ds. \end{aligned}$$

This implies the lemma since, evidently, $v(t,x) = u_2(t,x) - u_1(t,x) \geqslant 0$. □

Denote by $\alpha(x,g)$ the velocity field

$$\alpha(x,g) = \|A^{-1}(x)g\|^{-1}\sqrt{2c(x)}, \qquad \|g\| = 1,$$

and set

$$\begin{aligned} \tau_{F_0}(x) &= \inf\Big\{ \int_0^t \frac{\|\dot\varphi(s)\|}{\alpha\big(\varphi(s), \frac{\dot\varphi(s)}{\|\dot\varphi(s)\|}\big)}\, ds : \varphi \in C([0,T],H), \varphi(0) = x,\ \varphi(t) \in F_0 \Big\} \\ &= \inf\Big\{ \int_0^t \|A^{-1}(\varphi(s))\dot\varphi(s)\| \big(2c(x)\big)^{-1/2}\, dx : \varphi \in C([0,T],H), \\ &\qquad\qquad \varphi(0) = x,\ \varphi(t) \in F_0 \Big\}. \end{aligned}$$

LEMMA 4.4. *Let all the assumptions of Theorem* 4.1 *be fulfilled except* (4.12). *Then* $\lim_{\varepsilon\to 0} u_\varepsilon(t,x) = 1$ *for* $x \in F_t = \{ x \in H : \tau_{F_0}(x) < t \}$.

PROOF. Choose $x \in F_t$ and let $\varphi_1 \in C([0,t],h)$, $\varphi_1(0) = x$, $\varphi_1(t) \in F_0$, be such that

$$\tau_{F_0}(x) = \int_0^t \|A^{-1}(\varphi_1(s))\dot\varphi_1(s)\| \big(2c(x)\big)^{-1/2}\, dx.$$

Given a small number $\delta > 0$, consider a finite covering of the curve φ_1 by Riemannian open balls $U_1, \dots, U_n$ of radius δ. Let $A_1 = \varphi_1(t)$ be the center of U_1 and choose the center of U_k at the point A_k where ∂U_{k-1} intersects φ_1. Along with the solution $u_\varepsilon(t,x)$ of (4.1), consider the solution $u_1^\varepsilon(t,x)$ of (4.21) with

$$G = U_1, \quad c_1(u) = \inf_{x\in U_1} c(x,u), \quad f_1(x) = f(x), \quad \psi_1(t,x) = \psi(t,x)$$

and a solution $u_2^\varepsilon(t,x)$ of (4.21) with

$$G = U_1, \quad f_2(x) = f(x), \quad \psi_2(t,x) = 0, \quad c_2(x,u) = c(x,u).$$

Finally let $v_\varepsilon(t,x)$ be a solution of (4.1) with $c(x,u) = c_1(u)$ and initial function $f(x)$ for $x \in F_0 \cap U_1$ and $f(x) = 0$ for $x \notin F_0 \cap U_1$.

Since $u_\varepsilon(t,x)$ is nonnegative, it follows from Lemma 4.3 that $u_\varepsilon(t,x) \geqslant u_1^\varepsilon(t,x)$ for $x \in U_1$. Next, Lemma 4.3 implies that $u_2^\varepsilon(t,x) \geqslant u_1^\varepsilon(t,x)$ for $x \in U_1$. Thus the above arguments imply

$$u_1^\varepsilon(t,x) \geqslant v_\varepsilon(t,x) - o_\varepsilon(1) \quad \text{as } \varepsilon \to 0,\ x \in U_1.$$

Finally, by Theorem 4.2 we conclude that $\lim v_\varepsilon(t,x) = 1$ for $d\big(x, F_0 \cap U_1\big) < t\big(2c_1(0)\big)^{1/2}$, where $c_1(0) = \inf_{x\in U_1} c(x,0)$. This implies the inequality $\lim u_\varepsilon(t,x) \geqslant 1$ on the set $\{x \in U_1 : d(x,F_0) < t\sqrt{2c}\}$. Repeating the procedure in U_k, we see that if $\tau_{F_0}(z) < t$, then for a sufficiently small δ, the point z will be reached by the excitation spreading in U_k with velocity corresponding to $\inf_{x\in U_k} c(x,0)$ before the time t. Thus $\lim u_\varepsilon(t,z) \geqslant 1$. By the argument used in proof of Theorem 2.1, it follows that $\lim_{\varepsilon\to 0} u_\varepsilon(t,z) \leqslant 1$, and so $\lim_{\varepsilon\to 0} u_\varepsilon(t,z) = 1$ for $z \in F_t$. □

Now we are ready to investigate the case when (4.12) does not hold.

THEOREM 4.3. *Assume that all the conditions of Theorem* 2.1 *are fulfilled except the condition* (4.12). *Denote*

$$F_t = \{x \in H : \tau_{F_0}(x) < t\}, \qquad K = \{(t,x), t > 0, x \in F_t\}$$

and let $\overline{K}$ *be the closure of* K *in* $[0,\infty) \wedge H$. *Put*

$$\widetilde{S}(t,x) = \sup\{J_s(\varphi) : \varphi \in C([0,s],H),\ s \leqslant T,\ \varphi(0) = x,\\ (s,\varphi(s)) \in K,\ (\tau,\varphi(\tau)) \notin \overline{K},\ \tau \in [0,s)\}.$$

1. *Assume that* $\widetilde{S}(t,x) < 0$ *for* $(t,x) \notin \overline{K}$. *Then*

$$\lim_{\varepsilon\to 0} u_\varepsilon(t,x) = 1 \quad \textit{for } (t,x) \in K,$$
$$\lim_{\varepsilon\to 0} u_\varepsilon(t,x) = 0 \quad \textit{for } (t,x) \notin \overline{K}.$$

2. *Denote* $R_k = K \cup \{(t,x) : \widetilde{S}(t,x) > 0\}$ *and assume that for* $(t,x) \notin \overline{R}$

$$\widetilde{S}(t,x) = \sup\{J_s(\varphi) : \varphi \in C([0,s],H),\ \varphi(0) = x,\ (s,\varphi(s)) \in \overline{R},\\ (\tau,\varphi(\tau)) \notin \overline{R},\ \tau \in [0,s),\ s \leqslant t\}. \tag{4.23}$$

Then

$$\lim_{\varepsilon\to 0} u_\varepsilon(t,x) = 1 \quad \textit{for } (t,x) \in R,$$
$$\lim_{\varepsilon\to 0} u_\varepsilon(t,x) = 0 \quad \textit{for } (t,x) \notin \overline{R}.$$

PROOF. First, let us check that $\lim u_\varepsilon(t,x) = 0$ if $(t,x) \in \overline{R}$, $\widetilde{S}(t,x) < 0$. Put $\tau_t = \inf\{s : (s,\xi_\varepsilon(s)) \in \overline{K}\}$ and note that this random variable is a Markov time. Therefore

$$\begin{aligned} u_\varepsilon(t,x) &= \mathbf{E}u\big(t-\tau_t,\xi_\varepsilon(\tau_t)\big)\exp\Big\{\varepsilon^{-1}\int_0^{\tau_t} c\big(\xi_\varepsilon(s), u_\varepsilon(\tau_t - s, \xi_\varepsilon(s))\big)\,ds\Big\} \\ &\leqslant \mathbf{E}\Pi\big(t-\tau_t,\xi_\varepsilon(\tau_t)\big)\exp\Big\{\varepsilon^{-1}\int_0^{\tau_t} c\big(\xi_\varepsilon(s)\big)\,ds\Big\} = \eta_\varepsilon, \end{aligned} \tag{4.24}$$

where $c(x) = c(x,0)$ and $\Pi(t,x) = 0$ for $t = 0$, $x \in H \setminus F_0$ while $\Pi(t,x) = 1 \vee \sup\{f(x), x \in H\}$ on the lateral boundary of the domain G. As in the proof of Lemma 4.1 one may check that $\lim_{\varepsilon\to 0}\varepsilon\ln\eta_\varepsilon = \widetilde{S}(t,x)$. This relation along with (4.24) yields $\lim_{\varepsilon\to 0}u_\varepsilon(t,x) = 0$ provided $\widetilde{S}(t,x) < 0$. Finally, Lemma 4.4 implies that $\lim_{\varepsilon\to 0}u_\varepsilon(t,x) = 1$ for $(t,x)\in K$. The proof of the second assertion of the theorem is similar to that of Theorem 4.1 provided (4.12) is replaced by (4.23). □

References

[BD1] Ya. Belopol'skaya and Yu. Daletskiĭ, *Diffusion processes in Banach spaces and manifolds*, Trudy Moskov. Mat. Obshch. **37** (1978), 113–150; English transl in Trans. Moscow Math. Soc. **1980**, no. 2.

[BD2] ———, *Cauchy problem for a nonlinear parabolic system via operator-valued multiplicative functionals of Markov processes*, J. Soviet Math. **21** (1983), 653–679.

[BD3] ———, *Investigation of a Cauchy problem for quasilinear parabolic system with finite and infinite-dimensional arguments with the help of Markov random processes*, Izv. Vyssh. Uchebn. Zaved. Mat. **1978**, no. 6 (193), 5–17; English transl in Societ Math. (Iz. VUZ) **22** (1978).

[BD4] ———, *Stochastic equations and differential geometry*, Kluwer, Dorderecht, 1990.

[D] Yu. Daletskiĭ, *Infinite-dimensional elliptic operators and parabolic equations associated with them*, Russian Math. Surveys **22** (1967), no. 4, 1–53.

[F1] M. Freidlin, *Functional integration and partial differential equations*, Princeton Univ. Press, Princeton, NJ, 1985.

[F2] ———, *Quasilinear parabolic equations and measure in function space*, Functional Anal. Appl. **1** (1967), 237–240.

[F3] ———, *Propagation of concentration waves by a random motion connected with growth*, Dokl. Akad. Nauk SSSR **246** (1979), 544–548; English transl in Soviet Math. Dokl. **20** (1979).

[FW] M. Freidlin and A. Wentzell, *Random perturbations of dynamical systems*, "Nauka", Moscow, 1970; English transl, Springer-Verlag, New York, 1984.

[GF] J. Gärtner and M. Freidlin, *On the propagation of concentration waves in periodic and random media*, Dokl. Akad. Nauk SSSR **249** (1979), 521–525; English transl in Soviet Math. Dokl. **20** (1979).

[McK] H. P. McKean, *A class of Markov processes associated with nonlinear parabolic equations*, Proc. Nat. Acad. Sci. U.S.A. **59** (1966), 1907–1911.

[S] D. W. Stroock, *An introduction to the theory of large deviations*, Springer-Verlag, New York, 1984.

[V] S. R. S. Varadhan, *Diffusion processes in a small time interval*, Comm. Pure Appl. Math. **20** (1967), 659–685.

St. Petersburg Academy of Civil Engineering

Amer. Math. Soc. Transl.
(2) Vol. **164**, 1995

On the Stability of Solitary Waves for Nonlinear Schrödinger Equations

V. S. Buslaev and G. S. Perelman

§1. Main results

1.1. Solitary waves. The results of this paper can be considered as a development of some results of the authors' previous papers **[BP1–3]**. All three of those papers were devoted to the stability problem for the general nonlinear Schrödinger equation. In order to describe and compare the main results of this paper with the previous ones we must introduce the initial notions.

Consider the nonlinear Schrödinger equation:

$$
\begin{gathered}
i\vec{\psi}_t = \left[-\partial_x^2 + V(\psi_1\psi_2)\right]\sigma_3\vec{\psi}, \qquad \vec{\psi} = \vec{\psi}(x,t) = \begin{pmatrix} \psi_1(x,t) \\ \psi_2(x,t) \end{pmatrix} \in \mathbb{C}^2, \\
x \in \mathbb{R}, \qquad t \in \mathbb{R}_+, \qquad \sigma_3 = \begin{pmatrix} 1 & 0 \\ 0 & -1 \end{pmatrix}.
\end{gathered} \tag{1.1}
$$

Here for simplicity we shall deal only with the invariant class of even solutions: $\vec{\psi}(x,t) = \vec{\psi}(-x,t)$. This restriction can be surmounted with the help of some tricks proposed in **[BP2, BP3]**.

The second restriction of the class of solutions is somewhat more essential: without any special further comments we shall consider only solutions of the form

$$
\vec{\psi} = \begin{pmatrix} \psi \\ \overline{\psi} \end{pmatrix}, \qquad \psi_2 = \overline{\psi}_1 = \overline{\psi},
$$

also invariant with respect to the dynamics.

Assume that

(i) V is a smooth ($\in C^\infty$) real-valued function obeying the inequality

$$
V(\xi) \geqslant -V_1\xi^q, \qquad V_1 > 0,\ \xi \geqslant 1,\ q < 2;
$$

(ii) the point $\xi = 0$ is a root of V:

$$
V(\xi) = V_2\xi^p\big(1 + O(\xi)\big), \qquad p > 0.
$$

Further assumptions on V will be given in terms of the function

$$
U(\varphi, \alpha) = -\frac{1}{8}\alpha^2\varphi^2 - \frac{1}{2}\int_0^{\varphi^2} V(\xi)\,d\xi.
$$

1991 *Mathematics Subject Classification*. Primary 35Q55.

It is assumed that

(iii) for α from some interval, $\alpha \in A \subset \mathbb{R}_+$, the function $\varphi \mapsto U(\varphi,\alpha)$ has a positive root and the smallest positive root $\varphi_0 = \varphi_0(\alpha)$ is simple: $U_\varphi(\varphi_0,\alpha) \neq 0$.

Under assumptions (ii) and (iii) there exists an unique even positive solution $\varphi(y)$ of the equation

$$\varphi_{yy} = -U_\varphi,$$

vanishing at infinity:

$$\varphi(y,\alpha) \sim \varphi_0 \exp(-\alpha/2|y|).$$

The functions $\vec{w}(x,\vec{\sigma}) = e^{-i\beta\sigma_3}\vec{\varphi}(x,\alpha)$, $\vec{\varphi}(x,\alpha) = \varphi(x,\alpha)\binom{1}{1}$, $\vec{\sigma} = (\beta,\omega)$, $\omega = -\frac{1}{4}\alpha^2$, $\beta,\omega \in \mathbb{R}$, $\alpha \in A$, will be called *soliton states*. The set of the admissible $\vec{\sigma}$ will be denoted by Σ.

If $\vec{\sigma} = \vec{\sigma}(t)$ is a solution of the Hamilton system

$$\beta' = \omega, \qquad \omega' = 0, \tag{1.2}$$

then the function $\vec{w}(x,\sigma(t))$ is a solution of equation (1.1) called the *solitary wave*, or simply the *soliton*.

1.2. Linearization on the solitons. Consider the linearization of equation (1.1) on the soliton $\vec{w}(x,\vec{\sigma}(t))$:

$$\vec{\psi} \sim \vec{w} + \vec{\chi},$$
$$i\vec{\chi}_t = \left[-\partial_x^2 + V(\varphi^2)\right]\sigma_3\vec{\chi} + V'(\varphi^2)\varphi^2\left(\sigma_3 - ie^{2i\beta\sigma_3}\sigma_2\right)\vec{\chi},$$

where $\sigma_2 = \begin{pmatrix} 0 & i \\ -i & 0 \end{pmatrix}$. Instead of $\vec{\chi}$ introduce the function $\vec{f}$:

$$\vec{\chi}(x,t) = \exp(-i\beta(t)\sigma_3)\vec{f}(x,t).$$

This function satisfies the equation

$$i\vec{f}_t = H(\alpha)\vec{f}, \tag{1.3}$$

where

$$H(\alpha) = H_0(\alpha) + W(\alpha), \qquad H_0(\alpha) = (-\partial_x^2 + \alpha^2/4)\sigma_3,$$
$$W(\alpha) = \left(V(\varphi^2) + V'(\varphi^2)\varphi^2\right)\sigma_3 - iV'(\varphi^2)\varphi^2\sigma_2.$$

Further consider $H(\alpha)$ as a linear operator in $L_2(\mathbb{R} \to \mathbb{C}^2)$ defined on the domain where $H_0(\alpha)$ is selfadjoint.[1] It is clear that $H(\alpha)$ is real.

The continuous spectrum of $H(\alpha)$ lies on two half-axes $(-\infty,-E_0]$, $[E_0,\infty)$, $E_0 = \alpha^2/4$, and is simple. In addition, the operator $H(\alpha)$ can have a finite and finite-dimensional discrete spectrum on the real axis and on the imaginary one.

Moreover, $\sigma_1 H = -H\sigma_1$, $\sigma_2 H = -H^*\sigma_2$, and $\sigma_3 H = H^*\sigma_3$, where $\sigma_1 = \begin{pmatrix} 0 & 1 \\ 1 & 0 \end{pmatrix}$; therefore the spectrum is symmetric with respect to the transformations $E \to -E$ and $E \to \overline{E}$.

Let us indicate some more subtle spectral properties of $H(\alpha)$.

[1] Here (and later) we mean that L_2 is the subspace of the standard L_2 consisting of even functions.

1) The point $E = 0$ is a point of the spectrum, and

$$\vec{\xi}_1 = \begin{pmatrix} u_1 \\ \overline{u}_1 \end{pmatrix}, \qquad u_1 = -i\varphi,$$

is its corresponding eigenfunction: $H\vec{\xi}_1 = 0$; the corresponding invariant subspace is at least two-dimensional and

$$\vec{\xi}_2 = \begin{pmatrix} u_2 \\ \overline{u}_2 \end{pmatrix}, \qquad u_2 = -\frac{2}{\alpha}\varphi_\alpha,$$

is a function adjoint to $\vec{\xi}_1$: $H\vec{\xi}_2 = i\vec{\xi}_1$.

2) The invariant subspaces corresponding to the nonzero eigenvalues are eigenspaces.

3) The spectrum of $H(\alpha)$ is real, and the invariant subspace corresponding to the point $E = 0$ is generated by $\vec{\xi}_1$, $\vec{\xi}_2$ if and only if

$$\frac{d}{d\alpha}\|\varphi\|_2^2 > 0.$$

4) Let

$$\vec{\xi} = \vec{\xi}(x, \alpha) = \begin{pmatrix} v_1(x, \alpha) \\ v_2(x, \alpha) \end{pmatrix}$$

be an eigenvector corresponding to some eigenvalue $\lambda = \lambda(\alpha)$:

$$H(\alpha)\vec{\xi}(\cdot, \alpha) = \lambda(\alpha)\vec{\xi}(\cdot, \alpha),$$

and $0 < \lambda(\alpha) < E_0(\alpha)$. Then $\vec{\xi}$ can be normalized so that it becomes real and obeys the condition

$$\langle \vec{\xi}, \sigma_3 \vec{\xi} \rangle = \|v_1\|_2^2 - \|v_2\|_2^2 = 1.$$

Generically the equations $H(\alpha)\vec{\psi} = \pm E_0(\alpha)\vec{\psi}$ do not have solutions bounded at infinity. If, nevertheless, such bounded solutions exist, the points $\pm E_0$ are called *resonances*.

1.3. Description of the problem. Consider the Cauchy problem for equation (1.1) with the initial data

$$\vec{\psi}\big|_{t=0} = \vec{\psi}_0 \tag{1.4}$$

of the form

$$\vec{\psi}_0 = \vec{w}(\cdot, \vec{\sigma}_0) + \vec{\chi}_0, \tag{1.5}$$

where $\vec{\sigma}_0 \in \Sigma$. Assume that the norm

$$N = N(\chi_0) = \left\|(1 + x^2)\chi_0\right\|_2 + \|\chi_0'\|_2$$

is sufficiently small.

In **[BP2, BP3]** we considered this problem under the following conditions:

1) $p \geqslant 4$.

2) The operator $H(\alpha_0)$ does not have any eigenvalues except $E = 0$.

3) The invariant subspace corresponding to $E = 0$ is generated by the vectors $\vec{\xi}_1$, $\vec{\xi}_2$.

4) The points $\pm E_0(\alpha_0)$ are not resonances.

We were able to prove that the solution $\vec{\psi}$ admits the following asymptotic representation:

$$\vec{\psi}(\cdot, t) = \vec{w}\big(\cdot, \vec{\sigma}_+(t)\big) + e^{-il\sigma_3 t}\vec{f}_+ + o(1), \qquad t \to +\infty, \tag{1.6}$$

where $o(1)$ implies L_2-norm, $l = -\partial_x^2$, $\vec{f}_+ \in L_2$ and is small, $\vec{\sigma}_+(t)$ obeys equation (1.2), and $\vec{\sigma}_+(0) - \vec{\sigma}_0$ is also small.

In the present paper instead of condition 2) we shall also assume the existence of real nonzero eigenvalues of $H(\alpha_0)$ on the interval $\big(-E_0(\alpha_0), E_0(\alpha_0)\big)$. For simplicity we suppose that

2′) the interval $\big(0, E_0(\alpha_0)\big)$ contains exactly one eigenvalue $\lambda_0 = \lambda(\alpha_0)$, and $2\lambda_0 > E(\alpha_0)$.

Additionally we must replace condition 1) by the slightly stronger condition

1′) $p > 4$

and introduce the new condition

5′)

$$\langle \vec{F}, \sigma_3 \vec{\Psi} \rangle \neq 0,$$

where $\vec{\Psi}$ is the eigenfunction of the continuous spectrum of $H(\alpha_0)$ with the eigenvalue $2\lambda_0$,

$$\vec{F} = c_1 \begin{pmatrix} v_1(v_1 + 2v_2) \\ -v_2(v_2 + 2v_1) \end{pmatrix} + c_2 \begin{pmatrix} v_2^2 \\ -v_1^2 \end{pmatrix}, \tag{1.7}$$

$$c_1 = V'(\varphi^2)\varphi + \tfrac{1}{2}V''(\varphi^2)\varphi^3, \qquad c_2 = \tfrac{1}{2}V''(\varphi^2)\varphi^3. \tag{1.8}$$

The meaning of the last condition is rather simple: it means that the interaction of the eigenspace corresponding to the eigenvalue λ_0, via the term of double frequency $2\lambda_0$ generated by the nonlinearity of equation (1.1), with the subspace of the continuous spectrum, is nontrivial.

Under these broader conditions, the main result (1.6) remains valid with a slight modification: the curve $\vec{\sigma}_+(t)$ ceases to obey (1.2). In fact, asymptotically one only has the equations

$$\beta'_+ = \omega_+ + O(t^{-1}), \qquad \omega'_+ = 0,$$

although $\omega_+(0)$ is close to ω_0 again.

One can say that the previous behavior is reproduced owing to the interaction between the contribution of the additional discrete spectrum and the contribution of the continuous spectrum. By virtue of the dispersive swimming about, the contribution of the continuous spectrum locally vanishes as $t \to +\infty$ and forces the contribution of the additional discrete spectrum to disappear. Of course, in the following we shall control the velocity of this disappearance. More detailed, but only heuristic information on this account can be extracted from formulas of **[BP1]**.

1.4. Outline of the paper. The main idea of the proof will repeat the main idea of **[BP2, BP3]**, but some essential technical modifications will appear. Following **[BP1, BP2]**, we introduce some new coordinates for the description of the solutions with initial data (1.5). According to the language of **[BP2, BP3]**, we split motions. The new coordinates possess the following important property: for all time t they admit only small deviations from their initial values. We consider these coordinates in §2. We conclude the section by a system of equations for the new coordinates. The equations contain the contribution corresponding to the additional discrete spectrum in rather implicit form, and we must transform them in order to ensure the possibility of estimating the order of decrease of this contribution at infinity. Such a transformation is described in §4. In §5 we prove that the new coordinates indeed admit only small deviations from their initial values. As in **[BP2, BP3]**, for this purpose we use the method of majorants. In the last §6 we derive the asymptotic formula (1.6). The very short §3 contains a list of some auxiliary estimates used in §4 and very extensively in §5. They were either proved in **[GV1]** or can be proved fairly easily by the same methods

The list of references is essentially the same as in **[BP2]**, so we again refer our readers to that paper.

§2. The splitting of motions

2.1. Orthogonality condition. Write the solution $\vec{\psi}$ of the Cauchy problem (1.1), (1.4) as the sum

$$\vec{\psi} = e^{-i\beta\sigma_3}(\vec{\varphi} + \vec{\zeta} + \vec{f}), \tag{2.1}$$

where

$$\beta = \beta(t), \qquad \vec{\varphi} = \begin{pmatrix} \varphi \\ \varphi \end{pmatrix}, \qquad \varphi = \varphi(x, \alpha(t)),$$
$$\vec{\zeta} = \vec{\zeta}(x,t) = z(t)\vec{\xi}(x,\alpha(t)) + \overline{z(t)}\sigma_1\vec{\xi}(x,\alpha(t)).$$

Here

$$\vec{\sigma} = \vec{\sigma}(t) = (\beta(t), \omega(t)) \in \Sigma$$

is a trajectory in the set of parameters of the soliton. The vector $\vec{\xi}(x,\alpha)$ is the eigenvector of $H(\alpha)$ corresponding to the eigenvalue $\lambda(\alpha)$, and $\sigma_1\vec{\xi}(x,\alpha)$ corresponds to $-\lambda(\alpha)$. Finally, $z(t)$ is a complex-valued function.

The decomposition (2.1) can be fixed by the condition(s) of orthogonality:

$$\langle \vec{f}(t), \sigma_3\vec{\xi}_j(\cdot, \alpha(t))\rangle = 0, \qquad j = 1, 2, \tag{2.2}$$

$$\langle \vec{f}(t), \sigma_3\vec{\xi}(\cdot, \alpha(t))\rangle = 0, \tag{2.3}$$

$$\langle \vec{f}(t), \sigma_3\sigma_1\vec{\xi}(\cdot, \alpha(t))\rangle = 0. \tag{2.4}$$

Geometrically these conditions mean that for each t the component $\vec{f}(t)$ belongs to the subspace of the continuous spectrum of $H(\alpha(t))$.

Conditions (2.2) can be rewritten in a different form. Let

$$\vec{\psi}(t) = \vec{w}(\cdot, \vec{\sigma}(t)) + \vec{\chi}(t);$$

then

$$\langle \vec{\chi}(t), \sigma_3\vec{w}_\sigma(\cdot, \vec{\sigma}(t))\rangle = 0. \tag{2.5}$$

Here $\vec{w}_\sigma$ is the derivative of $\vec{w}(\cdot, \vec{\sigma})$ with respect to the parameter $\vec{\sigma}$:

$$\vec{w}_\beta = e^{-i\beta\sigma_3}\vec{\xi}_1, \qquad \vec{w}_\omega = e^{-i\beta\sigma_3}\vec{\xi}_2.$$

Condition (2.5) implies that $\exp(i\beta\sigma_3)\vec{\chi}$ belongs to the subspace generated by the continuous spectrum and by the eigenvalues $\pm\lambda(\alpha)$. The further decomposition of $\exp(i\beta\sigma_3)\vec{\chi}$ can be obtained immediately in terms of the eigenfunction $\vec{\xi}(\cdot,\alpha(t))$:

$$z = \langle e^{i\beta\sigma_3}\vec{\chi}, \sigma_3\vec{\xi}\rangle.$$

2.2. Initial data. Let the initial data have the form

$$\vec{\psi}_0 = \vec{w}(\cdot,\vec{\sigma}_0) + \vec{\chi}_0 \tag{2.6}$$

and let the N-norm of $\vec{\chi}_0$ be sufficiently small. Then the decomposition (2.6) of $\vec{\psi}_0$ can be transformed in a decomposition of the same form subjected to the orthogonality condition

$$\langle\vec{\chi}_0, \sigma_3\vec{w}_\sigma(\cdot,\vec{\sigma}_0)\rangle = 0,$$

and the N-norm of the new $\vec{\chi}_0$ will be small again.

More precisely, the function $\vec{\psi}_0$ of the form (2.6) can be represented as the sum

$$\vec{\psi}_0 = \vec{w}(\cdot,\vec{\sigma}_1) + \vec{\chi}_1,$$

where

$$\langle\vec{\chi}_1, \sigma_3\vec{w}_\sigma(\cdot,\vec{\sigma}_1)\rangle = 0,$$

the N-norm of χ_1 and the difference $\vec{\sigma}_1 - \vec{\sigma}_0$ being arbitrarily small if the N-norm of $\vec{\chi}_0$ is sufficiently small.

This fact is a consequence of the unique solvability of the equation

$$\langle\vec{\psi}_0 - \vec{w}(\cdot, v\sigma_1), \sigma_3\vec{w}_\sigma(\cdot,\vec{\sigma}_1)\rangle = 0.$$

In its turn this solvability is a consequence of the nondegeneration of the corresponding Jacobian

$$\begin{aligned} -\langle\vec{w}_{\sigma_k}(\cdot,\vec{\sigma}_0), \sigma_3\vec{w}_{\sigma_l}(\cdot,\vec{\sigma}_0)\rangle &= -2iA_{kl}(\alpha_0),\\ A(\alpha) = \begin{pmatrix} 0 & e\\ -e & 0\end{pmatrix}, \qquad & e = \frac{1}{\alpha}\frac{d}{d\alpha}\|\varphi\|_2^2. \end{aligned} \tag{2.7}$$

As a consequence of the above assumptions on $H(\alpha)$ we have $\det A(\alpha_0) \neq 0$ (see property 3) of $H(\alpha)$).

So one can assume that $\vec{\psi}_0$ admits the representation (2.6) or even the representation

$$\vec{\psi}_0 = e^{-i\beta_0\sigma_3}\big(\vec{\psi}(\cdot,\alpha_0) + z_0\vec{\xi}(\cdot,\alpha_0) + \bar{z}_0\sigma_1\vec{\xi}(\cdot,\alpha_0) + \vec{f}_0\big),$$

where $\vec{\chi}_0$ and $\vec{f}_0$ respectively satisfy conditions (2.5) and (2.2)–(2.4) for $t = 0$, and the N-norms of $\vec{\chi}_0$, $\vec{f}_0$ and z_0 are small enough.

2.3. Differential equations. Now we can replace the splitting conditions by differential equations. Jointly with the splitting conditions for the initial data, these equations will be equivalent to the initial splitting conditions. Combining these equations with equation (1.1), we shall arrive a complete system of differential equations for the components of decomposition (2.1), i.e., for the parameters $\vec{\sigma}$ of the soliton, for the amplitude z of the discrete spectrum's contribution and for the contribution $\vec{f}$ of the continuous spectrum.

Instead of β it is convenient to introduce a new parameter γ:

$$\beta(t) = \int_0^t \omega(\tau)\, d\tau + \gamma(t).$$

Equations (1.2) in terms of (γ, ω) acquire the form

$$\gamma' = 0, \qquad \omega' = 0.$$

Below, $\vec{\sigma}$ will also denote the vector (γ, ω).

Now rewrite (1.1) as an equation for $\vec{f}$:

$$i\vec{f}_t = H(\alpha)\vec{f} + \vec{N}_1(\alpha, z, \vec{f}), \tag{2.8}$$

where

$$\vec{N}_1(\alpha, z, \vec{f}) = \vec{N}(\varphi, \vec{f} + \vec{\zeta}) - \gamma'\sigma_3\vec{f} + l(\sigma)(\sigma_3 i\vec{\xi}_1 + \vec{\zeta}) + l_1(z)\vec{\xi},$$

$$\vec{N}(\varphi, \vec{f}) = V\left(|\varphi + f|^2\right)\sigma_3\vec{f} - V(\varphi^2)\sigma_3\vec{\varphi} - \left[V(\varphi^2) + V'(\varphi^2)\varphi^2\right]\sigma_3\vec{f} + iV'(\varphi^2)\varphi^2\sigma_2\vec{f},$$

$$l(\sigma) = -\gamma'\sigma_3 + i\omega'\frac{2}{\alpha}\partial_\alpha, \quad l_1(z) = (\lambda z - iz') - (\lambda\bar{z} + i\bar{z}')\sigma_1, \quad \vec{\zeta} = z\vec{\xi} + \bar{z}\sigma_1\vec{\xi}.$$

Substitute the expression for f_t from (2.8) into the derivative of the orthogonality conditions:

$$A\vec{\sigma}' = \vec{g}, \qquad \vec{\sigma} = (\gamma, \omega), \tag{2.9}$$

$$iz' - \lambda z = g. \tag{2.10}$$

The matrix A here is given by (2.7),

$$g_j = \langle \vec{N} + l(\sigma)\vec{\zeta}, \sigma_3\vec{\xi}_j\rangle + \langle \vec{f}, l(\sigma)\sigma_3\vec{\xi}_j\rangle,$$

$$g = \langle \vec{N} + l(\sigma)\vec{\zeta}, \sigma_3\vec{\xi}\rangle + \langle \vec{f}, l(\sigma)\sigma_3\vec{\xi}\rangle.$$

The expressions $\vec{g}$ and g also contain the derivative $\vec{\sigma}'$, which enters linearly in $l(\sigma)$.

In principle the system (2.9), (2.10) can be solved with respect to the derivatives and, together with equation (2.8), constitutes a complete system for $\vec{\sigma}$, z, $\vec{f}$:

$$\vec{\sigma}' = F(\alpha, z, \vec{f}), \tag{2.11}$$

$$iz' - \lambda z = G(\alpha, z, \vec{f}), \tag{2.12}$$

$$i\vec{f}_t = H(\alpha)\vec{f} + \vec{N}_1(\alpha, z, \vec{f}). \tag{2.13}$$

2.4. Equations on a finite interval. We hope that α has a finite limit α_+, as $t \to +\infty$, so that the operator $H(\alpha_+)$ could be a good approximation of $H(\alpha)$ in equation (2.13). Unfortunately, at the moment we do not know if α has a limit at all. This forces us to consider equations (2.13) on some finite interval $[0, t_1]$ and later investigate the limit as $t_1 \to \infty$. As an approximation of $H(\alpha)$ on this interval we can use the operator $H_1 = H(\alpha_1)$, $\alpha_1 = \alpha(t_1)$. This idea follows the papers **[BP2, BP3]**.

It is expedient to represent $\vec{f}$ as the sum $\vec{f} = \vec{k} + \vec{h}$ of the projections on the subspaces corresponding to the discrete and the continuous spectra of H_1. In particular, $\vec{f}(t_1) = \vec{h}(t_1)$.

Owing to the orthogonality conditions (2.2–4), the 4-dimensional component $\vec{k}$ can be expressed in terms of $\vec{h}$:

$$\begin{aligned} \langle \vec{k}, \sigma_3 \vec{\xi}_j \rangle + \langle \vec{h}, \sigma_3 \vec{\xi}_j \rangle &= 0, \qquad j = 1, 2, \\ \langle \vec{k}, \sigma_3 \vec{\xi} \rangle + \langle \vec{h}, \sigma_3 \vec{\xi} \rangle &= 0, \\ \langle \vec{k}, \sigma_3 \sigma_1 \vec{\xi} \rangle + \langle \vec{h}, \sigma_3 \sigma_1 \vec{\xi} \rangle &= 0. \end{aligned}$$

Since

$$\vec{k} = \sum_{j=1}^{4} k_j \vec{\xi}_j(\alpha_1), \qquad \vec{\xi}_3 = \vec{\xi}, \qquad \vec{\xi}_4 = \sigma_1 \vec{\xi},$$

we obtain a system for the coefficients k_j of the form

$$B(\alpha)\mathbf{k} = \mathbb{I}, \qquad \mathbf{k} = (k_1, k_2, k_3, k_4). \tag{2.14}$$

Here

$$B(\alpha) = \{ \langle \vec{\xi}_i(\alpha_1), \sigma_3(\vec{\xi}_j(\alpha)) \rangle_{i,j=1} \}^4, \qquad \mathbb{I}_j = -\langle \vec{h}, \sigma_3 \vec{\xi}_j(\alpha) \rangle.$$

It is clear that

$$B(\alpha) = B(\alpha_1) + B_1(\alpha), \tag{2.15}$$

where

$$B(\alpha_1) = \begin{pmatrix} -2iA(\alpha_1) & 0 \\ 0 & \sigma_3 \end{pmatrix},$$

$$B_1(\alpha) = \{ \langle \vec{\xi}_i(\alpha_1), \sigma_3(\vec{\xi}_j(\alpha) - \vec{\xi}_j(\alpha_1)) \rangle \}_{i,j=1}^4. \tag{2.16}$$

Since $\mathbb{I}(\alpha_1) = 0$, we have

$$\mathbb{I}_j = \langle \vec{h}, \sigma_3(\vec{\xi}_j(\alpha) - \vec{\xi}_j(\alpha_1)) \rangle. \tag{2.17}$$

In order to fix the component $\vec{h}$, it is sufficient to consider the projection of equation (2.13) on the subspace of the continuous spectrum of $H(\alpha_1)$. Separating as the main term of $H(\alpha)\vec{f}$ the operator $H(\alpha_1)\vec{f}$, one gets

$$i\vec{h}_t = H_1 \vec{h} - \Delta P_1 \sigma_3 \vec{f} + \vec{N}_2. \tag{2.18}$$

In this equation

$$\Delta = \beta' - \omega_1, \qquad \vec{N}_2 = P_1 (W(\alpha) - W(\alpha_1)) \vec{f} + P_1 (\vec{N}_1 + \gamma' \sigma_3 \vec{f}),$$

and $P_1 = P(\alpha_1)$, where $P(\alpha)$ is the spectral projection operator on the subspace of the continuous spectrum of $H(\alpha)$.

The component $-\Delta P_1 \sigma_3 \vec{f}$ dominates in the correction and affects the structure of the asymptotic behavior of $\vec{h}$, as $t \to \infty$, even in the main term. Fortunately, it is possible to replace this expression by another one which commutes with H_1, but has the same influence on the main term of the asymptotic representation:

$$P_1 \sigma_3 \vec{f} \to (P_+ - P_-) \vec{h}.$$

Here $P_\pm = P_\pm(\alpha_1)$, $P_\pm(\alpha)$ are the projection operators on the subspaces corresponding to the positive and negative parts of the continuous spectrum of $H(\alpha)$, respectively, $P(\alpha) = P_+(\alpha) + P_-(\alpha)$. The difference $P_1 \sigma_3 \vec{f} - (P_+ - P_-)\vec{h}$ has the same order as some terms already included in $\vec{N}_2$. This difference will be estimated in §4.

Thus, we rewrite equation (2.18) in the following final form:

$$i\vec{h}_t = H_1\vec{h} - \Delta(P_+ - P_-)\vec{h} + \vec{D}, \tag{2.19}$$

where

$$\vec{D} = \vec{N}_2 + \Delta\big[(P_+ - P_-) - P_1\sigma_3\big]\vec{f}.$$

Introduce the operator $\mathscr{U}(t,\tau)$:

$$i\mathscr{U}_t(t,\tau) = \big[H_1 - \Delta(P_+ - P_-)\big]\mathscr{U}(t,\tau), \qquad \mathscr{U}(\tau,\tau) = I.$$

It is simple to check that

$$\mathscr{U}(t,\tau)P_\pm = \exp\left[-iH_1(t-\tau) \pm i\int_\tau^t \Delta(s)\,ds\right]P_\pm.$$

The differential equation (2.19) can be rewritten as an equivalent integral equation:

$$\vec{h} = \mathscr{U}(t,0)P_1\vec{f}_0 - i\int_0^t \mathscr{U}(t,\tau)\vec{D}(\tau)\,d\tau. \tag{2.20}$$

§3. Preliminary estimates

Here we present without proofs a list of estimates for solutions of the nonlinear equation (1.1) and the linear equation (1.3). These estimates for nonlinear evolution can be found in **[GV1–2]**. For linear evolution, most of them were proved in **[BP2]**; the others can be obtained by the same methods.

Let H^1 be the standard Sobolev space with the norm

$$\|f\|_{H^1} = \|f\|_2 + \|f'\|_2.$$

Let the initial data ψ_0 belong to H^1 (see (1.4)). Then the Cauchy problem has a unique solution $\psi(\cdot,t) \in C(\mathbb{R}_+ \to H^1)$, which obeys the estimate

$$\|\psi(\cdot,t)\|_{H^1} \leqslant c(\|\psi_0\|_{H^1})\|\psi_0\|_{H^1}, \tag{3.1}$$

where c is a locally bounded function.

Moreover, if $\|(1+|x|)\psi_0\|_2 < \infty$, then

$$\big\|(1+|x|)\psi(x,t)\big\|_2 \leqslant c(\|\psi_0\|_{H^1})\big[\|(1+|x|)\psi_0\|_2 + |t|\|\psi_0\|_{H^1}\big]. \tag{3.2}$$

For the linear evolution $\mathscr{U}(t) = \exp(-itH(\alpha))$, the following estimates hold:

$$\big\|\mathscr{U}(t)P_\pm\vec{h}\big\|_2 \leqslant c\|h\|_2,$$

$$\big\|\mathscr{U}(t)P_\pm\vec{h}\big\|_\infty \leqslant c\begin{cases} t^{-1/2}\big[\|h\|_2 + \|h\|_W\big], \\ (1+t)^{-1/2}\big[\|h\|_{H^1} + \|h\|_W\big], \end{cases} \tag{3.3}$$

$$\big\|\rho\mathscr{U}(t)P_\pm\vec{h}\big\|_2 \leqslant c(1+|t|)^{-3/2}\big[\|h\|_2 + \|h\|_W\big]. \tag{3.4}$$

Here $P_\pm$ are the spectral projection operators corresponding to the positive $[E_0,\infty)$

and negative $(-\infty, -E_0]$ parts of the continuous spectrum of $H(\alpha)$, and $\|h\|_W$ means either $\|(1+x^2)h\|_1$ or $\|(1+x^2)h\|_2$ (the choice is arbitrary). Finally

$$\rho(x) = \left(1+|x|\right)^{-\nu}, \qquad \nu > 3.5. \tag{3.5}$$

All these estimates were almost proved in **[BP2]**, where instead of $P_\pm$ in the estimates we considered $P_+ + P_-$.

By the same methods one can prove also that

$$\|\rho\mathscr{U}(t)H^{-1}P_\pm\vec{h}\|_2 \leqslant c(1+t)^{-3/2}\|(1+x^2)h\|_2, \tag{3.6}$$

$$\|\rho\mathscr{U}(t)(H-2\lambda-i0)^{-1}P_\pm\vec{h}\|_2 \leqslant c(1+t)^{-3/2}\left[\|h\|_2 + \|(1+|x|)^3h\|_1\right]. \tag{3.7}$$

The sign "$-$" (in $-i0$) is essential: it corresponds to the sign of t. We remark that in the sense of distributions

$$e^{-i\lambda t}\frac{1}{\lambda - i0} \to \begin{cases} 0, & t\to+\infty, \\ 2\pi i\delta(\lambda), & t\to-\infty. \end{cases}$$

A specific property of the projection operators $P_\pm$ is that

$$\|(1+x^2)(P_1\sigma_3 - P_+ + P_-)\vec{h}\|_2 \leqslant c\|\rho h\|_2. \tag{3.8}$$

§4. Effective equations

4.1. Main nonlinear terms of $\vec{\sigma}'$ and z'. Let us return to the approach initiated in §2. One understands that the asymptotic vanishing of z is induced by the nonlinear terms of (2.11) and (2.12). In order to separate the effective equations, consider the main nonlinear terms of (2.11)–(2.13).

Below it will become clear that the contribution $\vec{f}$ of the continuous spectrum has asymptotically the same order (in the uniform norm) as z^2. In its turn, $z(t) = O(t^{-1/2})$.

We shall use these facts while deriving the equations. At this stage we are not worrying about formal rigor.

Consider (2.9) and (2.10). The main terms of their expansions arise from the main terms of the expansion of the vector $\vec{N} = \vec{N}(\varphi, \vec{f}+\vec{\zeta})$ in the expressions for $\vec{g}$ and g. By the definition of $\vec{N}$, the main order of $\vec{N}$ is quadratic in $f+\zeta$. It is necessary to consider also the third order terms:

$$\vec{N}(\varphi, \vec{f}) = \vec{L}_2(f) + \vec{L}_3(f) + \dots,$$

$$\vec{L}_2(f) = c_1\begin{pmatrix} f_1(f_1+2f_2) \\ -f_2(f_2+2f_1)\end{pmatrix} + c_2\begin{pmatrix} f_2^2 \\ -f_1^2\end{pmatrix},$$

where c_1 and c_2 are given by (1.8), and

$$\begin{aligned}\vec{L}_3(f) &= V'(\varphi^2)f_1f_2\begin{pmatrix} f_1 \\ -f_2\end{pmatrix} \\ &\quad + \frac{1}{2}V''(\varphi^2)\varphi^2(f_1+f_2)\left[2f_1f_2\begin{pmatrix} 1 \\ -1\end{pmatrix} + (f_1+f_2)\begin{pmatrix} f_1 \\ f_2\end{pmatrix}\right] \\ &\quad + \frac{1}{6}V'''(\varphi^2)\varphi^4(f_1+f_2)^3\begin{pmatrix} 1 \\ -1\end{pmatrix}.\end{aligned}$$

Remember now that the operator $l(\sigma)$ in the expressions for $\vec{g}$, g also contains the derivatives γ', ω'. Express them in terms of $\vec{g}$, g and in the resulting expression leave only the main second order terms generated by the vector N. As a result one can say that the main terms in $\vec{g}$, g are generated by the replacement, in (2.9) and (2.10), of $l(\sigma)\vec{\zeta} + \vec{N}(\varphi, \vec{f} + \vec{\zeta})$ by the explicit expression

$$l(\sigma)\vec{\zeta} + \vec{N}(\varphi, \vec{f} + \vec{\zeta}) \sim \vec{L}_2(\zeta) + \vec{L}_3(\zeta) + \vec{L}_3'(\zeta) + C(\zeta)\vec{f},$$

where

$$\vec{L}_3'(\zeta) = e^{-1}\Big(\langle \vec{L}_2(\zeta), \sigma_3\vec{\xi}_2\rangle\sigma_3 + i\langle \vec{L}_2(\zeta), \sigma_3\vec{\xi}_1\rangle \frac{2}{\alpha}\partial_\alpha\Big)\vec{\zeta},$$

$$C(\zeta) = 2c_1\begin{pmatrix} \zeta_1+\zeta_2 & \zeta_1 \\ -\zeta_2 & -\zeta_1-\zeta_2 \end{pmatrix} + 2c_2\begin{pmatrix} 0 & \zeta_2 \\ -\zeta_1 & 0 \end{pmatrix}.$$

Now one can write out the result, i.e., the main terms of the expansion of the nonlinear terms in equations (2.9) and (2.10):

$$e\omega' = \langle \vec{L}(\zeta) + C(\zeta)\vec{f}, \sigma_3\vec{\xi}_1\rangle + \cdots,$$

$$-e\gamma' = \langle \vec{L}(\zeta) + C(\zeta)\vec{f}, \sigma_3\vec{\xi}_2\rangle + \cdots,$$

$$iz' - \lambda z = \langle \vec{L}(\zeta) + C(\zeta)\vec{f}, \sigma_3\vec{\xi}\rangle + \cdots,$$

where $\vec{L}(\zeta) = \vec{L}_2(\zeta) + \vec{L}_3(\zeta) + \vec{L}_3'(\zeta)$.

It is also worthwhile to rewrite the main terms as explicit polylinear forms in z and $\bar{z}$. For $\vec{L}_2$:

$$\vec{L}_2(\zeta) = z^2\vec{F}(\alpha) - \bar{z}^2\sigma_1\vec{F}(\alpha) + |z|^2F_0(\alpha)\begin{pmatrix} 1 \\ -1 \end{pmatrix},$$

where $\vec{F}(\alpha) = \vec{L}_2(\vec{\xi})$ (see also (1.7)), and

$$F_0(\alpha) = 2c_1(\varphi)(v_1^2 + v_2^2 + v_1v_2) + 2c_2(\varphi)v_1v_2, \qquad \vec{v} = \vec{\xi} = \begin{pmatrix} v_1 \\ v_2 \end{pmatrix}.$$

For $\vec{L}_3 + \vec{L}_3'$:

$$\vec{L}_3 + \vec{L}_3' = z|z|^2\vec{F}_1(\alpha) - \bar{z}|z|^2\sigma_1\vec{F}_1(\alpha) + z^3\vec{F}_2(\alpha) - \bar{z}^3\sigma_1\vec{F}_2(\alpha);$$

here $\vec{F}_1(\alpha)$, $\vec{F}_2(\alpha)$ are real and have the same structure as $\vec{F}(\alpha)$, $F_0(\alpha)$: they are smooth functions of x and α, exponentially decreasing as $x \to \infty$. For $\vec{F}_1(\alpha)$, $\vec{F}_2(\alpha)$ one can obtain explicit expressions in terms of $\xi(\alpha)$, $\varphi(\alpha)$, $\varphi_\alpha(\alpha)$, but they are not essential for our further considerations.

Finally,

$$C(\zeta) = zC(\vec{\xi}) - \bar{z}\sigma_1C(\vec{\xi})\sigma_1.$$

4.2. The main terms of $\vec{h}$. The derivatives ω' and γ' are of second order with respect to z; this means that the deviation $\alpha - \alpha_1$ and the coefficient $\Delta = \gamma' - \omega_1$ are also of second order. Therefore the main terms of $\vec{D}$ are of second order and are generated by the expression

$$\vec{D} \approx \vec{N}_2 \approx P_1(\vec{N}_1 + \gamma'\sigma_3\vec{f}) \approx P_1\vec{N}_1 \approx P_1\vec{N} \approx P_1\vec{L}_2.$$

There is only one unclear point in this reduction: the difference $\vec{N}_1 - \vec{N}$ also contains a second order term $l(\sigma)\sigma_3 i\vec{\xi}_1 + l_1(z)\vec{\xi}$. But, in fact, one must estimate the difference $P_1(\vec{N}_1 - \vec{N})$, which contains $P_1\big(l(\sigma)\sigma_3 i\vec{\xi}_1 + l_1(z)\vec{\xi}\big)$. Since the projection operator $P_1 \approx P(\alpha)$ almost annuls the vectors $l(\sigma)\sigma_3\vec{\xi}_1$ and $\vec{\xi}$, the real order of the omitted terms is greater than 2. Thus, $\vec{D} = P_1\vec{L}_2 + \vec{D}_1$, where the order of $\vec{D}_1$ is greater than 2.

Introduce the new notation

$$\vec{h} = \vec{h}_1 + \vec{h}_2 + \vec{h}_3, \qquad \vec{h}_1 = \mathscr{U}(t,0)P_1\vec{f}_0,$$

$$\vec{h}_2 = -i\int_0^t \mathscr{U}(t,\tau)P_1L_2\,d\tau, \qquad \vec{h}_3 = -i\int_0^t \mathscr{U}(t,\tau)D_1\,d\tau.$$

In accordance with (3.3), the term $\vec{h}_1$ is locally of order $t^{-3/2}$ as $t\to\infty$. It is not the main asymptotic term. The main asymptotic term is contained in $\vec{h}_2$:

$$\vec{h}_2 = -i\int_0^t \mathscr{U}(t,\tau)P_1\left[z^2\vec{F}(\alpha) - \bar{z}^2\sigma_1\vec{F}(\alpha) + |z|^2F_0(\alpha)\begin{pmatrix}1\\-1\end{pmatrix}\right]d\tau.$$

The integral $\vec{h}_2$ allows some asymptotic simplification. Let

$$z(t) = e^{-i\lambda_1 t}z_1(t), \qquad \lambda_1 = \lambda(\alpha_1), \qquad \mathscr{U}(t,\tau) = e^{-iH_1(t-\tau)}\mathscr{U}_1(t,\tau).$$

After a natural integration by parts one obtains

$$\begin{aligned}
\vec{h}_2(t) &= \vec{h}_0(t) - \mathscr{U}(t,0)\vec{h}_0(0) + \vec{h}_2'(t),\\
\vec{h}_0(t) &= -P_1\Big[\big(H_1 - 2\lambda_1 - i0\big)^{-1}z^2(t)\vec{F}(\alpha)\\
&\qquad - \big(H_1 + 2\lambda_1 - i0\big)^{-1}\bar{z}^2(t)\sigma_1\vec{F}(\alpha) + H_1^{-1}|z(t)|^2F_0(\alpha)\begin{pmatrix}1\\-1\end{pmatrix}\Big]\\
&= |z|^2\vec{h}_{00}(\alpha) + z^2\vec{h}_{01}(\alpha) + \bar{z}^2\sigma_1\vec{h}_{01}(\alpha),\\
\vec{h}_2'(t) &= \int_0^t \mathscr{U}(t,\tau)\Big\{P_1\Big[\big(H_1 - 2\lambda_1 - i0\big)^{-1}e^{-i2\lambda_1\tau}\frac{d}{d\tau}\big(z_1^2\vec{F}\big)\\
&\qquad - \big(H_1 + 2\lambda_1 - i0\big)^{-1}e^{i2\lambda_1\tau}\frac{d}{d\tau}\big(\bar{z}_1^2\sigma_1\vec{F}\big)\\
&\qquad + H_1^{-1}\frac{d}{d\tau}|z_1|^2F_0\begin{pmatrix}1\\-1\end{pmatrix}\Big] + i\Delta(P_+ - P_-)\vec{h}_0(\tau)\Big\}\,d\tau.
\end{aligned}$$

Locally $\mathscr{U}(t,0)\vec{h}_0(0)$ is of order $t^{-3/2}$, and the integral $\vec{h}_2'$ is submitted to z^4. So the main order term of $\vec{h}_2$ is generated by $\vec{h}_0(t)$: $\vec{h}_2 \approx \vec{h}_0$, and consequently $\vec{h} \approx \vec{h}_0$.

Let us estimate the main order term of $\vec{f}$. Consider system (2.14) for the component $\vec{k}$, i.e., $\vec{f} = \vec{k} + \vec{h}$. It is clear from the system that $\vec{k}$ has higher (submitted) order than $\vec{h}$, so asymptotically

$$\vec{f} \approx \vec{h}_0. \tag{4.1}$$

For the following it is convenient to write $\vec{h} = \vec{h}_0 + \vec{h}'$ and $\vec{f} = \vec{h}_0 + \vec{f}'$, $\vec{f}' = \vec{k} + \vec{h}'$. We could control the order of the remainder $\vec{f}'$ more carefully; in fact, locally

$$\vec{f}' = O\big(Nt^{-3/2} + z^3\big),$$

where N is the N-norm of $\vec{\chi}_0$.

4.3. The system of effective equations. Now one can exclude $\vec{f}$ from the main terms of the equations for $\vec{\sigma}$ and z. Owing to (4.1),

$$l(\sigma)\vec{\zeta} + \vec{N}(\varphi, \vec{f} + \vec{\zeta}) = \vec{L}(\zeta) + C(\zeta)\vec{h}_0 + \cdots.$$

Here the remainder has order $NO(z)t^{-3/2} + O(z^4)$.

Consider the equations for ω and z:

$$e\omega' = \mathcal{F}, \qquad iz' - \lambda z = \mathcal{G}. \tag{4.2}$$

Separate explicitly the terms of the second and third orders:

$$\mathcal{F} = \mathcal{F}_2 + \mathcal{F}_3 + \mathcal{F}_R, \qquad \mathcal{G} = \mathcal{G}_2 + \mathcal{G}_3 + \mathcal{G}_R.$$

Here

$$\mathcal{F}_2 = 2a_0(\alpha)\operatorname{Im} z^2, \qquad \mathcal{F}_3 = 2\operatorname{Im}\left[a_1(\alpha)z|z|^2 + a_2(\alpha)z^3\right],$$

$a_0 = i\langle \vec{F}, \sigma_3\vec{\xi}_1\rangle$, a_0 is real, and a_1 and a_2 are smooth complex-valued functions of α, which can be expressed in terms of φ and $\vec{\xi}$.

Similarly,

$$\mathcal{G}_2 = b_1(\alpha)z^2 + b_2(\alpha)\bar{z}^2 + b_3(\alpha)|z|^2,$$
$$\mathcal{G}_3 = b_4(\alpha)|z|^2 z + b_5(\alpha)|z|^2\bar{z} + b_6(\alpha)z^3 + b_7(\alpha)\bar{z}^3.$$

All the coefficients b_j possess the same general properties as a_j, and it is not difficult to get the following explicit formulae:

$$b_1 = \langle \vec{F}, \sigma_3\vec{\xi}\rangle, \qquad b_2 = \langle \vec{F}, \sigma_3\sigma_1\vec{\xi}\rangle, \qquad b_3 = \langle F_0\begin{pmatrix}1\\1\end{pmatrix}, \vec{\xi}\rangle,$$
$$b_4 = \langle \vec{F}_1, \sigma_3\vec{\xi}\rangle + \langle C(\vec{\xi})\vec{h}_{00}, \sigma_3\vec{\xi}\rangle - \langle \sigma_1 C(\vec{\xi})\sigma_1\vec{h}_{01}, \sigma_3\vec{\xi}\rangle.$$

The influence of the different terms of $\mathcal{F}_j$, $\mathcal{G}_j$, even of the same order, on the asymptotic behavior of the solutions are different because of the oscillating character of z. In order to take this character into account, it is expedient to introduce the following new variables:

$$\widetilde{\omega} = \omega + 2c_0(\alpha)\operatorname{Re} z^2 + 2\operatorname{Re}\left(c_1(\alpha)z|z|^2 + c_2(\alpha)z^3\right),$$
$$\widetilde{z} = z + d_1(\alpha)z^2 + d_2(\alpha)\bar{z}^2 + d_3(\alpha)|z|^2 + d_4(\alpha)z|z|^2$$
$$+ d_5(\alpha)\bar{z}|z|^2 + d_6(\alpha)z^3 + d_7(\alpha)\bar{z}^3.$$

The coefficients c_j, d_j can be defined from the conditions

$$e\widetilde{\omega}' = \widetilde{\mathcal{F}}_R(\alpha, z), \tag{4.3}$$

$$i\widetilde{z}' - \lambda\widetilde{z} = b(\alpha)|\widetilde{z}|^2\widetilde{z} + \widetilde{\mathcal{G}}_R(\alpha, z), \tag{4.4}$$

and $\widetilde{\mathcal{F}}_R$, $\widetilde{\mathcal{G}}_R$ contain only terms whose orders are greater than 3.

The derivatives of $\widetilde{\omega}$, $\widetilde{z}$ can be expressed in terms of z, z' (and α, α', but this is not essential at the moment). Exclude z' using (4.2), so $\widetilde{\omega}'$ and $\widetilde{z}'$ will be expressed in terms of z. Now the undefined coefficients c_j, d_j can be found from the conditions $\widetilde{\mathcal{F}}_R, \widetilde{\mathcal{G}}_R = O(z^4)$.

In particular,

$$c_0 = -(2\lambda)^{-1}a_0, \qquad b = b_4 - \lambda^{-1}(2b_1b_3 + \tfrac{2}{3}b_2^2 - b_2b_3 + b_3^2). \tag{4.5}$$

4.4. The asymptotic behavior of z**.** In order to estimate qualitatively the behavior of z, or $\tilde{z}$, as $t \to \infty$, consider equations (4.4), neglect the remainder $\widetilde{\mathcal{G}}_R$ and replace $b(\alpha)$ by $b(\alpha_1)$:

$$i\tilde{z}' - \lambda\tilde{z} = b(\alpha_1)|\tilde{z}|^2\tilde{z}.$$

Let $r = \exp\Bigl(i\int_0^t \lambda\,d\tau\Bigr)\cdot\tilde{z}$; then $ir' = b(\alpha_1)|r|^2 r$. The general solution is given by

$$r(t) = r_0(1+\tau)^{-\frac{1}{2}+i\gamma}, \tag{4.6}$$

where $r_0 = r(0) = \tilde{z}(0)$, $\gamma = \operatorname{Re} b/2\operatorname{Im} b$, and $\tau = -2\operatorname{Im} b|r_0|^2 t$.

By virtue of the above assumptions we have $\operatorname{Im} b < 0$. Let us show this. Taking expression (4.5) for b into account, we get

$$\operatorname{Im} b(\alpha_1) = -\operatorname{Im}\langle\sigma_1 C(\vec{\xi})\sigma_1\vec{h}_{01}, \sigma_3\vec{\xi}\rangle = 2\langle\vec{h}_{01}, \sigma_3\vec{F}\rangle = 2\langle\vec{h}_{01}, \sigma_3 P_1\vec{F}\rangle.$$

Furthermore,

$$\begin{aligned}\operatorname{Im} b &= -\lim_{R\to\infty} 2\operatorname{Im}\int_{-R}^{R}\langle\sigma_3\vec{h}_{01}, (H - 2\lambda_1)\vec{h}_{01}\rangle_{\mathbb{C}^2}\,dx\\ &= \lim_{R\to\infty} 2\operatorname{Im}\langle\vec{h}_{01}, \frac{d}{dx}\vec{h}_{01}\rangle\Big|_{-R}^{R} = -4|\varkappa|^2 k.\end{aligned}$$

Here $k = \sqrt{2\lambda_1 - \alpha_1^2/4} > 0$ and the parameter $\varkappa$ can be characterized by the asymptotic representation

$$\vec{h}_{01} = \varkappa e^{ik|x|}\begin{pmatrix}1\\0\end{pmatrix} + o(1), \qquad x \to \infty.$$

In accordance with assumption 5′), we have $\varkappa \neq 0$, so $\operatorname{Im} b < 0$.

§5. Estimates of the majorants

5.1. The set of majorants. Following the approach of **[BP2, BP3]**, we proceed now to implement the method of majorants. The basis of the approach is formed by the system of equations (2.9), (2.10), and (2.20). Some improved versions of these equations, obtained in §4, will be also quite essential.

Introduce a natural system of norms for the components of the solution $\vec{\psi}$:

$$M_0(t) = |\alpha(t) - \alpha_1|, \qquad M_1(t) = |z(t)|,$$
$$M_2(t) = \|\rho f'\|_2, \qquad M_3(t) = \|f\|_\infty.$$

The weight function ρ was already introduced via formula (3.5).

These norms generate the system of the majorants:

$$\mathbb{M}_0(t) = \sup_{\tau\leqslant t}(1+\varepsilon\tau)M_0(\tau), \qquad \mathbb{M}_1(t) = \sup_{\tau\leqslant t}(1+\varepsilon\tau)^{1/2}M_1(\tau),$$

$$\mathbb{M}_2(t) = \sup_{\tau\leqslant t}(1+\varepsilon\tau)^{3/2}M_2(\tau), \qquad \mathbb{M}_3(t) = \sup_{\tau\leqslant t}\frac{(1+\varepsilon\tau)^{1/2}}{\ln(2+\varepsilon\tau)}M_3(\tau).$$

Finally, let $\mathbb{M}_j = \mathbb{M}_j(t_1)$.

The small parameter ε in the above definitions reflects the fact that the problem contains two time scales: one is defined by the time itself and the other is defined by the rate of decay of $z(t)$ as $t \to +\infty$. This rate is described by the variable $|\tilde{z}(0)|^2 t \sim N^2 t$ (see (4.6)). We shall fix a compromise scale parameter ε later.

Since the set of majorants contains both the functions $\vec{f}$ and $\vec{f}'$, it is worthwhile to estimate their difference: $\vec{f} = \vec{h}_0 + \vec{f}'$. It is clear from the definition of h_0 that

$$
\begin{aligned}
\|\rho h_0\|_2 &\leqslant \mu_0(\mathbb{M}_0)|z|^2, \\
\|(1+x^2)P_{\pm}\vec{h}_0\|_2 &\leqslant \mu_0(\mathbb{M}_0)|z|^2, \qquad (5.1)\\
\|h_0\|_\infty &\leqslant \mu_0(\mathbb{M}_0)|z|^2.
\end{aligned}
$$

Here $\mu_0(\mathbb{M}_0)$ is a function which is bounded in some finite neighborhood of 0 and may acquire the infinite value $+\infty$ outside some larger neighborhood.

All these estimates can be obtained by using the properties of the spectral resolution of H investigated in [**BP2**]. Since the necessary properties are quite common for all the Schrödinger type operators with potential quickly vanishing at infinity, we shall not enter into details.

Finally, let us introduce some important notation. We use $\mu(\mathbb{M})$ as a general notation for some functions of $\mathbb{M}_0, \dots, \mathbb{M}_3$, defined on $\mathbb{R}^4$, which are bounded in some finite neighborhood of 0 and may acquire the infinite value $+\infty$ outside some larger neighborhood. It will be assumed also that μ does not depend on ε. In all the formulas where $\mu(\mathbb{M})$ will appear, it would have been possible to give explicit expressions for them, but, in fact, such expressions are useless for our aims.

5.2. Estimates of the soliton's parameters. Consider equations (2.9) and (2.10). It is clear that

$$
\begin{aligned}
\|\vec{\sigma}'\|, \ |iz' - \lambda z| &\leqslant \mu(\mathbb{M}) \max_{j=1,2,3} \left|\langle \vec{N}(\varphi, \vec{f} + \vec{\zeta}), \sigma_3 \vec{\xi}_j \rangle\right| \qquad (5.2)\\
&\leqslant \mu(\mathbb{M}) \|\rho(f+\zeta)\|_2^2 \leqslant \mu(\mathbb{M})(1+\varepsilon t)^{-1}(\mathbb{M}_1 + \mathbb{M}_2)^2.
\end{aligned}
$$

Now consider the expressions $\mathscr{F}_R = \mathscr{F} - \mathscr{F}_2 - \mathscr{F}_3$ and $\mathscr{G}_R = \mathscr{G} - \mathscr{G}_2 - \mathscr{G}_3$, which, in fact, define the derivatives $\tilde{\omega}'$ and $i\tilde{z}' - \lambda\tilde{z}$. For example,

$$
\begin{aligned}
\mathscr{F}_R = \langle \vec{N}(\varphi, \vec{f} + \vec{\zeta}) - \vec{L}_2(\zeta) - \vec{L}_3(\zeta) - C(\zeta)\vec{f}, \sigma_3\vec{\xi}_1 \rangle + \langle C(\zeta)\vec{f}', \sigma_3\xi_1 \rangle \\
+ \langle l(\sigma)\vec{\zeta} - \vec{L}_3'(\zeta), \sigma_3\vec{\xi}_1 \rangle + \langle \vec{f}, l(\sigma)\sigma_3\vec{\xi}_1 \rangle.
\end{aligned}
$$

All the terms here can be estimated directly:

$$
\begin{aligned}
|\mathscr{F}_R| &\leqslant \mu(\mathbb{M})\left(\|\rho f\|_2^2 + \|\rho f\|_2 \cdot |z| + |z|^4\right) \qquad (5.3)\\
&\leqslant \mu(\mathbb{M})(1+\varepsilon t)^{-2}(\mathbb{M}_1^4 + \mathbb{M}_1\mathbb{M}_2 + \mathbb{M}_2^2).
\end{aligned}
$$

The function $\mathscr{G}_R$ has the same estimate:

$$
|\mathscr{G}_R| \leqslant \mu(\mathbb{M})(1+\varepsilon t)^{-2}(\mathbb{M}_1^4 + \mathbb{M}_1\mathbb{M}_2 + \mathbb{M}_2^2). \qquad (5.4)
$$

5.3. Estimates of $\vec{D}$. First of all, we estimate the component $\vec{k}$ of the decomposition $\vec{f} = \vec{k} + \vec{h}$. Consider the system (2.14) and take (2.15)–(2.17) into account:

$$
\|\mathbf{k}\|_{\mathbb{C}^4} \leqslant \mu(\mathbb{M})|\alpha - \alpha_1|\|\rho h\|_2.
$$

From the explicit expression for the spectral projection operator $P_d(\alpha) = I - P(\alpha)$ on the subspace of the discrete spectrum of $H(\alpha)$, it is clear that

$$\|\rho h\|_2 \leqslant \mu_0(\mathbb{M})\|\rho f\|_2.$$

Therefore

$$\|\vec{k}\|_\infty \leqslant \mu(\mathbb{M})(1+\varepsilon t)^{-2}\mathbb{M}_0\big(\mathbb{M}_2 + \mathbb{M}_1^2\big), \tag{5.5}$$
$$\|\rho\vec{k}\|_2, \|\vec{k}\|_2 \leqslant \mu(\mathbb{M})(1+\varepsilon t)^{-2}\mathbb{M}_0\big(\mathbb{M}_2 + \mathbb{M}_1^2\big).$$

The structure of the function $\vec{D}$ is rather complicated: $\vec{D} = P_1\vec{L}_2 + \vec{D}_1$, where $\vec{D}_1 = \vec{D}_2 + \vec{D}_3$, with

$$\vec{D}_2 = P_1\big[\vec{N}(\varphi, \vec{f}+\vec{\zeta}) - L_2(\zeta) - \widetilde{N}(\vec{f}+\vec{\zeta})\big] + P_1 l(\sigma)\big(\sigma_3 i\vec{\xi}_1 + \vec{\zeta}\big)$$
$$+ P_1 l_1(z)\vec{\xi} + P_1\big[W(\alpha) - W(\alpha_1)\big]\vec{f} + \Delta\big[(P_+ - P_-) - P_1\sigma_3\big]\vec{f}$$

and $\vec{D}_3 = P_1\widetilde{N}(\vec{f}+\vec{\zeta})$. Here

$$\widetilde{N}(\vec{f}) = V\big(|f|^2\big)\sigma_3\vec{f}.$$

The meaning of $\widetilde{N}$ is clear; it is the component of $\vec{N}(\varphi, \vec{f}+\vec{\zeta})$ that does not contain the function φ at all and, consequently, does not contain an exponentially decreasing (as $x \to \infty$) external factor.

Let us estimate the individual terms of $\vec{D}$ separately.

First of all,

$$\|(1+x^2)P_1\vec{L}_2\|_2 \leqslant \mu(\mathbb{M})(1+\varepsilon t)^{-1}\mathbb{M}_1^2.$$

For the first term of $\vec{D}_2$ we have

$$\big|\vec{N}(\varphi, \vec{f}+\vec{\zeta}) - \vec{L}_2(\zeta) - \widetilde{N}(\vec{f}+\vec{\zeta})\big| \leqslant \mu(\mathbb{M})\varphi\big[|f|(|f|+|z|)+|z|^3\big]$$

and

$$\big\|(1+x^2)P_1\big[\vec{N}(\varphi, \vec{f}+\vec{\zeta}) - \vec{L}_2(\zeta) - \widetilde{N}(\vec{f}+\vec{\zeta})\big]\big\|_2$$
$$\leqslant \mu(\mathbb{M})(1+\varepsilon t)^{-3/2}\big[\mathbb{M}_1^3 + (\mathbb{M}_1 + \mathbb{M}_3)(\mathbb{M}_2 + \mathbb{M}_1^2)\big].$$

The second term

$$P_1 l(\sigma)(\sigma_3 i\vec{\xi}_1 + \vec{\zeta}) = P_1\big[l(\sigma)\sigma_3 i\vec{\xi}_1 - l(\sigma)\sigma_3 i\vec{\xi}_1\big|_{t=t_1} + l(\sigma)\vec{\zeta}\big]$$

admits the estimate

$$\big\|(1+x^2)P_1 l(\sigma)(i\vec{\xi}_1 + \vec{\zeta})\big\|_2 \leqslant \mu(\mathbb{M})(1+\varepsilon t)^{-3/2}(\mathbb{M}_0 + \mathbb{M}_1)(\mathbb{M}_1 + \mathbb{M}_2)^2.$$

Here we have used the estimate (5.2) for $\vec{\sigma}'$. The third term

$$P_1 l_1(z)\vec{\xi} = P_1 l(z)\vec{\xi} - P_1 l(z)\vec{\xi}\big|_{t=t_1}$$

admits the estimate

$$\big\|(1+x^2)P_1 l_1(z)\vec{\xi}\big\|_2 \leqslant \mu(\mathbb{M})(1+\varepsilon t)^{-3/2}\mathbb{M}_0(\mathbb{M}_1 + \mathbb{M}_2)^2.$$

For the fourth term we have

$$\big\|(1+x^2)P_1\big[W(\alpha) - W(\alpha_1)\big]\vec{f}\big\|_2 \leqslant \mu(\mathbb{M})(1+\varepsilon t)^{-2}\mathbb{M}_0(\mathbb{M}_2 + \mathbb{M}_1^2).$$

In order to estimate the fifth term, take (3.8) and (5.2) into account:

$$\begin{aligned}\big\|(1+x^2)\Delta\big[(P_+-P_-)-P_1\sigma_3\big]\vec{f}\big\|_2 &\leqslant c|\Delta|\big(\|\rho h\|_2+\|(1+x^2)P_1\sigma_3\vec{k}\|_2\big)\\ &\leqslant \mu(\mathbb{M})(1+\varepsilon t)^{-2}(\mathbb{M}_0+\mathbb{M}_1^2+\mathbb{M}_2^2)(\mathbb{M}_1^2+\mathbb{M}_2).\end{aligned}$$

Finally, for $\vec{D}_2$ as a whole,

$$\big\|(1+x^2)\vec{D}_2\big\|_2 \leqslant \mu(\mathbb{M})(1+\varepsilon t)^{-3/2}(\mathbb{M}_0+\mathbb{M}_1+\mathbb{M}_2+\mathbb{M}_3)(\mathbb{M}_1^2+\mathbb{M}_2).$$

Moreover,

$$\begin{aligned}\big\|(1+x^2)(P_1\vec{L}_2+\vec{D}_2)\big\|_2&\\ \leqslant \mu(\mathbb{M})(1+\varepsilon t)^{-1}&\big[\mathbb{M}_1^2+\mathbb{M}_2(\mathbb{M}_0+\mathbb{M}_1+\mathbb{M}_2+\mathbb{M}_3)\big].\end{aligned}\tag{5.6}$$

The expression $\vec{D}_3 = P_1\widetilde{N}(\vec{f}+\vec{\zeta})$ is of different nature. The initial estimate is

$$|\widetilde{N}(\vec{f}+\vec{\zeta})| \leqslant c|f+\zeta|^{2p+1}.$$

Thus

$$\begin{aligned}\|\vec{D}_3\|_2 &\leqslant \mu_0(\mathbb{M}_0)\|f+\zeta\|_\infty^{2p}\|f+\zeta\|_2,\\ \|(1+x^2)\vec{D}_3\|_1 &\leqslant \mu_0(\mathbb{M}_0)\|f+\zeta\|_\infty^{2p-1}\|(1+|x|)(f+\zeta)\|_2^2.\end{aligned}$$

Here the presence of the function $\mu_0(\mathbb{M}_0)$ reflects the fact that $P_1 = I - P_d$ and the estimation of the projection operator P_d depend on the deviation $\alpha-\alpha_1$.

From the general estimates of the solutions of nonlinear Schrödinger equation, see (3.1) and (3.2), one derives the following inequalities:

$$\|f+\zeta\|_2 \leqslant \mu_0(\mathbb{M}_0), \qquad \|(1+|x|)(f+\zeta)\|_2 \leqslant \mu_0(\mathbb{M}_0)(1+t).$$

Consequently,

$$\begin{aligned}\|\vec{D}_3\|_2+\|(1+x^2)\vec{D}_3\|_1 &\leqslant \mu_0(\mathbb{M}_0)(1+|t|)^2\|f+\zeta\|_\infty^{2p-1}\\ &\leqslant \mu(\mathbb{M})\frac{(1+t)^2\ln^{2p-1}(2+\varepsilon t)}{(1+\varepsilon t)^{p-1/2}}(\mathbb{M}_1^{2p-1}+\mathbb{M}_3^{2p-1}).\end{aligned}$$

Provided that $p>4$, we have

$$\|\vec{D}_3\|_2+\|(1+x^2)\vec{D}_3\|_1 \leqslant \mu(\mathbb{M})(1+\varepsilon t)^{-3/2}\varepsilon^{-2}(\mathbb{M}_1^{2p-1}+\mathbb{M}_3^{2p-1}).\tag{5.7}$$

5.4. Estimates of $\vec{h}$. Here we derive an estimate for the L_∞-norm of $\vec{h}$ and, consequently, for the majorant $\mathbb{M}_3$. Consider the integral equation (2.20). By (3.3),

$$\|\mathscr{U}(t,0)P_1\vec{f}_0\|_\infty \leqslant c(1+t)^{-1/2}(\|f_0\|_{H^1}+\|(1+x^2)f_0\|_2) \leqslant \mu_0(N)N(1+t)^{-1/2},$$

where $N = \|(1+x^2)\chi_0\|_2+\|\chi_0'\|_2$. From (5.6) and (5.7), for the integrand in (2.20) we get

$$\begin{aligned}\|\mathscr{U}(t,\tau)(P_1\vec{L}_2+\vec{D}_2)\|_\infty &\leqslant \mu_0(\mathbb{M}_0)|t-\tau|^{-1/2}\|(1+x^2)(P_1\vec{L}_2+\vec{D}_2)\|_2\\ &\leqslant \mu(\mathbb{M})|t-\tau|^{-1/2}(1+\varepsilon\tau)^{-1}\big[\mathbb{M}_1^2+\mathbb{M}_2(\mathbb{M}_0+\mathbb{M}_1+\mathbb{M}_2+\mathbb{M}_3)\big],\\ \|\mathscr{U}(t,\tau)\vec{D}_3\|_\infty &\leqslant \mu_0(\mathbb{M}_0)|t-\tau|^{-1/2}(\|\vec{D}_3\|_2+\|(1+x^2)\vec{D}_3\|_1)\\ &\leqslant \mu(\mathbb{M})|t-\tau|^{-1/2}(1+\varepsilon\tau)^{-3/2}\varepsilon^{-2}(\mathbb{M}_1^{2p-1}+\mathbb{M}_3^{2p-1}).\end{aligned}$$

Combine the three last estimates:

$$\|h\|_\infty \leqslant \mu_0(N)N(1+t)^{-1/2} + \mu(\mathbb{M})\big[\mathbb{M}_1^2 + \mathbb{M}_2(\mathbb{M}_0+\mathbb{M}_1+\mathbb{M}_2+\mathbb{M}_3) \\ +\varepsilon^{-2}\big(\mathbb{M}_1^{2p-1}+\mathbb{M}_3^{2p-1}\big)\big]\int_0^t \frac{d\tau}{|t-\tau|^{1/2}(1+\varepsilon\tau)}.$$

The integral admits the estimate

$$\int_0^t \frac{d\tau}{\sqrt{t-\tau}(1+\varepsilon\tau)} \leqslant c\varepsilon^{-1/2}\frac{\ln(2+\varepsilon t)}{(1+\varepsilon t)^{1/2}}.$$

Now apply (5.5):

$$\mathbb{M}_3 \leqslant \mu(\mathbb{M})\big\{N + \varepsilon^{-1/2}\big[\mathbb{M}_1^2+\mathbb{M}_2(\mathbb{M}_0+\mathbb{M}_1+\mathbb{M}_2+\mathbb{M}_3)\big] \\ +\varepsilon^{-5/2}\big(\mathbb{M}_1^{2p-1}+\mathbb{M}_3^{2p-1}\big)\big\}.$$

So we have obtained the first inequality for the majorants. The second inequality for $\|\rho h'\|_2$ is based on the representation

$$\vec{h}' = \mathcal{U}(t,0)P_1\vec{f}_0 - \mathcal{U}(t,0)\vec{h}_0(0) \\ + \int_0^t \mathcal{U}(t,\tau)\big\{-i\vec{D}_1 + i\Delta(P_+-P_-)\vec{h}_0 - i\vec{D}_4\big\}\,d\tau,$$

$$\vec{D}_4 = iP_1\Big[(H_1-2\lambda-i0)^{-1}e^{-i2\lambda_1\tau}\frac{d}{d\tau}(z_1^2\vec{F}) \\ -(H_1+2\lambda_1-i0)^{-1}e^{i2\lambda_1\tau}\frac{d}{d\tau}(\bar{z}_1^2\sigma_1\vec{F}) + H_1^{-1}\frac{d}{d\tau}(|z_1|^2F_0)\begin{pmatrix}1\\-1\end{pmatrix}\Big].$$

Let us estimate the separate terms of this representation:

$$\|\rho\mathcal{U}(t,0)P_1\vec{f}_0\|_2 \leqslant \mu_0(\mathbb{M}_0)(1+t)^{-3/2}\|(1+x^2)\vec{f}_0\|_2 \leqslant \mu_0(\mathbb{M}_0)N(1+t)^{-3/2},$$

$$\|\rho\mathcal{U}(t,0)\vec{h}_0(0)\|_2 \\ \leqslant \mu_0(\mathbb{M}_0)(1+t)^{-3/2}|z(0)|^2\big[\|(1+x^2)F_0\|_2 + \|\vec{F}\|_2 + \|(1+|x|^3)\vec{F}\|_1\big] \\ \leqslant \mu(\mathbb{M})(1+t)^{-3/2}|z(0)|^2 \leqslant \mu(\mathbb{M})(1+t)^{-3/2}N^2,$$

$$\|\rho\mathcal{U}(t,\tau)\vec{D}_1\|_2 \leqslant \mu_0(\mathbb{M}_0)\big(1+|t-\tau|\big)^{-3/2}\big[\|(1+x^2)\vec{D}_2\|_2 + \|\vec{D}_3\|_2 + \|(1+x^2)\vec{D}_3\|_1\big] \\ \leqslant \mu(\mathbb{M})\big(1+|t-\tau|\big)^{-3/2}(1+\varepsilon\tau)^{-3/2} \\ \times\big[(\mathbb{M}_0+\mathbb{M}_1+\mathbb{M}_2+\mathbb{M}_3)(\mathbb{M}_1^2+\mathbb{M}_2) + \varepsilon^{-2}\big(\mathbb{M}_1^{2p-1}+\mathbb{M}_3^{2p-1}\big)\big].$$

In order to consider the contributions of $\vec{D}_4$ and $\Delta(P_+-P_-)\vec{h}_0$ we apply (3.6) and (3.7):

$$\|\rho\mathcal{U}(t,\tau)\vec{D}_4\|_2 \leqslant \mu_0(\mathbb{M}_0)\big(1+|t-\tau|\big)^{-3/2}\Big[\Big\|(1+x^2)\frac{d}{d\tau}(|z_1|^2F_0)\begin{pmatrix}1\\-1\end{pmatrix}\Big\|_2 \\ + \Big\|\frac{d}{d\tau}(z_1^2\vec{F})\Big\|_2 + \Big\|(1+|x|)^3\frac{d}{d\tau}(z_1^2\vec{F})\Big\|_1 \\ + \Big\|\frac{d}{d\tau}(\bar{z}_1^2\sigma_1\vec{F})\Big\|_2 + \Big\|(1+|x|)^3\frac{d}{d\tau}(\bar{z}_1^2\sigma_1\vec{F})\Big\|_1\Big] \\ \leqslant \mu(\mathbb{M})\big(1+|t-\tau|\big)^{-3/2}|z|\big(|z||\lambda-\lambda_1| + |iz'-\lambda z| + |z||w'|\big) \\ \leqslant \mu(\mathbb{M})\big(1+|t-\tau|\big)^{-3/2}(1+\varepsilon\tau)^{-3/2}\mathbb{M}_1\big(\mathbb{M}_0\mathbb{M}_1+\mathbb{M}_1^2+\mathbb{M}_2^2\big),$$

$$\begin{aligned}
&\|\rho\mathscr{U}(t,\tau)\Delta(P_+-P_-)\vec{h}_0\|_2 \\
&\quad \leqslant \mu_0(\mathbb{M}_0)(1+|t-\tau|)^{-3/2}|\Delta||z|^2 \\
&\qquad \times \Big[\|(1+x^2)F_0\begin{pmatrix}1\\-1\end{pmatrix}\|_2+\|\vec{F}\|_2+\|(1+|x|^3)\vec{F}\|_1\Big] \\
&\quad \leqslant \mu(\mathbb{M})(1+|t-\tau|)^{-3/2}|\Delta||z|^2 \\
&\quad \leqslant \mu(\mathbb{M})(1+|t-\tau|)^{-3/2}(1+\varepsilon\tau)^{-2}\mathbb{M}_1^2(\mathbb{M}_0+\mathbb{M}_1^2+\mathbb{M}_2^2).
\end{aligned}$$

In these chains of inequalities we used the estimate

$$|\Delta|=|\gamma'+\omega-\omega_1|\leqslant|\gamma'|+|\omega-\omega_1|\leqslant\mu(\mathbb{M})(1+\varepsilon t)^{-1}\big(\mathbb{M}_0+\mathbb{M}_1^2+\mathbb{M}_2^2\big).$$

Now we can combine the estimates for all the components of $\vec{h}'$:

$$\begin{aligned}
\|\rho h'\|_2 \leqslant \mu(\mathbb{M})(1+\varepsilon t)^{-3/2} & \\
\times\Big\{N+\Big[(\mathbb{M}_0+\mathbb{M}_1+\mathbb{M}_2+\mathbb{M}_3)(\mathbb{M}_1^2+\mathbb{M}_2)+\varepsilon^{-2}\big(\mathbb{M}_1^{2p-1}+\mathbb{M}_3^{2p-1}\big)\Big] & \\
\times\int_0^t\frac{d\tau}{(1+t-\tau)^{-3/2}}\Big(\frac{1+\varepsilon t}{1+\varepsilon\tau}\Big)^{3/2}\Big\}. &
\end{aligned}$$

The integral here is bounded.

Combining (5.5) and the last inequality, we obtain

$$\mathbb{M}_2\leqslant\mu(\mathbb{M})\Big[N+(\mathbb{M}_0+\mathbb{M}_1+\mathbb{M}_2+\mathbb{M}_3)(\mathbb{M}_1^2+\mathbb{M}_2)+\varepsilon^{-2}\big(\mathbb{M}_1^{2p-1}+\mathbb{M}_3^{2p-1}\big)\Big].$$

5.5. Closing the estimates. Here we shall deduce estimates for $\mathbb{M}_0$ and $\mathbb{M}_1$, and they will close the system of the inequalities for the majorants.

The derivation of the inequality for $\mathbb{M}_0$ is rather direct. First of all, note that as a consequence of (5.3):

$$|\widetilde{\omega}'|=\big|e^{-1}\widetilde{\mathscr{F}}_R\big|\leqslant\mu(\mathbb{M})(1+\varepsilon t)^{-2}\big(\mathbb{M}_1^4+\mathbb{M}_1\mathbb{M}_2+\mathbb{M}_2^2\big).$$

Furthermore,

$$|\omega-\widetilde{\omega}|\leqslant\mu(\mathbb{M})(1+\varepsilon t)^{-1}\mathbb{M}_1^2.$$

Finally,

$$\begin{aligned}
|\omega(t)-\omega(t_1)| &\leqslant|\widetilde{\omega}(t)-\widetilde{\omega}(t_1)|+|\omega(t)-\widetilde{\omega}(t)|+|\omega(t_1)-\widetilde{\omega}(t_1)| \\
&\leqslant\int_t^{t_1}|\widetilde{\omega}'(\tau)|\,d\tau+\mu(\mathbb{M})(1+\varepsilon t)^{-1}\mathbb{M}_1^2 \\
&\leqslant\mu(\mathbb{M})(1+\varepsilon t)^{-1}\big[\mathbb{M}_1^2+\varepsilon^{-1}\big(\mathbb{M}_1^4+\mathbb{M}_1\mathbb{M}_2+\mathbb{M}_2^2\big)\big].
\end{aligned}$$

This leads directly to the estimate for $\mathbb{M}_0$:

$$\mathbb{M}_0\leqslant\mu(\mathbb{M})\big[\mathbb{M}_1^2+\varepsilon^{-1}\big(\mathbb{M}_1^4+\mathbb{M}_1\mathbb{M}_2+\mathbb{M}_2^2\big)\big].$$

Considering $\mathbb{M}_1$, we must perform a somewhat more complicated computation. From (4.4) it follows that

$$(|r|^2)'=2\operatorname{Im}b(\alpha)|r|^4+2\operatorname{Im}(\widetilde{\mathscr{G}}_R\bar{\widetilde{z}}),$$

or

$$(|r|^2)'=2\operatorname{Im}b(\alpha_1)|r|^4+R,$$

where $R = 2\,\mathrm{Im}\big[\widetilde{\mathcal{G}}_R\bar{\tilde{z}} + (b(\alpha) - b(\alpha_1))|r|^4\big]$. By (5.4) we have $|R| \leqslant (1+\varepsilon t)^{-5/2}\mathcal{M}$, where

$$\mathcal{M} = \mu(\mathbb{M})\mathbb{M}_1\left(\mathbb{M}_0\mathbb{M}_1^3 + \mathbb{M}_1^4 + \mathbb{M}_1\mathbb{M}_2 + \mathbb{M}_2^2\right).$$

Let$\rho = |r|^2$, $A = -2\,\mathrm{Im}\,b(\alpha_1) > 0$; then

$$\rho' = -A\rho^2 + R. \tag{5.8}$$

It can be verified directly that $\rho_0 = B(1+\varepsilon t)^{-1}$, $B > 0$, satisfies

$$\rho_0' = -A\rho_0^2 + (1+\varepsilon t)^{-2}\mathcal{M}, \tag{5.9}$$

where $\varepsilon B = AB^2 - \mathcal{M}$.

Comparison of (5.8) and (5.9) implies that under the assumption $\rho(0) \leqslant B$ the solution ρ satisfies $\rho(t) \leqslant B(1+\varepsilon t)^{-1}$. In other words,

$$\rho(0) \leqslant \frac{\varepsilon + \sqrt{4\mathcal{M}A + \varepsilon^2}}{2A} \implies \rho(t) \leqslant \frac{\varepsilon + \sqrt{4\mathcal{M}A + \varepsilon^2}}{2A}(1+\varepsilon t)^{-1}.$$

A bit more roughly,

$$\rho(0) \leqslant \varepsilon A^{-1} \implies \rho(t) \leqslant \frac{\varepsilon + \sqrt{\mathcal{M}A}}{A}(1+\varepsilon t)^{-1}.$$

This means that, if in the sense of order

$$N^2 \leqslant \varepsilon, \tag{5.10}$$

then

$$|r(t)| \leqslant \mu(\mathbb{M})(1+\varepsilon t)^{-1/2}\left[\varepsilon^{1/2} + \sqrt[4]{\mathcal{M}}\right].$$

Since $|z| \leqslant |\tilde{z}| + \mu(\mathbb{M})(1+\varepsilon t)^{-1}\mathbb{M}_1^2$, this finally implies

$$\mathbb{M}_1 \leqslant \mu(\mathbb{M})\left[\varepsilon^{1/2} + \mathbb{M}_1^{1/4}\left(\mathbb{M}_0\mathbb{M}_1^3 + \mathbb{M}_1^4 + \mathbb{M}_1\mathbb{M}_2 + \mathbb{M}_2^2\right)^{1/4}\right].$$

So, under condition (5.10), we have closed the system of inequalities for the majorants.

5.6. Estimates of the majorants. We collect the inequalities we have obtained:

$$\mathbb{M}_0 \leqslant \mu(\mathbb{M})\left[\mathbb{M}_1^2 + \varepsilon^{-1}\left(\mathbb{M}_1^4 + \mathbb{M}_1\mathbb{M}_2 + \mathbb{M}_2^2\right)\right],$$

$$\mathbb{M}_1 \leqslant \mu(\mathbb{M})\left[\varepsilon^{1/2} + \mathbb{M}_1^{1/4}\left(\mathbb{M}_0\mathbb{M}_1^3 + \mathbb{M}_1^4 + \mathbb{M}_1\mathbb{M}_2 + \mathbb{M}_2^2\right)^{1/4}\right],$$

$$\mathbb{M}_2 \leqslant \mu(\mathbb{M})\left[N + (\mathbb{M}_0 + \mathbb{M}_1 + \mathbb{M}_2 + \mathbb{M}_3)(\mathbb{M}_1^2 + \mathbb{M}_2) + \varepsilon^{-2}\left(\mathbb{M}_1^{2p-1} + \mathbb{M}_3^{2p-1}\right)\right],$$

$$\mathbb{M}_3 \leqslant \mu(\mathbb{M})\Big[N + \varepsilon^{-1/2}\left[\mathbb{M}_1^2 + \mathbb{M}_2(\mathbb{M}_0 + \mathbb{M}_1 + \mathbb{M}_2 + \mathbb{M}_3)\right] + \varepsilon^{-5/2}\left(\mathbb{M}_1^{2p-1} + \mathbb{M}_3^{2p-1}\right)\Big].$$

The majorant $\mathbb{M}_0$ can be excluded from the inequalities for $\mathbb{M}_1$, $\mathbb{M}_2$, $\mathbb{M}_3$:

$$\mathbb{M}_1 \leqslant \mu(\mathbb{M})\Big[\varepsilon^{1/2} + \mathbb{M}_1^{1/4}\big(1+\varepsilon^{-1}\mathbb{M}_1^3\big)^{1/4}\big(\mathbb{M}_1^4 + \mathbb{M}_1\mathbb{M}_2 + \mathbb{M}_2^2\big)^{1/4}\Big],$$

$$\mathbb{M}_2 \leqslant \mu(\mathbb{M})\Big[N + \big[\varepsilon^{-1}\big(\mathbb{M}_1^4 + \mathbb{M}_1\mathbb{M}_2 + \mathbb{M}_2^2\big) + \mathbb{M}_1 + \mathbb{M}_2 + \mathbb{M}_3\big]\big(\mathbb{M}_1^2 + \mathbb{M}_2\big) + \varepsilon^{-2}\big(\mathbb{M}_1^{2p-1} + \mathbb{M}_3^{2p-1}\big)\Big],$$

$$\mathbb{M}_3 \leqslant \mu(\mathbb{M})\Big[N + \varepsilon^{-1/2}\mathbb{M}_1^2 + \varepsilon^{-1/2}\big[\varepsilon^{-1}\big(\mathbb{M}_1^4 + \mathbb{M}_1\mathbb{M}_2 + \mathbb{M}_2^2\big) + \mathbb{M}_1 + \mathbb{M}_2 + \mathbb{M}_3\big]\mathbb{M}_2 + \varepsilon^{-5/2}\big(\mathbb{M}_1^{2p-1} + \mathbb{M}_3^{2p-1}\big)\Big].$$

Introduce new scales:

$$\mathbb{M}_1 = \varepsilon^{1/2-\gamma_1}\widehat{\mathbb{M}}_1, \qquad \mathbb{M}_2 = \varepsilon^{3/2-\gamma_2}\widehat{\mathbb{M}}_2, \qquad \mathbb{M}_3 = \varepsilon^{1/2-\gamma_3}\widehat{\mathbb{M}}_3,$$

where γ_1, γ_2, γ_3 are small positive numbers that will be fixed later. Let $N = \varepsilon^{3/2}$ and $14\gamma_1 \leqslant 7\gamma_3 \leqslant \gamma_2 \leqslant 1/2$. Then

$$\widehat{\mathbb{M}}_1 \leqslant \mu(\mathbb{M})\Big[\varepsilon^{\gamma_1} + \widehat{\mathbb{M}}_1^{1/4}\big(1+\widehat{\mathbb{M}}_1^3\big)^{1/4}\big(\widehat{\mathbb{M}}_1^4 + \widehat{\mathbb{M}}_1\widehat{\mathbb{M}}_2 + \widehat{\mathbb{M}}_2^2\big)^{1/4}\Big],$$

$$\widehat{\mathbb{M}}_2 \leqslant \mu(\mathbb{M})\Big[\varepsilon^{\gamma_2} + \big[\big(\widehat{\mathbb{M}}_1^4 + \widehat{\mathbb{M}}_1\widehat{\mathbb{M}}_2 + \widehat{\mathbb{M}}_2^2\big) + \widehat{\mathbb{M}}_1 + \widehat{\mathbb{M}}_2 + \widehat{\mathbb{M}}_3\big]\big(\widehat{\mathbb{M}}_1^2 + \widehat{\mathbb{M}}_2\big) + \big(\widehat{\mathbb{M}}_1^{2p-1} + \widehat{\mathbb{M}}_3^{2p-1}\big)\Big],$$

$$\widehat{\mathbb{M}}_3 \leqslant \mu(\mathbb{M})\Big[\varepsilon^{1+\gamma_3} + \widehat{\mathbb{M}}_1^2 + \big(\widehat{\mathbb{M}}_1^4 + \widehat{\mathbb{M}}_1\widehat{\mathbb{M}}_2 + \widehat{\mathbb{M}}_2^2 + \widehat{\mathbb{M}}_1 + \widehat{\mathbb{M}}_2 + \widehat{\mathbb{M}}_3\big)\widehat{\mathbb{M}}_2 + \big(\widehat{\mathbb{M}}_1^{2p-1} + \widehat{\mathbb{M}}_3^{2p-1}\big)\Big].$$

One can take the function μ to be spherically symmetric and monotone. Then the three above relations can be simplified:

$$\widehat{\mathbb{M}}_1 \leqslant \mu(\widehat{\mathbb{M}})\Big[\varepsilon^{\gamma_1} + \widehat{\mathbb{M}}_1^{5/4} + \widehat{\mathbb{M}}_1^{1/4}\big(\widehat{\mathbb{M}}_1\widehat{\mathbb{M}}_2 + \widehat{\mathbb{M}}_2^2\big)^{1/4}\Big],$$

$$\widehat{\mathbb{M}}_2 \leqslant \mu(\widehat{\mathbb{M}})\Big[\varepsilon^{\gamma_2} + \widehat{\mathbb{M}}_1^3 + \widehat{\mathbb{M}}_1\widehat{\mathbb{M}}_2 + \widehat{\mathbb{M}}_2^2 + \widehat{\mathbb{M}}_3^2\Big],$$

$$\widehat{\mathbb{M}}_3 \leqslant \mu(\widehat{\mathbb{M}})\Big[\varepsilon^{1+\gamma_3} + \widehat{\mathbb{M}}_1^2 + \widehat{\mathbb{M}}_2^2 + \widehat{\mathbb{M}}_3^2\Big].$$

Fix the ball $\|\widehat{\mathbb{M}}\| \leqslant R$ where $\mu(\widehat{\mathbb{M}})$ is bounded by a constant μ_1, and consider the set of solutions of the above system in this ball for small ε.

Replace $\widehat{\mathbb{M}}_2$ in the first inequality by the estimate from the second one:

$$\widehat{\mathbb{M}}_1 \leqslant \mu_2\Big[\varepsilon^{\gamma_1} + \widehat{\mathbb{M}}_1^{1/4}(\widehat{\mathbb{M}}_1 + \widehat{\mathbb{M}}_2 + \widehat{\mathbb{M}}_3) + \widehat{\mathbb{M}}_1^{1/2}\big(\widehat{\mathbb{M}}_1\widehat{\mathbb{M}}_2 + \widehat{\mathbb{M}}_2^2 + \widehat{\mathbb{M}}_3^2\big)^{1/4}\Big];$$

here μ_2 can be expressed explicitly in terms of R and μ_1. Continue the procedure and the third step gives

$$\begin{aligned}\widehat{\mathbb{M}}_1 &\leqslant \mu_3\big[\varepsilon^{\gamma_1} + \widehat{\mathbb{M}}_1^{1/4}(\widehat{\mathbb{M}}_1 + \widehat{\mathbb{M}}_2 + \widehat{\mathbb{M}}_3) + \widehat{\mathbb{M}}_1\big(\widehat{\mathbb{M}}_1\widehat{\mathbb{M}}_2 + \widehat{\mathbb{M}}_2^2 + \widehat{\mathbb{M}}_3^2\big)^{1/4}\big]\\ &\leqslant \mu_3\big[\varepsilon^{\gamma_1} + (\widehat{\mathbb{M}}_1 + \widehat{\mathbb{M}}_2 + \widehat{\mathbb{M}}_3)^{5/4}\big].\end{aligned}$$

So we have the system

$$\widehat{\mathbb{M}}_1 \leqslant \mu_4\big[\varepsilon^{\gamma_1} + (\widehat{\mathbb{M}}_1 + \widehat{\mathbb{M}}_2 + \widehat{\mathbb{M}}_3)^{5/4}\big],$$
$$\widehat{\mathbb{M}}_2 \leqslant \mu_4\big[\varepsilon^{\gamma_2} + (\widehat{\mathbb{M}}_1 + \widehat{\mathbb{M}}_2 + \widehat{\mathbb{M}}_3)^{2}\big],$$
$$\widehat{\mathbb{M}}_3 \leqslant \mu_4\big[\varepsilon^{1+\gamma_3} + (\widehat{\mathbb{M}}_1 + \widehat{\mathbb{M}}_2 + \widehat{\mathbb{M}}_3)^{2}\big].$$

This means that

$$\widehat{\mathbb{M}}_1, \widehat{\mathbb{M}}_2, \widehat{\mathbb{M}}_3 \leqslant \mu_5\varepsilon^{\gamma_1}. \tag{5.11}$$

Alternatively,

$$\|\widehat{\mathbb{M}}\| > R. \tag{5.12}$$

Since all the norms $\widehat{\mathbb{M}}_j$ are continuous in time t and for $t = 0$ obey inequality (5.11), the variant (5.12) is impossible.

As a simple consequence one finally obtains

$$\mathbb{M}_0 \leqslant cN^{2/3(1+\gamma_1-\gamma_2)} \leqslant cN^{1/3}, \tag{5.13}$$

$$\mathbb{M}_1 \leqslant cN^{2/3\cdot 1/2} \leqslant cN^{1/3}, \tag{5.14}$$

$$\mathbb{M}_2 \leqslant cN^{2/3(3/2+\gamma_1-\gamma_2)} \leqslant cN^{2/3}, \tag{5.15}$$

$$\mathbb{M}_3 \leqslant cN^{2/3(1/2+\gamma_1-\gamma_3)} \leqslant cN^{2/3(1/2-1/14)}. \tag{5.16}$$

The constant c here does not depend on N.

§6. Scattering

6.1. Asymptotic behavior of the parameters. All the estimates of the majorants found in §4 are valid for arbitrary t_1, and the boundaries for $\mathbb{M}_j$ do not depend on t_1. Therefore one can consider the limit transition $t_1 \to \infty$, and as a result one gets estimates which are valid for $t \in \mathbb{R}_+$. In the following formulas the dependence of the constants on the initial data will not be controlled.

From estimate (5.14) for $\mathbb{M}_1$ it follows that $|z(t)| \leqslant c(1+t)^{-1/2}$, $t \in \mathbb{R}_+$. Estimate (5.13) for $\mathbb{M}_0$ gives $|\omega(t) - \omega(t_1)| \leqslant c(1+t)^{-1}$, where $0 \leqslant t \leqslant t_1 < \infty$. So the finite limit $\omega_+ = \lim_{t\to\infty} \omega(t)$ exists, and the estimate

$$|\omega(t) - \omega_+| \leqslant c(1+t)^{-1}$$

is valid for $t \in \mathbb{R}_+$. Estimate (5.2) leads to the inequality $|\gamma'| \leqslant c(1+t)^{-1}$, $t \in \mathbb{R}_+$, and, therefore,

$$\beta(t) = \omega_+ t + O(\ln t), \qquad t \to \infty.$$

6.2. The dispersive remainder. Introduce the integral equation (2.20) with $t_1 = \infty$. The operator H_1 is transformed into the following operator:

$$H_+ = H(\alpha_+), \qquad \alpha_+ = \alpha(\omega_+).$$

Owing to (5.6), the component $\vec{k}$ in the corresponding decomposition $\vec{f} = \vec{k} + \vec{h}$ can be estimated as $\|\vec{k}\|_2 \leqslant c(1+t)^{-2}$, $t \in \mathbb{R}_+$.

Rewrite the integral equation for the component $\vec{h}$:

$$\vec{h} = \mathscr{U}_+(t,0)P(\alpha_+)\vec{f}_0 - i\int_0^t \mathscr{U}_+(t,\tau)\vec{D}\,d\tau,$$

where

$$\begin{aligned}\mathscr{U}_+(t,\tau)P_\pm(\alpha_+) &= \exp\bigl[-iH_+(t-\tau) \mp i\omega_+(t-\tau) \pm i\bigl(\beta(t)-\beta(\tau)\bigr)\bigr]P_\pm(\alpha_+)\\ &= \mathscr{U}_+(t,0)\mathscr{U}_+(0,\tau)P_\pm(\alpha_+).\end{aligned}$$

All the other notation also has the natural meaning.

As in §5 we must split D into two (three) terms: $\vec{D} = P(\alpha_+)\vec{L}_2 + \vec{D}_1$, $\vec{D}_1 = \vec{D}_2 + \vec{D}_3$. Apply (5.6) and (5.7) for $\vec{D}_1 = \vec{D}_2 + \vec{D}_3$. They give

$$\begin{aligned}\|(1+x^2)\vec{D}_2(t)\|_2 &\leqslant c(1+t)^{-3/2},\\ \|\vec{D}_2(t)\|_2 + \|(1+x^2)\vec{D}_3(t)\|_1 &\leqslant c(1+t)^{-3/2}.\end{aligned}$$

Consequently,

$$\begin{aligned}\|\mathscr{U}(t,\tau)\vec{D}_1(t)\|_2 &\leqslant \|\vec{D}_1(\tau)\|_2 \leqslant c(1+\tau)^{-3/2},\\ \|\mathscr{U}(t,\tau)\vec{D}_1(\tau)\|_\infty &\leqslant c|t-\tau|^{-1/2}(1+\tau)^{-3/2}.\end{aligned}$$

Using these estimates, one can represent the component $\vec{h}$ in the following way:

$$\begin{aligned}\vec{h}(t) = \mathscr{U}_+(t,0)\biggl[P(\alpha_+)\vec{f}_0 &- i\int_0^\infty \mathscr{U}_+(0,\tau)\vec{D}_1\,d\tau\\ &- i\int_0^t \mathscr{U}_+(0,\tau)P(\alpha_+)\vec{L}_2\,d\tau\biggr] + \vec{h}_R(t),\end{aligned}$$

$$\vec{h}_R(t) = i\int_t^\infty \mathscr{U}_+(t,\tau)\vec{D}_1\,d\tau.$$

It is clear that $\|h_R\|_\infty = O\bigl(t^{-1}\bigr)$ and $\|h_R\|_2 = O\bigl(t^{-1/2}\bigr)$.

The integral

$$\int_0^\infty \mathscr{U}_+(0,\tau)P(\alpha_+)\vec{L}_2\,d\tau$$

converges only conditionally and must be investigated more carefully. This is not too difficult, but for this purpose it would be necessary to restore in detail the structure of the spectral representation of the operator $H(\alpha)$ described in **[BP2]**. We cannot enter into these details here, and restrict ourselves to the formulation of the result:

$$\int_0^t \mathscr{U}_+(0,\tau)P(\alpha_+)\vec{L}_2\,d\tau = \int_0^\infty \mathscr{U}_+(0,\tau)P(\alpha_+)\vec{L}_2\,d\tau + \vec{h}_{OR},$$

where $\vec{h}_{OR} = o(1)$ in $L_2 \cap L_\infty$. So in $L_2 \cap L_\infty$ we have

$$\vec{h}(t) = \mathscr{U}_+(t,0)\biggl[P(\alpha_+)\vec{f}_0 - i\int_0^\infty \mathscr{U}_+(0,\tau)\vec{D}\,d\tau\biggr] + o(1).$$

The vector

$$\biggl[P(\alpha_+)\vec{f}_+ - i\int_0^\infty \mathscr{U}_+(0,\tau)\vec{D}\,d\tau\biggr]$$

belongs to the subspace of the continuous spectrum of H_+, such that in L_2

$$e^{-iH_+t}\left[P_+(\alpha_+)\vec{f}_0 - i\int_0^\infty \mathscr{U}_+(0,\tau)\vec{D}\,d\tau\right] = e^{-iH_0t}\vec{h}_+ + o(1),$$

where $\vec{h}_+ \in L_2$. For $\vec{h}$ this gives

$$\vec{h}(t) = e^{+i\beta(t)\sigma_3}e^{-ilt\sigma_3}\vec{h}_+ + o(1), \qquad l = -\partial_x^2.$$

Now the final asymptotic formula for the solution $\vec{\psi}$ acquires the form

$$\vec{\psi}(t) = e^{-i\beta(t)\sigma_3}\vec{\varphi}(\cdot,\alpha_+) + e^{-ilt\sigma_3}\vec{h}_+ + o(1).$$

References

[BP1] V. S. Buslaev and G. S. Perel′man, *Nonlinear scattering: states which are close to a soliton*, Zap. Nauchn. Sem. S.-Petersburg. Otdel. Mat. Inst. Steklov. (POMI) **200** (1992), 38–51; English transl. in J. Math. Sci. (to appear).

[BP2] ———, *Scattering for nonlinear Schrödinger equation: states which are close to a soliton*, Algebra i Analiz **4** (1992), no. 5, 63–102; English transl. in St. Petersburg Math. J. **4** (1993), no. 5.

[BP3] ———, *On nonlinear scattering of states which are close to a soliton*, Méthodes Semi-Classiques, Vol. II (Internat. Conf., Nantes, 1991), Astérisque, vol. 210, Soc. Math. France, Paris, 1993, pp. 49–63.

[GV1] J. Ginibre and G. Velo, *On a class of nonlinear equations.* I, II, J. Funct. Anal. **32** (1979), 1–71.

[GV2] ———, *On a class of nonlinear Schrödinger equations.* III: *Special theories in* dim 1, 2, 3, Ann. Inst. H. Poincaré Phys. Théor. **28** (1978), 287–316.

Department of Mathematics and Mechanics, St. Petersburg State University

Translated by the authors

Amer. Math. Soc. Transl.
(2) Vol. **164**, 1995

On Semigroups Generated by Initial-Boundary Value Problems Describing Two-Dimensional Visco-Plastic Flows

O. Ladyzhenskaya and G. Serëgin

§1. Introduction

We consider the following two-dimensional initial-boundary value problem (1.1):

$$(1.1_1)\qquad \begin{aligned}\partial_t v(x,t) + v_k(x,t)\partial_{x_k} v(x,t) - \operatorname{div} \frac{\partial D}{\partial \varepsilon}(\varepsilon(v(x,t)))\\ = -\nabla p(x,t) + g(x,t),\end{aligned}$$

and

$$(1.1_2)\qquad \operatorname{div} v(x,t) = 0 \qquad \text{for } (x,t) \in Q \equiv \Omega \times (0,+\infty),$$

$$(1.1_3)\qquad v(x,t) = 0 \qquad \text{for } (x,t) \in \partial\Omega \times [0,+\infty),$$

$$(1.1_4)\qquad v(x,0) = \varphi(x) \quad \text{for } x \in \Omega.$$

Here Ω is a bounded domain in $\mathbb{R}^2$, $v = (v_1, v_2)\colon Q \to \mathbb{R}^2$ is the velocity field, $\varepsilon(v) = (\varepsilon_{ij}(v))$, $i, j = 1, 2$, is the symmetric part of the gradient $\partial_x v$, called the *strain rate tensor* (so that $2\varepsilon_{ij}(v) = \partial_{x_i} v_j + \partial_{x_j} v_i$), the function $D\colon \mathbb{M}_s^{2\times 2} \to \mathbb{R}^+ = [0,+\infty)$ is the dissipative potential, $\mathbb{M}_s^{2\times 2}$ is the space of symmetric 2×2 matrices, $p\colon Q \to \mathbb{R}^1$ is the pressure field, and $g\colon Q \to \mathbb{R}^2$ is the field of external forces.

The system (1.1) determines the fields v and p if the fields g, φ and the function D are known and satisfy certain conditions (see **[L1–L4, MM]**).

For

$$D(\varepsilon) = \nu|\varepsilon|^2 = \nu \sum_{i,j=1}^{2} \varepsilon_{ij}^2, \qquad \nu = \text{const} > 0,$$

(1.1) is the *Navier-Stokes system*. Its global unique solvability has been proved in various phase spaces (**[L4, L5]**). If g does not depend on t, then the solution operators $V_t\colon \varphi \to v(t)$ form a semigroup $\{V_t, t \in \mathbb{R}^+, \overset{\circ}{J}(\Omega)\}$ of the first class in the phase space $\overset{\circ}{J}(\Omega) \subset L_2(\Omega; \mathbb{R}^2)$, and this semigroup has a minimal global B-attractor $\mathfrak{M}$. This attractor is a connected compact invariant subset of $\overset{\circ}{J}(\Omega)$ bounded in the

1991 *Mathematics Subject Classification*. Primary 35K60; Secondary 73E60.
Key words and phrases: Visco-plastic theory, initial-boundary value problem, semigroup, solution operators, attractor.

space $\overset{\circ}{J}{}_2^1(\Omega)$ ([**L5**], see also [**L6, L7**]). In [**L5**] some other properties of $\mathfrak{M}$ were also established. Later some estimates of the Hausdorff and fractal dimensions of $\mathfrak{M}$ were found (see [**L7, T, BV**] and the references given there).

In this paper we consider media for which

$$\frac{\partial D}{\partial \varepsilon}(\varepsilon) = \left(\frac{\partial D}{\partial \varepsilon_{ij}}(\varepsilon)\right), \quad i, j = 1, 2,$$

are nonlinear functions of ε, but $D(\varepsilon)$ is assumed to be a convex function of ε. First we investigate smooth functions D and, after that, the nonsmooth functions D of the form

$$D_0(\varepsilon) = \nu|\varepsilon|^2 + \sqrt{2}k_*|\varepsilon| \tag{1.3}$$

corresponding to the Bingham fluid (see [**MM**]). In (1.3) ν and k_* are positive parameters characterizing viscous and plastic properties of the fluid.

In the first case we shall suppose that D satisfies the following conditions:

$$D\colon \mathbb{M}_s^{2\times 2} \to \mathbb{R}^2 \quad \text{is a } C^2\text{-function and } D(0) = 0; \tag{1.4$_1$}$$

$$D(\varepsilon) \geqslant \nu|\varepsilon|^2, \quad \nu = \text{const} > 0, \quad \text{for any } \varepsilon \in \mathbb{M}_s^{2\times 2}; \tag{1.4$_2$}$$

$$\left(\frac{\partial^2 D(\varepsilon)}{\partial \varepsilon^2}\eta\right) : \eta = \frac{\partial^2 D(\varepsilon)}{\partial \varepsilon_{ij}\partial \varepsilon_{kl}}\eta_{ij}\eta_{kl} \geqslant 2\nu|\eta|^2 \quad \text{for any } \varepsilon \text{ and } \eta \in \mathbb{M}_s^{2\times 2}; \tag{1.4$_3$}$$

$$\left|\frac{\partial D}{\partial \varepsilon}(\varepsilon)\right| \leqslant \nu_1|\varepsilon| + \nu_2 \quad \text{for any } \varepsilon \in \mathbb{M}_s^{2\times 2} \text{ for positive constants } \nu_1,\ \nu_2. \tag{1.4$_4$}$$

From (1.4_k) it follows that $D(\varepsilon)$ is a convex function of ε and

$$\frac{\partial D}{\partial \varepsilon}(\varepsilon) : \varepsilon = \frac{\partial D}{\partial \varepsilon_{ij}}(\varepsilon)\varepsilon_{ij} \geqslant D(\varepsilon). \tag{1.4$_5$}$$

The unique solvability of the problem (1.1_k), $k = 1, 2, 3, 4$, with $D(\varepsilon)$ satisfying conditions (1.4_k), $k = 1, 2, 3, 4, 5$, and

$$\varphi \in \overset{\circ}{J}(\Omega), \qquad g \in L_{2,1}(Q_T), \qquad Q_T \equiv \Omega \times (0, T), \quad \forall T \in \mathbb{R}^+,$$

can be proved by the same method as in [**L2**] (see also [**L3**]) for the three-dimensional modifications of the Navier-Stokes equations. It is also true for $g \in L_2(0, T; H^{-1})$, where H^{-1} is the space dual to $H^1 \equiv \overset{\circ}{J}{}_2^1(\Omega)$ relative to $\overset{\circ}{J}(\Omega)$.

If additionally g has the derivative $\partial_t g \in L_2(0, T; H^{-1})$, the solution of the problem (1.1) will be smoother for $t > 0$ and the solution operators V_t will be compact for $t > 0$. In order to prove these facts, we shall use a new method suggested in [**L8**] (see also [**L9**]) for the Navier-Stokes equations. The result which we shall get in this way permits to use (for the case when g does not depend on t) some results from the theory of attractors developed in [**L5–L7**]. In particular, we shall prove the existence of a compact minimal global B-attractor for (1.1).

In §3 we consider problem (1.1) with D_0 from (1.3) and prove, for it also, the existence of a compact minimal global B-attractor. This fact was proved in [**S**] for the case when Ω is a square and the unknown functions v and p satisfy periodic boundary conditions.

We shall use the following Hilbert spaces:

- $L_2(\Omega; \mathbb{R}^2)$ with the standard inner product $(\cdot, \cdot)$ and norm $\|\cdot\|$.

• $H \equiv \overset{\circ}{J}(\Omega)$, the subspace of $L_2(\Omega; \mathbb{R}^2)$ defined as the closure in the norm $\|\cdot\|$ of the set

$$\dot{J}^\infty(\Omega) = \{u \parallel u \in C^\infty(\Omega; \mathbb{R}^2),\ \operatorname{div} u = 0,\ \operatorname{supp} u \text{ is compact in } \Omega\};$$

• P is the orthogonal projection in $L_2(\Omega; \mathbb{R}^2)$ on H.

• $H^1 \equiv \overset{\circ}{J}{}^1_2(\Omega)$ is a Hilbert space defined as the closure of the set $\dot{J}^\infty(\Omega)$ in the norm

$$\|u\|_{(1)} \equiv \|u\|_{H^1} \equiv \|\partial_x u\| \equiv \left(\int_\Omega |\partial_x u(x)|^2\, dx\right)^{1/2}.$$

• H^{-1} is the Hilbert space dual to H^1 relative to H, the norm in H^{-1} being defined as usual by

$$\|u\|_{(-1)} \equiv \|u\|_{H^{-1}} \equiv \sup_{\|v\|_{(1)} \leqslant 1} (u, v).$$

• The norm in $L_p(\Omega; \mathbb{R}^k)$ will be denoted by $\|u\|_{p,\Omega}$.

We restrict ourselves to the case of bounded domains Ω, although the results on global unique solvability of problem (1.1) formulated below are true for any Ω in $\mathbb{R}^2$.

Let us denote by $\{\lambda_k\}_{k=1}^\infty$ the spectrum of the Stokes operator $-\widetilde{\Delta} = -P\Delta$ under the Dirichlet condition, i.e., the λ_k are eigenvalues of the problem

$$\widetilde{\Delta} u = -\lambda u \quad \text{in } \Omega, \qquad u\big|_{\partial\Omega} = 0, \qquad u \in H,$$

and $\{\varphi_k\}_{k=1}^\infty$ are the corresponding eigenfunctions, so that $\widetilde{\Delta}\varphi_k = -\lambda_k\varphi_k$ and $(\varphi_k, \varphi_l) = \delta_{kl}$. It is known (see **[L4]**) that $\{\varphi_k\}_{k=1}^\infty$ forms a basis in H and H^1.

In accordance with the principal ideas of **[L4]**, we shall study only the velocity fields v and call them *solutions* of the problem (1.1_k), $k = 1, 2, 3, 4$. These fields v are determined by (1.1_k), $k = 2, 3, 4$, and by the projection of (1.1_1) on H. After that the fields p can be determined from the complete system of equations (1.1_k) and the fields v.

§2. Problem (1.1) with smooth D

The following theorem holds for any bounded domain Ω.

THEOREM 1. *Let D satisfy the conditions (1.4_k), $k = 1, 2, 3, 4$, of §1. Then for any $g \in L_2(0, T; H^{-1})$ and any $\varphi \in H$ there exists a unique solution v of the problem (1.1) having the following properties*:

$$v \in C([0, T]; H) \cap L_2(0, T; H^1), \qquad \partial_t v \in L_2(0, T; H^{-1}). \tag{2.1}$$

It satisfies the variational identity

$$\begin{aligned}(\partial_t v(t), w) + (v_k(t)\partial_{x_k} v(t), w) + \int_\Omega \frac{\partial D}{\partial \varepsilon}(\varepsilon(v(x,t))) : \varepsilon(w(x))\, dx \\ = (g(t), w) \quad \textit{for any } w \in H^1 \textit{ and for a.a. } t \in [0, T],\end{aligned} \tag{2.2_1}$$

the initial condition

$$v(0) = \varphi, \tag{2.2_2}$$

and the energy equality

$$\frac{1}{2}\|v(t)\|^2 + \int_0^t \int_\Omega \frac{\partial D}{\partial \varepsilon}\big(\varepsilon(v(x,\theta))\big) : \varepsilon(v(x,\theta))\,dx\,d\theta = \frac{1}{2}\|\varphi\|^2 + \int_0^t \big(g(\theta), v(\theta)\big)\,d\theta. \tag{2.3}$$

The latter implies two energy estimates:

$$\|v(t)\|^2 \leqslant \|\varphi\|^2 \exp\Big\{-\frac{\lambda_1 \nu}{2} t\Big\} + \frac{2}{\nu}\int_0^t \|g(\tau)\|^2_{(-1)} \exp\Big\{-\frac{\lambda_1 \nu}{2}(t-\tau)\Big\}\,d\tau, \tag{2.4$_1$}$$

$$\frac{\nu}{2}\int_0^t \|\partial_x v(\theta)\|^2\,d\theta \leqslant \int_0^t \int_\Omega D\big(\varepsilon(v(x,\theta))\big)\,dx\,d\theta \leqslant \|\varphi\|^2 + \frac{2}{\nu}\int_0^t \|g(\tau)\|^2_{(-1)}\,d\tau. \tag{2.4$_2$}$$

Moreover,

$$\int_0^t \|\partial_t v(\tau)\|^2_{(-1)}\,d\tau \leqslant \Phi_1\Big(t, \|\varphi\|, \int_0^t \|g(\tau)\|^2_{(-1)}\,d\tau, \nu^{-1}, \nu_1, \nu_2\Big). \tag{2.5}$$

The solution v continuously depends on φ in the norm of H.

If, in addition, g has the derivative $\partial_t g \in L_2(0,T;H^{-1})$, then v has the properties

$$v \in L_\infty(\gamma, T; H^1), \qquad \partial_t v \in L_\infty(\gamma, T; H) \cap L_2(\gamma, T; H^1) \tag{2.6}$$

for any $\gamma \in (0,T)$, and the estimate

$$\sup_{t\in[\gamma,T]} \big(\|\partial_x v(t)\| + \|\partial_t v(t)\|\big) + \|\partial^2_{xt} v\|_{2,Q_{\gamma T}} \leqslant \Phi_2\Big(T, \|\varphi\|, \int_0^T \big(\|g(t)\|^2_{(-1)} + \|\partial_t g(t)\|^2_{(-1)}\big)\,dt, \nu^{-1}, \gamma^{-1}\Big) \tag{2.7}$$

is valid. Here $Q_{\gamma T} \equiv \Omega \times (\gamma, T)$, and the functions Φ_1 and Φ_2 are continuous and increasing in each variable.

Proof of Theorem 2.1. We use the known relations

$$2\|\varepsilon(w)\|^2 = \|\partial_x w\|^2 + \|\operatorname{div} w\|^2, \tag{2.8}$$

$$\|w\|_{4,\Omega} \leqslant \frac{1}{\sqrt{2}}\|w\|\|\partial_x w\|, \tag{2.9}$$

which hold for any $w \in \overset{\circ}{W}{}^1_2(\Omega;\mathbb{R}^2)$ (for (2.9), see **[L4]**).

In order to prove Theorem 2.1, we apply the Galerkin-Faedo approximations $\{v^n\}_{n=1}^{\infty}$ calculated using the coordinate system $\{\psi_k\}_{k=1}^{\infty}$ in the space H^1. These approximations have the form

$$v^n(x,t) = \sum_{k=1}^{n} a_k^n(t)\psi_k(x)$$

and their coefficients $a_k^n(t)$ are determined by the following system of ordinary differential equations:

$$\begin{aligned}(2.11_1)\qquad &\left(\partial_t v^n(t), w^n\right) + \left(v_i^n(t)\partial_{x_i} v^n(t), w^n\right)\\ &+\left(\frac{\partial D}{\partial \varepsilon}(\varepsilon(v^n(t))), \varepsilon(w^n)\right) = \left(g(t), w^n\right), \quad \text{for any } w^n \in H_n^1\end{aligned}$$

and the initial condition

$$(2.11_2)\qquad v^n(0) = \varphi^n \in H_n^1,$$

where $H_n^1 = \operatorname{Span}\{\psi_1, \psi_2, \dots, \psi_n\}$ and

$$(2.12)\qquad \varphi^n \longrightarrow \varphi \quad \text{in } H.$$

Here and later we use the abbreviation

$$\left(\frac{\partial D}{\partial \varepsilon}(\varepsilon(v(t))), \varepsilon(w)\right) \equiv \int_\Omega \frac{\partial D}{\partial \varepsilon}(\varepsilon(v(x,t))) : \varepsilon(w(x))\, dx.$$

Now we shall prove some a priori estimates for $\{v^n\}_{n=1}^{\infty}$. We $w^n = v^n(t)$ in (2.11_1). Then, after elementary transformations, we get the relation

$$(2.13)\qquad \frac{1}{2}\partial_t\|v^n(t)\|^2 + \left(\frac{\partial D}{\partial \varepsilon}(\varepsilon(v^n(t))), \varepsilon(v^n(t))\right) = \left(g(t), v^n(t)\right).$$

By using (1.4_2), (1.4_5), and (2.8), the following inequalities may be deduced from (2.13):

$$\begin{aligned}&\frac{1}{2}\partial_t\|v^n(t)\|^2 + \int_\Omega D\big(\varepsilon(v^n(x,t))\big)\, dx\\ &\leqslant \|g(t)\|_{(-1)}\|\partial_x v^n(t)\| \leqslant \frac{1}{2}\int_\Omega D\big(\varepsilon(v^n(x,t))\big)\, dx + \frac{1}{\nu}\|g(t)\|_{(-1)}^2\end{aligned}$$

and, therefore,

$$\begin{aligned}(2.14_1)\qquad &\partial_t\|v^n(t)\|^2 + \frac{\nu}{2}\|\partial_x v^n(t)\|^2\\ &\leqslant \partial_t\|v^n(t)\|^2 + \int_\Omega D\big(\varepsilon(v^n(x,t))\big)\, dx \leqslant \frac{2}{\nu}\|g(t)\|_{(-1)}^2,\end{aligned}$$

$$(2.14_2)\qquad \partial_t\|v^n(t)\|^2 + \frac{\nu\lambda_1}{2}\|v^n(t)\|^2 \leqslant \frac{2}{\nu}\|g(t)\|_{(-1)}^2.$$

The integration of (2.14_1) gives

$$\begin{aligned}(2.15_1)\qquad &\frac{\nu}{2}\int_0^t \|\partial_x v^n(\theta)\|^2\, d\theta \leqslant \int_0^t\int_\Omega D\big(\varepsilon(v^n(x,\theta))\big)\, dx\, d\theta\\ &\leqslant \|\varphi^n\|^2 + \frac{2}{\nu}\int_0^t \|g(\theta)\|_{(-1)}^2\, d\theta\end{aligned}$$

and the integration of (2.14_2) gives

$$(2.15_2)\quad \|v^n(t)\|^2 \leqslant \|\varphi^n\|^2 \exp\Big\{-\frac{\lambda_1 \nu}{2}t\Big\} + \frac{2}{\nu}\int_0^t \|g(\tau)\|^2_{(-1)} \exp\Big\{-\frac{\lambda_1\nu}{2}(t-\tau)\Big\}\,d\tau.$$

Next, by (2.11_1) we have

$$\begin{aligned}(\partial_t v^n(t), w^n) &= (v_i^n(t)\partial_{x_i} w^n, v^n(t)) - \Big(\frac{\partial D}{\partial \varepsilon}(\varepsilon(v^n(t))), \varepsilon(w^n)\Big) + (g(t), w^n)\\ &\leqslant \|v^n(t)\|^2_{4,\Omega}\|\partial_x w^n\| + \int_\Omega \Big|\frac{\partial D}{\partial \varepsilon}(\varepsilon(v^n(x,t)))\Big| |\varepsilon(w^n)|\,dx\\ &\quad + \|g(t)\|_{(-1)}\|\partial_x w^n\|.\end{aligned}$$

Taking into account condition (1.4_4) and relations (2.8), (2.9), we obtain

$$\begin{aligned}(\partial_t v^n(t), w^n) \leqslant \Big\{&\frac{1}{\sqrt{2}}\|v^n(t)\|\|\partial_x v^n(t)\| + \|g(t)\|_{(-1)}\\ &+ \frac{1}{\sqrt{2}}\Big(\int_\Omega (\nu_1|\varepsilon(v^n(x,t))| + \nu_2)^2\,dx\Big)^{1/2}\Big\}\|\partial_x w^n\|\end{aligned}$$

and, therefore,

$$\begin{aligned}&\int_0^t \Big[\sup_{\substack{w^n\in H_n^1\\ w^n\neq 0}} \frac{|(\partial_t v^n(\theta), w^n)|}{\|\partial_x w^n\|}\Big]^2 d\theta\\ &\quad\leqslant 3\int_0^t \Big(\frac{1}{2}\|v^n(\theta)\|^2\|\partial_x v^n(\theta)\|^2 + \|g(t)\|^2_{(-1)}\\ &\qquad\qquad + \frac{1}{2}\int_\Omega (\nu_1|\partial_x v^n(x,\theta)| + \nu_2)^2\,dx\Big)\,d\theta.\end{aligned}$$

The estimates (2.15_k), $k = 1, 2$, give a majorant for the right-hand side of the last inequality. So, we have

$$(2.16)\quad \int_0^T \Big[\sup_{\substack{w^n\in H_n^1\\ w^n\neq 0}} \frac{|(\partial_t v^n(\theta), w^n)|}{\|\partial_x w^n\|}\Big]^2 \leqslant \Phi_1\Big(t, \|\varphi^n\|, \int_0^t \|g(\tau)\|^2_{(-1)}\,d\tau, \nu^{-1}, \nu_1, \nu_2\Big).$$

Now we must carry out a limiting procedure and show that $\{v^n\}_{n=1}^\infty$ has a limit point v that will be the solution of the problem (1.1_k), $k = 1, 2, 3, 4$. Such a procedure has often been carried out for various problems (for our problem $(1{,}1_k)$ see **[L2, L3]**). For the reader's convenience, we have decided to extract some general analytical facts generally used in such limiting procedures and to formulate them separately as Lemmas 2.1 and 2.2. It seems to us that the statement of Lemma 2.2, which concerns the inclusion $\partial_t u \in L_2(0, T; \mathscr{H}^{-1})$ for the limit function u, is new.

LEMMA 2.1. *Let $\mathscr{H}$ and $\mathscr{H}^1$ be separable Hilbert spaces with scalar products $(\cdot,\cdot)$ and $(\cdot,\cdot)_{(1)}$ and with corresponding norms $\|\cdot\|$ and $\|\cdot\|_{(1)}$. Suppose that $\mathscr{H}^1$ is imbedded* *in $\mathscr{H}$ and the imbedding is dense and compact. Let $\mathscr{H}^{-1}$ be the space dual to $\mathscr{H}^1$ relative to $\mathscr{H}$ with the norm*

$$\|u\|_{(-1)} = \sup_{\|v\|_{(-1)}\leqslant 1} (u, v).$$

Suppose that a function u has the properties

$$u \in L_2(0, T; \mathscr{H}^1) \quad \text{and} \quad \partial_t u \in L_2(0, T; \mathscr{H}^{-1}). \tag{2.17}$$

Then $u\colon [0, T] \to \mathscr{H}$ (perhaps after modifications on a set of zero Lebesgue measure) is continuous, i.e.,

$$u \in C\big([0, T]; \mathscr{H}\big). \tag{2.18}$$

Moreover, the function $t \mapsto \|u(t)\|^2$ is absolutely continuous on $[0, T]$, and

$$\|u(t)\|^2 - \|u(\tau)\|^2 = 2\int_\tau^t \big(\partial_t u(\theta), u(\theta)\big)\, d\theta, \quad \forall t, \tau \in [0, T]. \tag{2.19}$$

LEMMA 2.2. *Let the spaces $\mathscr{H}^1$, $\mathscr{H}$ and $\mathscr{H}^{-1}$ be the same as in Lemma 2.1. Suppose that $\{\mathscr{H}_n^1\}_{n=1}^\infty$ is a family of linear subspaces of $\mathscr{H}^1$ with the property*

(2.20) *for any $w \in \mathscr{H}^1$ there is a sequence $w^n \in \mathscr{H}_n^1$ such that $w^n \to w$ in $\mathscr{H}^1$.*

Let $\big\{t \mapsto u^n(t) \in \mathscr{H}_n^1,\ t \in [0, T]\big\}_{n=1}^\infty$ be a sequence satisfying the following inclusions and estimates:

$$u^n \in C\big([0, T]; \mathscr{H}\big) \cap L_2\big(0, T; \mathscr{H}^1\big), \qquad \partial_t u^n \in L_2\big(o, T; \mathscr{H}^{-1}\big), \tag{2.21}$$

$$\sup_{t\in[0,T]} \|u^n(t)\| \leqslant c_1, \tag{2.22}$$

$$\|u^n\|_{L_2(0,T;\mathscr{H}^1)} \leqslant c_2, \tag{2.23}$$

$$\|\alpha_n\|_{L_2(0,T)} \leqslant c_3, \tag{2.24}$$

where

$$\alpha_n(t) \equiv \sup_{\substack{w^n\in\mathscr{H}_n^1\\ w^n\neq 0}} \frac{\big|\big(\partial_t u^n(t), w^n\big)\big|}{\|w^n\|_{(1)}}$$

with positive constants c_k, $k = 1, 2, 3$, independent of n.

Then there exists an infinite subsequence (still denoted by $\{u^n\}_{n=1}^\infty$) which has the following properties:

$$u^n \rightharpoonup u \quad \text{in } L_2(0, T; \mathscr{H}^1), \tag{2.25}$$

$$\big(u^n(t), w\big) \to \big(u(t), w\big) \quad \text{in } C\big([0, T]\big) \text{ for any fixed } w \in \mathscr{H}, \tag{2.26}$$

$$u^n \to u \quad \text{in } L_2(0, T; \mathscr{H}). \tag{2.27}$$

Moreover, the limit function u satisfies the inclusions and estimates

$$u \in C\big([0, T]; \mathscr{H}\big) \cap L_2(0, T; \mathscr{H}^1), \qquad \partial_t u \in L_2(0, T; \mathscr{H}^{-1}) \tag{2.28$_1$}$$

$$\|u(t)\| \leqslant \liminf_{n\to\infty} \|u^n(t)\| \quad \text{for any } t \in [0, T], \tag{2.28$_2$}$$

$$\|u\|_{L_2(0,T;\mathscr{H}^1)} \leqslant \liminf_{n\to\infty} \|u^n\|_{L_2(0,T;\mathscr{H}^1)}, \tag{2.28$_3$}$$

$$\|\partial_t u\|_{L_2(0,T;\mathscr{H}^{-1})} \leqslant c_3. \tag{2.28$_4$}$$

Let us continue the proof of Theorem 2.1. If we set $\mathscr{H} = H$, $\mathscr{H}^1 = H^1$, $\mathscr{H}^{-1} = H^{-1}$, $\mathscr{H}^1_n = H^1_n$, then by (2.15_k), $k = 1, 2$, and (2.16) all the assumptions of Lemma 2.2 hold. For this reason we can state (after the selection of a subsequence) that

$$v^n \rightharpoonup v \quad \text{in } L_2(0, T; H^{-1}), \tag{2.29}$$

$$(v^n(t), w) \to (v(t), w) \quad \text{in } C([0, T]) \text{ for any fixed } w \in H, \tag{2.30}$$

$$v^n \to v \quad \text{in } L_2(Q_T). \tag{2.31}$$

Moreover, the limit function v satisfies the following inclusions:

$$v \in L_2(0, T; H^1), \quad \partial_t v \in L_2(0, T; H^{-1}), \tag{2.32}$$

$$v \in C([0, T]; H). \tag{2.33}$$

It should be noted that if $\psi_k = \varphi_k$, $k = 1, 2, \dots$ (about the system of functions $\{\varphi_k\}_{k=1}^{\infty}$ see the Introduction), then, in addition (see the Appendix),

$$\partial_t v^n \rightharpoonup \partial_t v \quad \text{in } L_2(0, T; H^{-1}).$$

So the inclusions (2.1) are proved. Relation (2.2_2) follows from (2.30), (2.33) and (2.11_2), (2.12). Using the same facts and also (2.29), we obtain the energy estimates (2.4_k), $k = 1, 2$, by taking the limit in (2.15_k), $k = 1, 2$.

In order to prove the variational equality (2.2_1), let us take an arbitrary non-negative function $\xi \in C_0^1(0, T)$. Using (2.11_1), we can write

$$\int_0^T \xi(t)\Big\{\big(\partial_t v^n(t), w^n - v^n(t)\big) + \big(v_i^n(t)\partial_{x_i} v^n(t), w^n - v^n(t)\big) + \Big(\frac{\partial D}{\partial \varepsilon}(\varepsilon(v^n(t))), \varepsilon(w^n) - \varepsilon(v^n(t))\Big) - \big(g(t), w^n - v^n(t)\big)\Big\}\, dt = 0,$$

for all $w^n \in H^1_n$. The convexity of D, the relation $\operatorname{div} v^n(t) = 0$ and the last equality imply

$$\begin{aligned} &-\int_0^T \partial_t\xi(t)\big(v^n(t), w^n\big)\, dt + \frac{1}{2}\int_0^T \partial_t \xi(t)\|v^n(t)\|^2\, dt \\ &+ \int_0^T \xi(t)\big(v_i^n(t)\partial_{x_i} v^n(t), w^n\big)\, dt \\ &+ \int_0^T \xi(t)\int_\Omega \big(D(\varepsilon(w^n(x))) - D(\varepsilon(v^n(x, t)))\big)\, dx\, dt \\ &- \int_0^T \xi(t)\big(g(t), w^n - v^n(t)\big)\, dt \geqslant 0. \end{aligned} \tag{2.34}$$

For any $w \in H^1$ let $\{w^n\}_{n=1}^{\infty}$, $w^n \in H^1_n$, be a sequence that converges to w in H^1. Then by (2.29), (2.31) and the convexity of D, the inequality (2.34) leads to the relation

$$-\int_0^T \partial_t\xi(t)\big(v(t), w\big)\, dt + \frac{1}{2}\int_0^T \partial_t\xi(t)\|v(t)\|^2\, dt + \int_0^T \xi(t)\big(v_i(t)\partial_{x_i} v(t), w\big)\, dt + \int_0^T \xi(t)\Big\{\int_\Omega \big(D(\varepsilon(w(x))) - D(\varepsilon(v(x, t)))\big)\, dx \,\big(g(t), w - v(t)\big)\Big\}\, dt \geqslant 0.$$

To take the limit in the third term of the left-hand side of (2.34) we used the same arguments as in **[L4,** Chapter VI, §7]. Thus, due to Lemma 2.1 and (2.32), we have

$$\int_0^T \xi(t)\Big\{\big(\partial_t v(t), w - v(t)\big) + \big(v_i(t)\partial_{x_i} v(t), w - v(t)\big) + \int_\Omega \big(D\big(\varepsilon(w(x))\big) - D\big(\varepsilon(v(x,t))\big)\big)\,dx - \big(g(t), w - v(t)\big)\Big\}\,dt \geqslant 0$$

for any $w \in H^1$ and any nonnegative $\xi \in C_0^1(0,T)$. This inequality gives

$$\big(\partial_t v(t), w - v(t)\big) + \big(v_i(t)\partial_{x_i} v(t), w - v(t)\big) + \int_\Omega \big(D\big(\varepsilon(w(x))\big) - D\big(\varepsilon(v(x,t))\big)\big)\,dx - \big(g(t), w - v(t)\big) \geqslant 0 \tag{2.35}$$

for any $w \in H^1$ and a.a. $t \in [0,T]$. If we set $w = v(t) + u$ in (2.35) for any $u \in H^1$ then, by the convexity of D, which implies the inequality

$$D\big(\varepsilon(u(x)) + \varepsilon(v(x,t))\big) - D\big(\varepsilon(v(x,t))\big) \geqslant \frac{\partial D}{\partial \varepsilon}\big(\varepsilon(v(x,t))\big) : \varepsilon(u(x)),$$

the following relation holds:

$$\big(\partial_t v(t), u\big) + \big(v_i(t)\partial_{x_i} v(t), u\big) + \left(\frac{\partial D}{\partial \varepsilon}\big(\varepsilon(v(t))\big), \varepsilon(u)\right) - \big(g(t), u\big) \geqslant 0.$$

Finally, the arbitrariness of $u \in H^1$ gives the desired equality (2.2_1).

The energy equality (2.3) is obtained by integrating (2.2_1) with respect to t with $w = u(t)$, and applying Lemma 2.1.

To show that the estimate (2.5) is true we can apply the same arguments as in the proof of (2.16). In fact, the variational equality (2.2_1) gives

$$\begin{aligned}\big(\partial_t v(t), w\big) &= \big(v_i(t)\partial_{x_i} w, v(t)\big) - \left(\frac{\partial D}{\partial \varepsilon}\big(\varepsilon(v(t))\big), \varepsilon(w)\right) + \big(g(t), w\big)\\ &\leqslant \|v(t)\|_{4,\Omega}^2 \|\partial_x w\| + \int_\Omega \left|\frac{\partial D}{\partial \varepsilon}\big(\varepsilon(v(x,t))\big)\right| \big|\varepsilon(w(x))\big|\,dx\\ &\quad + \|g(t)\|_{(-1)} \|\partial_x w\|\end{aligned}$$

and by (1.4_4), (2.8), and (2.9) we get

$$\begin{aligned}\|\partial_t v(t)\|_{(-1)} &\leqslant \frac{1}{\sqrt{2}} \|v(t)\| \|\partial_x v(t)\| + \|g(t)\|_{(-1)}\\ &\quad + \frac{1}{\sqrt{2}} \left(\int_\Omega \big(\nu_1 |\varepsilon(v(x,t))| + \nu_2\big)^2\,dx\right)^{1/2}.\end{aligned}$$

This and (2.4_k), $k = 1, 2$, lead to (2.5) with the same majorant as in (2.16).

In order to prove uniqueness and continuity of v with respect to φ, we consider the difference $u(t) \equiv v(t) - v'(t)$ between the two solutions $v(t) \equiv v(t, \varphi)$ and

$v'(t) \equiv v(t, \varphi')$ having the properties (2.1). Then, using (2.2_1) for $v(t)$ and $v'(t)$, and Lemma 2.1, we deduce that

$$\tag{2.36} \begin{aligned} &\frac{1}{2}\|u(t)\|^2 + \int_0^t \left(\frac{\partial D}{\partial \varepsilon}(\varepsilon(v(\theta))) - \frac{\partial D}{\partial \varepsilon}(\varepsilon(v'(\theta))), \varepsilon(u(\theta)) \right) d\theta \\ &\qquad = \frac{1}{2}\|u(0)\|^2 - \int_0^t \left(u_i(\theta)\partial_{x_i} v(\theta) + v'(\theta)\partial_{x_i} u(\theta), u(\theta) \right) d\theta. \end{aligned}$$

Due to (1.4_3), (2.8), and (2.9), it follows from (2.36) that

$$\begin{aligned} \frac{1}{2}\|u(t)\|^2 + \nu \int_0^t \|\partial_x u(\tau)\|^2 \, d\tau &\leqslant \frac{1}{2}\|u(0)\|^2 - \int_0^t \left(u_i(\theta)\partial_{x_i} v(\theta), u(\theta) \right) d\theta \\ &\leqslant \frac{1}{2}\|u(0)\|^2 + \int_0^t \|\partial_x v(\tau)\| \|u(\tau)\|_{4,\Omega}^2 \, d\tau \\ &\leqslant \frac{1}{2}\|u(0)\|^2 + \frac{1}{\sqrt{2}} \int_0^t \|\partial_x v(\tau)\| \|u(\tau)\| \|\partial_x u(\tau)\| \, d\tau \\ &\leqslant \frac{1}{2}\|u(0)\|^2 + \int_0^t \left(\nu \|\partial_x u(\tau)\|^2 + \frac{1}{8\nu} \|\partial_x v(\tau)\|^2 \|u(\tau)\|^2 \right) d\tau. \end{aligned}$$

Thus we have

$$\tag{2.37} \|u(t)\|^2 \leqslant \|u(0)\|^2 + \frac{1}{4\nu} \int_0^t \|\partial_x v(\tau)\|^2 \|u(\tau)\|^2 \, d\tau.$$

The estimate (2.4_2) gives a majorant for

$$\begin{aligned} \frac{1}{4\nu} \int_0^t \|\partial_x v(\tau)\|^2 \, d\tau &\leqslant \frac{1}{2\nu}\|\varphi\|^2 + \frac{1}{\nu^3} \int_0^t \|g(\tau)\|_{(-1)}^2 \, d\tau \\ &\equiv \Phi_3\left(\|\varphi\|, \int_0^t \|g(\tau)\|_{(-1)}^2 \, d\tau, \nu^{-1} \right) \end{aligned}$$

and, therefore, we can conclude from (2.37) that

$$\begin{aligned} \|u(t)\|^2 &\equiv \|v(t,\varphi) - v(t,\varphi')\|^2 \\ &\leqslant \|\varphi - \varphi'\|^2 \Phi_4\left(\|\varphi\|, \int_0^t \|g(\tau)\|_{(-1)}^2 \, d\tau, \nu^{-1} \right). \end{aligned}$$

This estimate shows that problem (1.1) with $\varphi \in H$ has not more than one solution in the class (2.1) and the solution $v(t, \varphi)$ of problem (1.1) is a uniformly continuous function of φ on any ball $B_R(H)$.

So, we shall establish (2.6) and (2.7). To do this, set $w^n = \partial_t v^n(t)$ in (2.11_1):

$$\tag{2.39} \begin{aligned} &\|\partial_t v^n(t)\|^2 + \partial_t \int_\Omega D(\varepsilon(v^n(x,t))) \, dx \\ &\qquad = -\left(v_i^n(t)\partial_{x_i} v^n(t), \partial_t v^n(t) \right) + \left(g(t), \partial_t v^n(t) \right). \end{aligned}$$

Using (2.9), from (2.39) we can derive

$$
\begin{aligned}
&\|\partial_t v^n(t)\|^2 + \partial_t \int_\Omega D\big(\varepsilon(v^n(x,t))\big)\,dx \\
&\quad = \|v^n(t)\|_{4,\Omega}\|\partial_x v^n(t)\|\|\partial_t v^n(t)\|_{4,\Omega} + \|g(t)\|_{(-1)}\|\partial^2_{xt} v^n(t)\| \\
&\quad \leqslant \frac{1}{\sqrt{2}}\|v^n(t)\|^{1/2}\|\partial_x v^n(t)\|^{3/2}\|\partial_t v^n(t)\|^{1/2}\|\partial^2_{xt} v^n(x,t)\|^{1/2} \\
&\qquad + \|g(t)\|_{(-1)}\|\partial^2_{xt} v^n(t)\|.
\end{aligned}
\tag{2.40}
$$

Next, let us differentiate (2.11_1) with respect to t and then take $w^n = \partial_t v^n(t)$. After a simple calculation we get

$$
\begin{aligned}
&\frac{1}{2}\partial_t\|\partial_t v^n(t)\|^2 + \left(\frac{\partial^2 D}{\partial \varepsilon^2}\big(\varepsilon(v^n(t))\big)\varepsilon\big(\partial_t v^n(t)\big), \varepsilon\big(\partial_t v^n(t)\big)\right) \\
&\quad = -\big(\partial_t v^n_k(t)\partial_{x_k} v^n(t), \partial_t v^n(t)\big) + \big(\partial_t g(t), \partial_t v^n(t)\big).
\end{aligned}
\tag{2.41}
$$

Taking (1.4_3) and (2.9) into account, we deduce from (2.41) that

$$
\begin{aligned}
&\frac{1}{2}\partial_t\|\partial_t v^n(t)\|^2 + \nu\|\partial^2_{xt} v^n(t)\|^2 \\
&\quad \leqslant \|\partial_x v^n(t)\|\|\partial_t v^n(t)\|^2_{4,\Omega} + \|\partial_t g(t)\|_{(-1)}\|\partial^2_{xt} v^n(t)\| \\
&\quad \leqslant \frac{1}{\sqrt{2}}\|v^n(t)\|_{(1)}\|\partial_t v^n(t)\|\|\partial_t v^n(t)\|_{(1)} + \|\partial_t g(t)\|_{(-1)}\|\partial_t v^n(t)\|_{(1)} \\
&\quad \leqslant \frac{\nu}{4}\|\partial_t v^n(t)\|^2_{(1)} + \frac{1}{2\nu}\|v^n(t)\|^2_{(1)}\|\partial_t v^n(t)\|^2 \\
&\qquad + \frac{\nu}{4}\|\partial_t v^n(t)\|^2_{(1)} + \frac{1}{\nu}\|\partial_t g(t)\|^2_{(-1)}.
\end{aligned}
$$

After combining similar terms and multiplying by 2, we get the desired relation

$$
\partial_t\|\partial_t v^n(t)\|^2 + \nu\|\partial_t v^n(t)\|^2_{(1)} \leqslant \frac{1}{\nu}\|v^n(t)\|^2_{(1)}\|\partial_t v^n(t)\|^2 + \frac{2}{\nu}\|\partial_t g(t)\|^2_{(-1)}.
\tag{2.42}
$$

Assuming that $\sup_n \|\partial_t v^n(0)\|$ is finite, we could derive the estimates for $\|\partial_t v^n(t)\|$ and $\|\partial^2_{xt} v^n\|_{2,Q_t}$, $Q_t \equiv \Omega \times (0,t)$ from (2.42). But we want to investigate solutions v with any $\varphi = v(0) \in H$. For this reason we consider the sum of (2.40) multiplied by $4t$ and (2.42) multiplied by t^2:

$$
\begin{aligned}
&\partial_t\left[4t\int_\Omega D\big(\varepsilon(v^n(x,t))\big)\,dx + t^2\|\partial_t v^n(t)\|^2\right] + 2t\|\partial_t v^n(t)\|^2 + \nu t^2\|\partial_t v^n(t)\|^2_{(1)} \\
&\quad \leqslant 4\int_\Omega D\big(\varepsilon(v^n(x,t))\big)\,dx + \frac{2}{\nu}t^2\|\partial_t g(t)\|^2_{(-1)} \\
&\qquad + 2\sqrt{2}\|v^n(t)\|^{1/2}\|v^n(t)\|^{3/2}_{(1)}\big(t\|\partial_t v^n(t)\|\big)^{1/2}\big(t\|\partial_t v^n(t)\|_{(1)}\big)^{1/2} \\
&\qquad + 4\|g(t)\|_{(-1)}\big(t\|\partial_t v^n(t)\|_{(1)}\big) + \nu^{-1}\|v^n(t)\|^2_{(1)}\big(t\|\partial_t v^n(t)\|\big)^2.
\end{aligned}
\tag{2.43}
$$

Setting

$$
y(t) \equiv 4t\int_\Omega D\big(\varepsilon(v^n(x,t))\big)\,dx + t^2\|\partial_t v^n(t)\|^2,
$$

we can get the following inequality from (2.43):

$$\partial_t y(t) + 2t\|\partial_t v^n(t)\|^2 + \nu t^2\|\partial_t v^n(t)\|^2_{(1)} \leqslant c_1(t)y(t) + c_2(t) \tag{2.44}$$

with $c_k(\cdot)$, $k = 1, 2$, for which the estimates (2.15_k), $k = 1, 2$, give some majorants of $\int_0^t c_k(\tau)\,d\tau$, $k = 1, 2$. From these facts we get (2.7) for $v = v^n$ and $\varphi = \varphi^n$ from (2.44). Finally, the desired estimate (2.7) follows from (2.29)–(2.31) and (2.12). Theorem 2.1 is proved. □

Estimate (2.7) is the new estimate mentioned in the Introduction.

Let us consider the case when g does not depend on t, i.e.,

$$g(t) = \mu f, \qquad \mu \in \mathbb{R}^+,\ f \in H^{-1}, \tag{2.45}$$

where μ is a positive parameter.

We fix f and μ, and from now on we shall not explicitly show the dependence of solutions and their majorants on f and μ. Let $v(t,\varphi)$ be a solution of problem (1.1) with $v(0) = \varphi \in H$. According to Theorem 2.1, only one solution $v(t,\varphi)$, $t \in \mathbb{R}^+$, corresponds to any $\varphi \in H$. It possesses all the properties enumerated in Theorem 2.1. In particular, $v \in C(\mathbb{R}^+; H)$ and $v(t,\varphi)$ depends on φ continuously. Thus the solution operators

$$V_t \colon \varphi \in H \longrightarrow v(t,\varphi) \in H, \quad t \in \mathbb{R}^+$$

of problem (1.1) are defined on the whole space H and are continuous in H. They have also the semigroup property, i.e.,

$$V_{t_1+t_2} = V_{t_1} V_{t_2}, \quad \forall t_1, t_2 \geqslant 0.$$

Moreover, the family $\{V_t, t \in \mathbb{R}^+, H\}$ is a continuous semigroup (i.e., the mapping $V_t \colon \mathbb{R}^+ \times H \longrightarrow H$ is continuous). This follows from (2.38) and the continuity of solutions $v(t,\varphi)$ in t. For this semigroup there exist B-absorbing bounded sets (see **[L5–L7]**)

$$B_R(H) = \{u \mid u \in H, \|u\| < R\}, \qquad R > R_0 \equiv \frac{2\mu}{\nu\sqrt{\lambda_1}\|f\|_{(-1)}}. \tag{2.46}$$

In fact, by (2.4_1) we can get the inequality

$$\|v(t,\varphi)\|^2 \leqslant \|\varphi\|^2 \exp\Big\{-\frac{\lambda_1\nu}{2}t\Big\} + \frac{4\mu^2\|f\|^2_{(-1)}}{\lambda_1\nu^2}\Big[1 - \exp\Big\{-\frac{\lambda_1\nu}{2}t\Big\}\Big],$$

which implies that for any bounded set B of the space H there is a number $t(B,R) > 0$ such that

$$V_t(B) \subset B_R(H) \quad \text{for all } t \geqslant t(B,R). \tag{2.47}$$

The inclusion (2.47) means that the ball $B_R(H)$ is a B-absorbing set for the semigroup $\{V_t, t \in \mathbb{R}^+, H\}$.

Another important property of this semigroup is that it is a semigroup of class 1 (see **[L6, L7]**). This follows from inequality (2.7) for solutions $v(t) \equiv v(t,\varphi)$. Namely, (2.7) implies the estimate

$$\|\partial_x v(t,\varphi)\| \leqslant \Phi_2(2t, \|\varphi\|, 2t\mu^2\|f\|^2_{(-1)}, \nu^{-1}, t^{-1}), \tag{2.48}$$

which guarantees the compactness of the operators $V_t \colon H \longrightarrow H$ for any $t > 0$.

Now, taking all these facts into account, we can apply results proved in **[L5–L7]** for semigroups of class 1, and so obtain the following theorem.

THEOREM 2.2. *Suppose that Ω is a bounded domain in $\mathbb{R}^2$ and condition* (2.45) *holds. Then the family of solution operators to the problem* (1.1) *is the continuous semigroup $\{V_t, t \in \mathbb{R}^+, H\}$ of class* 1 *having bounded B-absorbing sets* (2.46). *It possesses the compact minimal global B-attractor $\mathfrak{M}_\mu$, which is a connected invariant subset in H bounded in H^1. For any $\varphi \in \mathfrak{M}_\mu$ the solution $v(t,\varphi)$, $t \in \mathbb{R}^+$, may be extended for all $t \in \mathbb{R}^1$ continuously so that $v(t,\varphi)$ will be the solution of* (1.1) *on $\mathbb{R}^1$. Moreover, for all solutions $v(t,\varphi)$, $\varphi \in \mathfrak{M}_\mu$, $t \in \mathbb{R}^1$, there are derivatives $\partial_t v \in L_\infty(\mathbb{R}^1, H)$ and $\partial^2_{xt} v \in L_2(\Omega \times (-T,T))$, $\forall T < +\infty$, and*

$$\sup_{\varphi\in\mathfrak{M}_\mu}\sup_{t\in\mathbb{R}^1}\left\{\|v(t,\varphi)\|_{(1)}, \|\partial_t v(t,\varphi)\|, \int_t^{t+1}\|\partial_t v(\tau,\varphi)\|^2_{(1)}\,d\tau\right\} \leqslant \Phi_5\bigl(\mu\|f\|_{(-1)}, \nu^{-1}, \nu_1, \nu_2\bigr). \tag{2.49}$$

The number of determining modes on $\mathfrak{M}_\mu$ is finite and may be majorized by a value depending only on $\mu\|f\|_{(-1)}$, ν^{-1}, ν_1 and ν_2. The set $\mathfrak{M}_\mu$ consists of one point, namely the stationary solution of (1.1_k), $k = 1,2,3$, *if μ is sufficiently small.*

§3. Problem (1.1) for nonsmooth D

In the case when the dissipative potential has the form (1.3) we must rewrite the initial-boundary value problem (1.1) in the following way:

$$\partial_t v(x,t) + v_k(x,t)\partial_{x_k} v(x,t) - \operatorname{div}\sigma(x,t) = g(x,t), \tag{3.1$_1$}$$

$$\operatorname{div} v(x,t) = 0, \tag{3.1$_2$}$$

$$\sigma(x,t) + p(x,t)\mathbb{I} \in \partial D_0(\varepsilon(v(x,t))) \quad \text{for } (x,t) \in Q \equiv \Omega\times(0,+\infty), \tag{3.1$_3$}$$

$$v(x,t) = 0 \quad \text{for } (x,t) \in \partial\Omega\times[0,+\infty), \tag{3.1$_4$}$$

$$v(x,0) = \varphi(x) \quad \text{for } x \in \Omega, \tag{3.1$_5$}$$

where σ is the Cauchy stress tensor, $\mathbb{I}$ is the identity matrix in $\mathbb{M}^{2\times 2}_s$, and $\partial D_0(\varepsilon)$ is the subdifferential of D_0 at ε, i.e.,

$$\varkappa \in \partial D_0(\varepsilon) \iff D_0(\tau) - D_0(\varepsilon) \geqslant \varkappa : (\tau-\varepsilon), \quad \text{for all } \tau \in \mathbb{M}^{2\times 2}_s. \tag{3.2}$$

We shall prove the following theorem.

THEOREM 3.1. *For any $g \in L_2(0,T;H^{-1})$ and any $\varphi \in H$ there exists a unique solution v of problem* (3.1_k), $k = 1,2,3,4,5$, *having the properties* (2.1). *It satisfies the variational inequality*

$$\begin{gathered}\bigl(\partial_t v(t), w - v(t)\bigr) + \bigl(v_k(t)\partial_{x_k} v(t), w - v(t)\bigr) \\ + \int_\Omega \bigl(D_0(\varepsilon(w(x))) - D_0(\varepsilon(v(x,t)))\bigr)\,dx \geqslant \bigl(g(t), w - v(t)\bigr) \\ \text{for all } w \in H^1 \text{ and for a.a. } t \in [0,T],\end{gathered} \tag{3.3}$$

the initial condition (2.2_2), *and the energy inequality*

$$\frac{1}{2}\|v(t)\|^2 + \int_0^t\int_\Omega D_0(\varepsilon(v(x,\theta)))\,dx\,d\theta \leqslant \frac{1}{2}\|\varphi\|^2 + \int_0^t \bigl(g(\theta), v(\theta)\bigr)\,d\theta. \tag{3.4}$$

The estimates (2.4_k), $k = 1, 2$, (2.5) *with* $D = D_0$, $\nu_1 = 2\nu$, $\nu_2 = \sqrt{2}k_*$ *are true. The solution* v *continuously depends on* φ *in the norm of* H. *If, in addition,* $\partial_t g \in L_2(0, T; h^{-1})$, *then the solution* v *satisfies* (2.6) *and* (2.7).

REMARK 3.1. By Lemma 2.1 and (2.1), the variational inequality (3.3) is correct.

REMARK 3.2. The functions Φ_1 and Φ_2 are the same as in Theorem 2.1.

REMARK 3.3. Concerning the first part of Theorem 3.1, see also [**DL**].

PROOF OF THEOREM 3.1. Let us consider the regularized potential

$$D_0^\delta(\varepsilon) = \nu|\varepsilon|^2 + \sqrt{2}k_*\big(\sqrt{\delta^2 + |\varepsilon|^2} - \delta\big), \quad \forall \varepsilon \in \mathbb{M}_s^{2\times 2}$$

with a positive parameter δ. This function satisfies all conditions (1.4_k). For this reason we can apply Theorem 2.1, setting $D = D_0^\delta$, $v = v^\delta$, $\nu_1 = 2\nu$, $\nu_2 = \sqrt{2}k_*$. So, we have

$$v^\delta \in C\big([0, T]; H\big) \cap L_2\big(0, T; H^{-1}\big), \qquad \partial_t v^\delta \in L_2\big(0, T; H^{-1}\big); \tag{3.5}$$

$$\begin{gathered}\big(\partial_t v^\delta(t), w\big) + \big(v_k^\delta(t)\partial_{x_k} v^\delta(t), w\big) + \Big(\frac{\partial}{\partial\varepsilon} D_0^\delta\big(\varepsilon(v^\delta(t))\big), \varepsilon(w)\Big) = \big(g(t), w\big) \\ \forall w \in H^1 \text{ and for a.a. } t \in [0, T];\end{gathered} \tag{3.6$_1$}$$

$$v^\delta(0) = \varphi; \tag{3.6$_2$}$$

$$\|v^\delta(t)\|^2 \leqslant \|\varphi\|^2 \exp\Big\{-\frac{\lambda_1\nu}{2}t\Big\} + \frac{2}{\nu}\int_0^t \|g(\tau)\|_{(-1)}^2 \exp\Big\{-\frac{\lambda_1\nu}{2}(t-\tau)\Big\}\, d\tau; \tag{3.7$_1$}$$

$$\int_0^t \int_\Omega D_0^\delta\big(\varepsilon(v^\delta(x, t))\big)\, dx\, d\tau \leqslant \|\varphi\|^2 + \frac{2}{\nu}\int_0^t \|g(\tau)\|_{(-1)}^2\, d\tau; \tag{3.7$_2$}$$

$$\int_0^t \|\partial_t v^\delta(\tau)\|_{(-1)}^2\, d\tau \leqslant \Phi_1\Big(t, \|\varphi\|, \int_0^t \|g(\tau)\|_{(-1)}^2\, d\tau, \nu^{-1}, 2\nu, \sqrt{2}k_*\Big) \tag{3.7$_3$}$$

and if $\partial_t g \in L_2(0, T; H^{-1})$ then

$$\begin{aligned}&\sup_{t\in[\gamma, T]}\big(\|\partial_x v^\delta(t)\| + \|\partial_t v^\delta(t)\|\big) + \|\partial_{xt}^2 v^\delta\|_{2, Q_{\gamma T}} \\ &\quad \leqslant \Phi_2\Big(T, \|\varphi\|, \int_0^T \big(\|g(t)\|_{(-1)}^2 + \|\partial_t g(t)\|_{(-1)}^2\big)\, dt, \nu^{-1}, \gamma^{-1}\Big).\end{aligned} \tag{3.8}$$

So we can apply Lemma 2.2 for $\mathscr{H}^{-1} = H^{-1}$, $\mathscr{H} = H$, $\mathscr{H}^1 = \mathscr{H}_n^1 = H^1$ and assert that there is a sequence $\delta \to 0$ such that

$$v^\delta \rightharpoonup v \qquad \text{in } L_2(0, T; H^1), \tag{3.9$_1$}$$

$$\big(v^\delta(t), w\big) \to \big(v(t), w\big) \qquad \text{in } C([0, T]) \text{ for any fixed } w \in H, \tag{3.9$_2$}$$

$$v^\delta \to v \qquad \text{in } L_2(Q_T), \tag{3.9$_3$}$$

$$\partial_t v^\delta \rightharpoonup \partial_t v \qquad \text{in } L_2(0, T; H^{-1}). \tag{3.9$_4$}$$

By (3.9_k) and Lemma 2.1, the limit function v satisfies (2.1) and (3.1_k) for $k = 2, 4, 5$.

In order to prove that the variational inequality (3.3) holds for v let us take an arbitrary nonnegative function $\xi \in C_0^1(0,T)$ and consider the relation

$$\frac{\partial}{\partial \varepsilon} D_0^\delta(\varepsilon) : (\varkappa - \varepsilon) \leqslant D_0^\delta(\varkappa) - D_0^\delta(\varepsilon) \quad \forall \varkappa, \varepsilon \in \mathbb{M}_s^{2\times 2}. \tag{3.10}$$

Then the variational equality (3.6_1) gives

$$\begin{aligned} &\int_0^T \xi(t)\big(\partial_t v^\delta(t), w - v^\delta(t)\big)\, dt + \int_0^T \xi(t)\big(v_k^\delta(t)\partial_{x_k} v^\delta(t), w - v^\delta(t)\big)\, dt \\ &\quad + \int_0^T \xi(t) \int_\Omega \big(D_0^\delta(\varepsilon(w(x))) - D_0^\delta(\varepsilon(v^\delta(x,t)))\big)\, dx\, dt \\ &\quad \geqslant \int_0^T \big(g(t), w - v^\delta(t)\big)\, dt \end{aligned} \tag{3.11}$$

for any $w \in H^1$. Let us pass to the limit in (3.11). It is easily to see that for the first and second integrals on the left-hand side of (3.11) the limit can be determined in the same way as in the proof of Theorem 2.1. By the obvious identity

$$D_0^\delta(\varepsilon) - D_0(\varepsilon) = -2\delta \frac{|\varepsilon|}{\sqrt{\delta^2 + |\varepsilon|^2} + |\varepsilon| + \delta}, \tag{3.12}$$

the convexity of D_0 and (3.9_1), the inequality

$$\liminf_{\delta\to 0} \int_0^T \xi(t) \int_\Omega D_0^\delta(\varepsilon(v^\delta(x,t)))\, dx\, dt \geqslant \int_0^T \xi(t) \int_\Omega D_0(\varepsilon(v(x,t)))\, dx\, dt$$

is true. So, after passing to the limit in (3.11), we get the inequality

$$\begin{aligned} &\int_0^T \xi(t)\big(\partial_t v(t), w - v(t)\big)\, dt + \int_0^T \xi(t)\big(v_k(t)\partial_{x_k}(t), w - v(t)\big)\, dt \\ &\quad + \int_0^T \xi(t) \int_\Omega \big[D_0(\varepsilon(w(x))) - D_0(\varepsilon(v(x,t)))\big]\, dx\, dt \\ &\quad \geqslant \int_0^T \xi(t)\big(g(t), w - v(t)\big)\, dt \end{aligned}$$

for any $w \in H^1$. Due to the arbitrariness of ξ this implies inequality (3.3).

Using the same arguments, we can derive the following inequality from (3.6_1):

$$\begin{aligned} &\big(\partial_t v(t), w - v(t)\big) + \big(v_k(t)\partial_{x_k} v(t), w - v(t)\big) \\ &\quad + 2\nu\big(\varepsilon(v(t)), \varepsilon(w) - \varepsilon(v(t))\big) \\ &\quad + \sqrt{2}k_* \int_\Omega \big(|\varepsilon(w(x))| - |\varepsilon(v(x,t))|\big)\, dx \geqslant \big(g(t), w - v(t)\big) \end{aligned} \tag{3.13}$$

for any $w \in H^1$ and a.a. $t \in [0,T]$. Actually, (3.13) is equivalent to (3.3) (see **[DL]**). We use (3.13) in order to prove the uniqueness of v. Let $v(t)$ and $v'(t)$ satisfy (3.13), have the properties (2.1), and $v(0) = \varphi \in H$, $v'(0) = \varphi' \in H$. Setting $w = v'(t)$

in (3.13) and $w = v(t)$ in (3.13) for v', we add the resulting inequalities and, after elementary calculations, get an analog of (2.20) for $u \equiv v - v'$:

$$\begin{aligned} &\frac{1}{2}\|u(t)\|^2 + 2\nu \int_0^t \|\varepsilon(u(\tau))\|^2 \, d\tau \\ &\leqslant \frac{1}{2}\|u(0)\|^2 - \int_0^t \left(u_k(\theta)\partial_{x_k} v(\theta) + v_k'(\theta)\partial_{x_k} u(\theta), u(\theta)\right) d\theta, \end{aligned} \tag{3.14}$$

which implies (2.37). Then, taking the limit as $\delta \to 0$ in the inequality

$$\begin{aligned} \frac{1}{4\nu}\int_0^t \|v^\delta(\tau)\|_{(1)}^2 \, d\tau &\leqslant \frac{1}{2\nu}\|\varphi\|^2 + \frac{1}{\nu^3}\int_0^t \|g(\tau)\|_{(-1)}^2 \, d\tau \\ &\equiv \Phi_3\left(\|\varphi\|, \int_0^t \|g(\tau)\|_{(-1)}^2 \, d\tau, \nu^{-1}\right), \end{aligned}$$

we obtain (2.38) from (2.37), and this proves that v is unique and continuous on φ in the norm of H.

Finally, to prove (2.7), it suffices to take into account (3.9_k), $k = 1, 2, 3, 4$, and to pass to the limit in (3.8). Theorem 3.1 is proved. □

Let us estimate the difference $v^\delta - v$. For this purpose for v^δ we use the inequality

$$\begin{aligned} &\left(\partial_t v^\delta(t), w - v^\delta(t)\right) + \left(v_k^\delta(t)\partial_{x_k} v^\delta(t), w - v^\delta(t)\right) \\ &\quad + 2\nu\left(\varepsilon(v^\delta(t)), \varepsilon(w) - \varepsilon(v^\delta(t))\right) \\ &\quad + \sqrt{2}k_* \int_\Omega \left(\sqrt{\delta^2 + |\varepsilon(w(x))|^2} - \sqrt{\delta^2 + |\varepsilon(v^\delta(x,t))|^2}\right) dx \\ &\quad \geqslant \left(g(t), w - v^\delta(t)\right), \end{aligned} \tag{3.15}$$

which holds for any $w \in H^1$ and a.a. $t \in [0, T]$. This follows from (3.6_1) and (3.10). Setting $w = v^\delta(t)$ in (3.13), $w = v(t)$ in (3.15), and adding these inequalities, after elementary calculations and the application of (2.9) we get the inequalities

$$\begin{aligned} &\frac{1}{2}\partial_t\|v^\delta(t) - v(t)\|^2 + 2\nu\|\varepsilon(v^\delta(t)) - \varepsilon(v(t))\|^2 \\ &\quad \leqslant \left(v_k^\delta(t) - v_k(t)\right)\partial_{x_k} v(t), v^\delta(t) - v(t)\big) \\ &\quad\quad + \sqrt{2}k_*\delta \int_\Omega \left(\frac{1}{|\varepsilon(v)| + \sqrt{\delta^2 + |\varepsilon(v)|^2}} + \frac{1}{|\varepsilon(v^\delta)| + \sqrt{\delta^2 + |\varepsilon(v^\delta)|^2}}\right) dx \\ &\quad \leqslant \frac{1}{\sqrt{2}}\|v^\delta(t) - v(t)\|\|v^\delta(t) - v(t)\|_{(1)}\|v(t)\|_{(1)} + 2\sqrt{2}k_*\delta|\Omega| \\ &\quad \leqslant \frac{\nu}{2}\|v^\delta(t) - v(t)\|_{(1)}^2 + \frac{1}{\nu}\|v^d l(t) - v(t)\|^2\|v(t)\|_{(1)}^2 + 2\sqrt{2}k_*\delta|\Omega|. \end{aligned}$$

Since $v^\delta(0) - v(0) = 0$, we can derive the desired estimate from the last relation:

$$\|v^\delta(t) - v(t)\|^2 \leqslant 4\sqrt{2}k_* t\delta|\Omega| \exp\left\{\frac{2}{\nu}\int_0^t \|\partial_x v(\tau)\|^2 \, d\tau\right\}. \tag{3.16}$$

Let us suppose again that the function g satisfies condition (2.45) and let $v(t, \varphi)$ be the solution of problem (3.1) with $\varphi = v(0) \in H$. Now we can determine the solution operators

$$V_t^0 \colon \varphi \in H \to v(t, \varphi) \in H, \quad t \in \mathbb{R}^+.$$

Repeating the arguments used to prove Theorem 2.2, we get

THEOREM 3.2. *Suppose that* Ω *is a bounded domain in* $\mathbb{R}^2$ *and condition* (2.45) *holds. Then the family of solution operators of problem* (3.1) *forms the continuous semigroup* $\{V_t^0, t \in \mathbb{R}^+, H\}$ *of class* 1 *having bounded* B*-absorbing sets* (2.46). *This semigroup possesses a compact minimal global* B*-attractor* $\mathfrak{M}_\mu^0$, *which is a connected invariant subset in* H *bounded in* H^1. *For any* $\varphi \in \mathfrak{M}_\mu^0$ *the solution* $v(t,\varphi)$, $t \in \mathbb{R}^+$, *may be extended continuously so that* $v(t,\varphi)$ *will be the solution of* (3.1) *on the whole line* $\mathbb{R}^1$.

Moreover, for all solutions $v(t,\varphi)$, $\varphi \in \mathfrak{M}_\mu^0$, $t \in \mathbb{R}^1$, *there are derivatives* $\partial_t v$ *from* $L_\infty(\mathbb{R}^1, H)$ *and* $\partial_{xt}^2 v$ *from* $L_2(\Omega \times (-T,T))$, $\forall T < +\infty$, *and the estimates* (2.49) *are valid with* $\nu_1 = 2\nu$, $\nu_2 = \sqrt{2}k_*$.

The number of determining modes on $\mathfrak{M}_\mu^0$ *is finite and may be majorized by a value depending only on* $\mu\|f\|_{(-1)}$, ν, k_*.

The last statement of Theorem 3.2 is proved in the same way as Theorem 3 in [S].

Due to the plastic properties of Bingham fluids, the attractor $\mathfrak{M}_\mu^0$ is trivial for small values of parameter μ.

THEOREM 3.3. *Let*

$$\mu_* = \inf\left\{ \sqrt{2}k_* \int_\Omega |\varepsilon(w)|\,dx \;\|\; w \in H^1,\ \int_\Omega f \cdot w\,dx = 1 \right\}. \tag{3.17}$$

Then $\mathfrak{M}_\mu^0 = \{0\}$ *for any* $\mu \in [0, \mu_*]$. *Moreover, if* $0 \leqslant \mu < \mu_*$, *then for any set* B *bounded in* H *there is a number* $t(B,\mu)$ *such that* $V_t^0(\varphi) = 0$ *for all* $\varphi \in B$ *and all* $t > t(B,\mu)$.

When $\mu = \mu_*$, *the following exponential estimate holds*:

$$\|V_t^0(\varphi)\| \leqslant \|\varphi\| \exp\{-\lambda_1 \nu t\}, \quad t \geqslant 0.$$

PROOF OF THEOREM 3.3. We have the two estimates

$$\frac{1}{2}\partial_t \|v(t)\|^2 + 2\nu\|\varepsilon(v(t))\|^2 + \sqrt{2}k_*\left(1 - \frac{\mu}{\mu_*}\right) \int_\Omega |\varepsilon(v(x,t))|\,dx \leqslant 0,$$

$$\|w\| \leqslant c \int_\Omega |\varepsilon(w(x))|\,dx, \quad \forall w \in \overset{\circ}{W}{}_2^1(\Omega; \mathbb{R}^2).$$

The first of them follows from inequality (3.3) with $w = 0$ and the definition of the number μ_* (the limit load coefficient). The second one is proved in [**MM**]. Next, we repeat the arguments used for the proof of Theorem 2 in [S] and obtain all the assertions of Theorem 3.3. Theorem 3.3 is proved. □

§4. Appendix

In the proofs of Lemmas 2.1 and 2.2 we shall use a special basis in the triple $\mathscr{H}^1 \subset \mathscr{H} \subset \mathscr{H}^{-1}$. Let us describe this basis and a construction of the space $\mathscr{H}^{-1}$ which is dual to $\mathscr{H}^1$ relative to $\mathscr{H}^0 \equiv \mathscr{H}$. Simultaneously we construct a full scale of Hilbert spaces $\mathscr{H}^s$, $s \in \mathbb{R}^1$, in which $\mathscr{H}^0$ and $\mathscr{H}^1$ are given separable Hilbert spaces and $\mathscr{H}^{-s}$ is dual to $\mathscr{H}^s$ with respect to $\mathscr{H}^0$ for any $s \in \mathbb{R}^1$.

We shall denote the inner product in $\mathscr{H}^0$ by $(\cdot,\cdot)$ and the norm by $\|\cdot\|$; in $\mathscr{H}^1$ they are denoted by $(\cdot,\cdot)_{(1)}$ and $\|\cdot\|_{(1)}$ respectively.

Let us consider *problem* (I): to find an element $u \in \mathscr{H}^1$ for which the identity

$$(u,\eta)_{(1)} = (f,\eta) \tag{4.1}$$

holds for any $\eta \in \mathscr{H}^1$. Here f is a given element of $\mathscr{H}^0$.

PROPOSITION 4.1. *For any $f \in \mathscr{H}^0$ problem* (I) *has a unique solution u. The solution operator*

$$A\colon f \to u \tag{4.2}$$

is linear, selfadjoined, positive, and compact in $\mathscr{H}^0$. Its eigenvectors can be chosen so that they form an orthonormal basis $\{e_k\}_{k=1}^{\infty}$ in $\mathscr{H}^0$ and an orthobasis in $\mathscr{H}^1$. The corresponding eigenvalues $\{\mu_k\}$ form a monotonic sequence $\mu_1 \geqslant \mu_2 \geqslant \cdots > 0$ with $\mu_k \to 0$ as $k \to \infty$.

Proposition 4.1 is an abstract form of results proved by Friedrichs in 1934 for the Dirichlet problem in a bounded domain for elliptic equations ([**F**]). Later on these results were described in different forms in many publications (see, for example, [**LM,** Chapter II, §4]).

According to Proposition 4.1,

$$Ae_k = \mu_k e_k \quad \text{and} \quad (e_i, e_k) = \delta_{ik}. \tag{4.3}$$

The identity (4.1) can be rewritten in the form

$$(Af,\eta)_{(1)} = (f,\eta) \quad \text{for any } f \in \mathscr{H}^0 \text{ and any } \eta \in \mathscr{H}^1. \tag{4.4}$$

It follows from (4.3) and (4.4) that in $\mathscr{H}^1$ we have

$$(e_i, e_k)_{(1)} = \mu_i^{-1}\delta_{ik} \tag{4.5}$$

For any $f \in \mathscr{H}^0$ one can consider the Fourier series

$$f = \sum_{k=1}^{\infty} a_k(f) e_k \tag{4.6$_1$}$$

with

$$a_k(f) = (f, e_k) \tag{4.6$_2$}$$

The series (4.6$_1$) converges in the norm of $\mathscr{H}^0$, and

$$\|f\| = \left(\sum_{k=1}^{\infty} |a_k(f)|^2\right)^{1/2}. \tag{4.7$_1$}$$

The inner product for $f \in \mathscr{H}^0$ and $g \in \mathscr{H}^0$ is given by

$$(f,g) = \sum_{k=1}^{\infty} a_k(f) a_k(g) \tag{4.7$_2$}$$

On the other hand, any series

$$\sum_{k=1}^{\infty} a_k e_k \tag{4.8}$$

with $\sum_{k=1}^{\infty} |a_k|^2 < \infty$ represents the element f of $\mathscr{H}^0$ with $(f, e_k) = a_k$.

The space $\mathscr{H}^1$ is the set of all series (4.6_1) for which

$$(4.9) \qquad \sum_{k=1}^{\infty} \mu_k^{-1} |a_k(f)|^2 < +\infty, \qquad \|f\|_{(1)} = \left(\sum_{k=1}^{\infty} \mu_k^{-1} |a_k(f)|^2 \right)^{1/2}.$$

By definition, the elements of $\mathscr{H}^s$ arc the series (4.8) for which

$$\sum_{k=1}^{\infty} \mu_k^{-s} |a_k|^2 < +\infty.$$

The norm of the element $\sum_{k=1}^{\infty} a_k e_k$ in $\mathscr{H}^s$ is defined by

$$(4.10_1) \qquad \left\| \sum_{k=1}^{\infty} a_k e_k \right\|_{(s)} = \left(\sum_{k=1}^{\infty} \mu_k^{-s} |a_k|^2 \right)^{1/2}.$$

According to (4.10_1), the inner product in $\mathscr{H}^s$ is determined by

$$(4.10_2) \qquad \left(\sum_{k=1}^{\infty} a_k e_k, \sum_{k=1}^{\infty} b_k e_k \right)_{(s)} = \sum_{k=1}^{\infty} \mu_k^{-s} a_k b_k.$$

For $s = 0$ and 1 these definitions correspond to (4.7) and (4.9). The eigenvectors e_k and their sums $\sum_{k=1}^{m} a_k e_k$, $m < +\infty$, belong to any $\mathscr{H}^s$.

For any $\sum_{k=1}^{\infty} a_k e_k \equiv f \in \mathscr{H}^{-s}$ and $\sum_{k=1}^{\infty} b_k e_k \equiv g \in \mathscr{H}^s$, the numerical series $\sum_{k=1}^{\infty} a_k b_k$ converges since

(4.11)
$$\left| \sum_{k=1}^{\infty} a_k b_k \right| \leqslant \sum_{k=1}^{\infty} |a_k b_k| \leqslant \left(\sum_{k=1}^{\infty} \mu_k^{s} |a_k|^2 \right)^{1/2} \left(\sum_{k=1}^{\infty} \mu_k^{-s} |b_k|^2 \right)^{1/2} = \|f\|_{(-s)} \|g\|_{(s)}.$$

Due to (4.7_2), it is reasonable to denote $\sum_{k=1}^{\infty} a_k b_k$ by

$$(4.12_1) \qquad (f, g) \equiv \sum_{k=1}^{\infty} a_k b_k$$

for all $\sum_{k=1}^{\infty} a_k e_k \equiv f \in \mathscr{H}^{-s}$ and $\sum_{k=1}^{\infty} b_k e_k \equiv g \in \mathscr{H}^s$. Then for $\sum_{k=1}^{\infty} a_k e_k \equiv f \in \mathscr{H}^s$ and $g = e_k$ we get the relations

$$(4.12_2) \qquad a_k = (f, e_k).$$

They correspond to (4.6_2) and make it possible to retain the form (4.6_k), $k = 1, 2$, for any $f \in \mathscr{H}^s$.

A fixed element f of $\mathscr{H}^{-s}$ generates the linear functional l_f (i.e., additive and bounded) on $\mathscr{H}^s$ determined by the formula

$$(4.13) \qquad l_f(g) = (f, g) \equiv \sum_{k=1}^{\infty} a_k b_k.$$

It is easy to prove that the norm of l_f (as a linear functional on $\mathscr{H}^s$) is equal to $\|f\|_{(-s)}$, and that any linear functional l on $\mathscr{H}^s$ can be represented in the form (4.13) by using some element f of $\mathscr{H}^{-s}$.

The spaces $l_2(0,T;\mathscr{H}^s)$ can be determined as linear sets of the series

$$f(t)=\sum_{k=1}^{\infty}a_k(t;f)e_k,\quad a_k(\cdot;f)\in L_2(0,T),\tag{4.14$_1$}$$

with the norm

$$\|f\|_{L_2(0,T;\mathscr{H}^s)}\equiv\left(\int_0^T\sum_{k=1}^{\infty}|a_k(t;f)|^2\mu_k^{-s}\,dt\right)^{1/2}<+\infty.\tag{4.14$_2$}$$

Here $L_2(0,T;\mathscr{H}^s)$ is the Hilbert space with inner product

$$\begin{aligned}(f,g)_{L_2(0,T;\mathscr{H}^s)}&=\int_0^T\sum_{k=1}^{\infty}a_k(t;f)a_k(t;g)\mu_k^{-s}\,dt\\&=\sum_{k=1}^{\infty}\int_0^T a_k(t;f)a_k(t;g)\mu_k^{-s}\,dt.\end{aligned}\tag{4.14$_3$}$$

For any $f\in L_2(0,T;\mathscr{H}^s)$ and any $g\in L_2(0,T;\mathscr{H}^{-s})$ we have

$$(f,g)_{L_2(0,T;\mathscr{H})}\equiv\int_0^T(f(t),g(t))\,dt=\int_0^T\sum_{k=1}^{\infty}a_k(t;f)a_k(t;g)\,dt.\tag{4.15}$$

We shall also use the following known fact:

PROPOSITION 4.2. *For any $\varepsilon>0$ there is a number $N(\varepsilon)$ such that for any $u\in\mathscr{H}^1$*

$$\|u\|^2\leqslant\sum_{k=1}^{N(\varepsilon)}|(u,e_k)|^2+\varepsilon\|u\|_{(1)}^2.\tag{4.16}$$

Comparing (4.7_1) and (4.9) it is easy to see that (4.16) is true if $N(\varepsilon)$ satisfies the inequality $\mu_{N(\varepsilon)+1}<\varepsilon$.

PROOF OF LEMMA 2.1. Conditions (2.17) mean that for $u\in L_2(0,T;\mathscr{H}^1)$ there exists an element $w\in L_2(0,T;\mathscr{H}^{-1})$ for which the identity

$$\int_0^T(w(t),v)\xi(t)\,dt=-\int_0^T(u(t),v)\partial_t\xi(t)\,dt\tag{4.17}$$

holds for any $v\in\mathscr{H}^1$ and any $\xi\in C_0^1(0,T)$.

Let us consider the representations

$$u(t)=\sum_{k=1}^{\infty}a_k(t;u)e_k,\qquad a_k(t;u)=(u(t),e_k),$$

$$w(t)=\sum_{k=1}^{\infty}a_k(t;w)e_k,\qquad a_k(t;w)=(w(t),e_k).$$

Relations (4.17) with $v=e_k$ give us the following identities:

$$\int_0^T a_k(t;w)\xi(t)\,dt=-\int_0^T a_k(t;u)\partial_t\xi(t)\,dt,\quad\forall\xi\in C_0^1(0,T).\tag{4.18}$$

This means that $a_k(\cdot;u)$ is an absolutely continuous function of $t \in [0,T]$ and

$$\partial_t a_k(\cdot;u) = a_k(\cdot;w) \in L_2(0,T).$$

Hence

$$|a_k(t;u)|^2 - |a_k(\tau;u)|^2 = 2\int_\tau^t a_k(\theta;u)\partial_\theta a_k(\theta;u)\,d\theta$$
$$\leqslant 2\Big(\int_0^T |a_k(\theta;u)|^2\,d\theta\Big)^{1/2}\Big(\int_0^T |\partial_\theta a_k(\theta;u)|^2\,d\theta\Big)^{1/2}, \qquad 0 \leqslant \tau, t \leqslant T.$$

Integration of the last inequality over $\tau \in [0,T]$ yields

$$|a_k(t;u)|^2 \leqslant \frac{1}{T}\int_0^T |a_k(\tau;u)|^2\,d\tau$$
$$+\frac{2}{T}\Big(\int_0^T |a_k(\tau;u)|^2\,d\tau\Big)^{1/2}\Big(\int_0^T |\partial_\tau a_k(\tau;u)|^2\,d\tau\Big)^{1/2}.$$

Taking the sum of these inequalities over $k = m,\dots,m+p$, we get

$$\sup_{t\in[0,T]}\sum_{k=m}^{m+p}|a_k(t;u)|^2 \leqslant \frac{1}{T}\int_0^T\sum_{k=m}^{m+p}|a_k(\tau;u)|^2\,d\tau$$
$$+\frac{2}{T}\Big(\int_0^T\sum_{k=m}^{m+p}\mu_k^{-1}|a_k(\tau;u)|^2\,d\tau\Big)^{1/2}\Big(\int_0^T\sum_{k=m}^{m+p}\mu_k|\partial_\tau a_k(\tau;u)|^2\,d\tau\Big)^{1/2}.$$

The right-hand side is less than any $\varepsilon > 0$ if only $m \geqslant m(\varepsilon)$ and $p > 0$. This means that the continuous functions

$$u^n(t) \equiv \sum_{k=1}^n a_k(t;u)e_k, \qquad n = 1,2,\dots,$$

converge in $\mathscr{H}$ uniformly in $t \in [0,T]$, and therefore $u \in C([0,T];\mathscr{H})$.

The assertion (2.19) follows from the relation

$$\lim_{n\to\infty}\big(\|u^n(t)\|^2 - \|u^n(\tau)\|^2\big) = \lim_{n\to\infty} 2\int_\tau^t \big(\partial_\theta u^n(\theta), u^n(\theta)\big)\,d\theta$$

since each term here has a proper limit. □

Proof of Lemma 2.2. Assertions (2.25) and (2.28_3) follow directly from (2.23) since $L_2(0,T);\mathscr{H}^1)$ is a Hilbert space. In order to prove (2.26) and therefore (2.28_2), for each e_k we choose a sequence $w_k^n \in \mathscr{H}_n^1$, $n = 1,2,\dots$, such that

$$w_k^n \xrightarrow[n\to+\infty]{} e_k \quad \text{in } \mathscr{H}^1. \tag{4.19}$$

This is possible thanks to condition (2.20). Next, we have

$$
\begin{aligned}
&\big|(u^n(t+\Delta t)-u^n(t),e_k)\big| \\
&\quad\leqslant \big|(u^n(t+\Delta t),w_k^n-e_k)\big|+\big|(u^n(t),w_k^n-e_k)\big| \\
&\qquad+\big|(u^n(t+\Delta t)-u^n(t),w_k^n)\big| \\
&\quad\leqslant 2c_1\|w_k^n-e_k\|+\left|\int_t^{t+\Delta t}(\partial_t u^n(\theta),w_k^n)\,d\theta\right| \\
&\quad\leqslant 2c_1\|w_k^n-e_k\|+\sqrt{|\Delta t|}\left(\int_0^T\alpha_n^2(t)\,dt\right)^{1/2}\|w_k^n\|_{(1)} \\
&\quad\leqslant 2c_1\|w_k^n-e_k\|+\sqrt{|\Delta t|}c_3\|w_k^n\|_{(1)}.
\end{aligned}
\tag{4.20}
$$

Therefore each sequence $\{(u^n(t),e_k)\}_{n=1}^\infty$ is equicontinuous on $[0,T]$, and by (2.22) it is also bounded. Therefore it is compact in $C([0,T])$, and we can choose a subsequence converging to a continuous function $a_k\colon[0,T]\to\mathbb{R}^1$. Using the standard diagonal procedure, we can choose a subsequence $\{u^{\widetilde{n}_j}\}_{j=1}^\infty$ for which $\{(u^{\widetilde{n}_j}(t),e_k)\}_{j=1}^\infty$ converge to $a_k(t)$ for each $k\in\mathbb{N}$. We choose $\{\widetilde{n}_j\}_{j=1}^\infty$ from the subsequence $\{n_j\}_{j=1}^\infty$ so that the convergence (2.25) takes place. Thus $a_k(t)=(u(t),e_k)$, $a_k\in C([0,T])$, $u(t)=\sum_{k=1}^\infty(u(t),e_k)e_k$ belongs to $\mathscr{H}$, and $\|u(t)\|\leqslant c_1$ for any $t\in[0,T]$. It is easy to prove that $(u^{\widetilde{n}_j}(t),w)$, $\widetilde{n}_j\to+\infty$, converges uniformly in $t\in[0,T]$ for any $w\in\mathscr{H}$ and its limit is $(u(t),w)$. For convenience we now write $\{u^n\}_{n=1}^\infty$ instead of $\{u^{\widetilde{n}_j}\}_{j=1}^\infty$. The assertion (2.27) is true by the following arguments. Inequality (4.16), for $v=u^n(t)-u(t)$, gives the estimates

$$
\begin{aligned}
&\|u^n(t)-u(t)\|^2\leqslant\sum_{k=1}^{N(\varepsilon)}\big|(u^n(t)-u(t),e_k)\big|^2+2\varepsilon\big(\|u^n(t)\|_{(1)}^2+\|u(t)\|_{(1)}^2\big),\\
&\int_0^T\|u^n(t)-u(t)\|^2\,dt\leqslant\int_0^T\sum_{k=1}^{N(\varepsilon)}\big|(u^n(t)-u(t),e_k)\big|^2\,dt+4\varepsilon c_2^2\equiv\rho(\varepsilon,n).
\end{aligned}
\tag{4.21}
$$

The right-hand side of (4.21) will be less than an arbitrary $\delta>0$ if $n\geqslant n(\delta)$. In fact, first we take $\varepsilon<\delta/8c_2^2$ and after that choose $n=n(\delta)$ so that the first term of the right-hand side will be less than $\delta/2$ for any $n\geqslant n(\delta)$; thus $\rho(\varepsilon,n)<\delta$ for $n>n(\varepsilon)$.

Now we want to prove that $\partial_t u\in L_2(0,T;\mathscr{H}^{-1})$. Let us take an arbitrary $v\in\mathscr{H}^1$ and a sequence $\{w_v^n\in\mathscr{H}_n^1\}_{n=1}^\infty$ such that

$$
w_v^n\xrightarrow[n\to+\infty]{}v\quad\text{in }\mathscr{H}^1. \tag{4.22}
$$

It is easy to see that for $\beta_v^n(t)\equiv(\partial_t u^n(t),w_v^n)$ and any $\xi\in L_2(0,T)$ the inequalities

$$
\left|\int_0^T\xi(t)\beta_v^n(t)\,dt\right|\leqslant\int_0^T|\xi(t)|\alpha_n(t)\,dt\,\|w_v^n\|_{(1)}, \tag{4.23$_1$}
$$

$$
\left|\int_0^T\xi(t)\beta_v^n(t)\,dt\right|\leqslant\|\xi\|_{L_2(0,T)}\|\alpha_n\|_{L_2(0,T)}\|w_v^n\|_{(1)} \tag{4.23$_2$}
$$

hold. The last one and (4.22), (2.24) guarantee

$$
\|\beta_v^n\|_{L_2(0,T)}\leqslant c_3\|w_v^n\|_{(1)}\leqslant c_v. \tag{4.23$_3$}
$$

So, the sequence $\{\beta_v^n\}_{n=1}^\infty$ is bounded in $L_2(0,T)$ and therefore weakly compact in

$L_2(0,T)$. Let us show that $\{\beta_v^n\}_{n=1}^\infty$ has only one limit point, denoted by β_v (in the weak topology of $L_2(0,T)$), and

$$\beta_v(\cdot) = (\partial_t u(\cdot), v) \quad \text{as elements of } L_2(0,T). \tag{4.24}$$

Suppose that $\{\beta_v^{n_j}\}_{j=1}^\infty$ weakly converges to β_v. For $\beta_v^{n_j}$ we have the relations

$$\int_0^T \xi(t)\beta_v^{n_j}(t)\,dt \equiv \int_0^T \xi(t)\big(\partial_t u^{n_j}(t), w_v^{n_j}\big)\,dt = -\int_0^T \partial_t\xi(t)\big(u^{n_j}(t), w_v^{n_j}\big)\,dt$$

with arbitrary $\xi \in C_0^1(0,T)$. In the limit as $n_j \to +\infty$ they give the identity

$$\int_0^T \xi(t)\beta_v(t)\,dt = -\int_0^T \partial_t\xi_t(u(t),v)\,dt, \quad \forall \xi \in C_0^1(0,T), \tag{4.25}$$

which implies (4.24) and

$$\beta_v = \operatorname*{w-lim}_{n_j\to+\infty} \beta_v^{n_j}.$$

Let us note that β_v depends on v linearly and

$$\|\beta_v\|_{L_2(0,T)} \leqslant c_3\|v\|_{(1)} \tag{4.26}$$

Next, by using (2.24) we can choose a subsequence $\{\alpha_{n_j}\}_{j=1}^\infty$ such that

$$\alpha_{n_j} \rightharpoonup \alpha \quad \text{in } L_2(0,T) \quad \text{and} \quad \|\alpha\|_{L_2(0,T)} \leqslant c_3. \tag{4.27}$$

For this sequence $\{\alpha_{n_j}\}_{j=1}^\infty$ we can pass to the limit in (4.23_1), obtaining

$$\left|\int_0^T \xi(t)\beta_v(t)\,dt\right| \leqslant \int_0^T |\xi(t)|\alpha(t)\,dt\|v\|_{(1)}, \quad \forall \xi \in L_2(0,T). \tag{4.28}$$

Let us denote by $\mathscr{M}_v$ the common set of Lebesgue points of the functions $t \mapsto \alpha(t)$ and $t \mapsto \beta_v(t)$. It is known that

$$\operatorname{meas} \mathscr{M}_v = T \quad \text{for any } v \in \mathscr{H}^1. \tag{4.29}$$

Inequalities (4.28) guarantee that

$$|\beta_v(t)| \leqslant \alpha(t)\|v\|_{(1)} \quad \text{for } t \in \mathscr{M}_v. \tag{4.30}$$

Let us take a countable set $\{v_k\}_{k=1}^\infty$ which is dense in $\mathscr{H}^1$. For it the intersection $\mathscr{M} \equiv \bigcap_{k=1}^\infty \mathscr{M}_{v_k}$ also has measure T, and

$$|\beta_{v_k}(t)| \leqslant \alpha(t)\|v_k\|_{(1)} \quad \text{for } t \in \mathscr{M}. \tag{4.31}$$

Let $\mathscr{H}_0^1$ be the linear span of the set $\{v_k\}_{k=1}^\infty$. For any element v of $\mathscr{H}_0^1$ the set $\mathscr{M}_v$

includes M because $\beta_v(t)$ is linear in v (see (4.24)) and the inequalities (4.28) hold for any $v \in \mathscr{H}^1$. Therefore

$$|\beta_v(t)| \leqslant \alpha(t)\|v\|_{(1)} \quad \text{for any } v \in \mathscr{H}_0^1 \text{ and any } t \in \mathscr{M}. \tag{4.32}$$

This allows us to extend $\beta_v(t)$ (for $t \in \mathscr{M}$) continuously from $\mathscr{H}_0^1$ to the entire space $\mathscr{H}^1$. Let us denote this extension by $\widetilde{\beta}_v(t)$. So

$$\widetilde{\beta}_v(t) \equiv \lim_{v_{k_i} \to v} \beta_{v_{k_i}}(t), \qquad v \in \mathscr{H}^1,\ t \in \mathscr{M}, \tag{4.33}$$

where the v_{k_i} belong to $\mathscr{H}_0^1$ and converge to v in $\mathscr{H}^1$, and

$$|\widetilde{\beta}_v(t)| \leqslant \alpha(t)\|v\|_{(1)} \quad \text{for any } v \in \mathscr{H}^1 \text{ and any } t \in \mathscr{M}. \tag{4.34}$$

The extension $\widetilde{\beta}_v(t)$ is also linear in $v \in \mathscr{H}^1$, and as a linear functional on $\mathscr{H}^1$ can be represented in the form

$$\widetilde{\beta}_v(t) = (w(t), v), \qquad v \in \mathscr{H}^1,\ t \in \mathscr{M}, \tag{4.35}$$

where $w(t) \in \mathscr{H}^{-1}$ and $\|w(t)\|_{(-1)} \leqslant \alpha(t)$. The latter and (4.27) imply that $w \in L_2(0,T;\mathscr{H}^{-1})$ and $\|w\|_{L_2(0,T;\mathscr{H}^{-1})} \leqslant c_3$.

On the other hand, $\widetilde{\beta}_v$ is equal to β_v as an element of $L_2(0,T)$. Indeed, for $v \in \mathscr{H}_0^1$ this is obvious, and for $v \in \mathscr{H}^1 \setminus \mathscr{H}_0^1$ (4.26) gives

$$\|\beta_v - \beta_{v_k}\|_{L_2(0,T)} \leqslant c_3\|v - v_{k_i}\|_{(1)} \longrightarrow 0, \tag{4.36}$$

where $v_{k_i} \in \mathscr{H}_0^1$ and $\lim_{k_i \to +\infty} v_{k_i} = v$ in $\mathscr{H}^1$, so that

$$|\beta_v(t) - \beta_{v_k}(t)| \longrightarrow 0 \quad \text{for } t \in \widehat{\mathscr{M}_v},\ \operatorname{meas}\widehat{\mathscr{M}_v} = T;$$

this and (4.33) imply that $\widetilde{\beta}_v(t) = \beta_v(t)$ for $t \in \mathscr{M}_v \cap \mathscr{M}$, and hence $\widetilde{\beta}_v = \beta_v$ in $L_2(0,T)$.

Now we have reached the following conclusions

$$(\partial_t u(\cdot), v) = \beta_v = \widetilde{\beta}_v = (w(\cdot), v) \quad \text{in } L_2(0,T) \text{ for any } v \in \mathscr{H}^1,$$
$$w = \partial_t u \in L_2(0,T;\mathscr{H}^{-1}) \quad \text{and} \quad \|\partial_t u\|_{L_2(0,T;\mathscr{H}^{-1})} \leqslant c_3.$$

So, Lemma 2.2 is proved. □

The comparatively long proof of the statement $\partial_t u \in L_2(0,T;\mathscr{H}^{-1})$ is not necessary when $\mathscr{H}_n^1 = \operatorname{Span}\{e_1, e_2, \dots, e_n\}$. In this case for any $w \in \mathscr{H}^1$ we have

$$\|w\|_{(1)} \geqslant \|w^n\|_{(1)} \quad \text{and} \quad (\partial_t u^n(t), w) = (\partial_t u^n(t), w^n),$$

where $w^n = \sum_{k=1}^n (w, e_k)e_k$. Hence

$$\|\partial_t u^n(t)\|_{(-1)} = \sup_{\substack{w \in \mathscr{H}^1 \\ w \neq 0}} \frac{|(\partial_t u^n(t), w)|}{\|w\|_{(1)}} \leqslant \sup_{\substack{v^n \in \mathscr{H}_n^1 \\ v^n \neq 0}} \frac{|(\partial_t u^n(t), v^n)|}{\|v^n\|_{(1)}} = \alpha_n(t)$$

and

$$\|\partial_t u^n\|_{L_2(0,T;\mathscr{H}^{-1})} \leqslant c_3, \qquad n = 1, 2, \dots.$$

Since $L_2(0,T;\mathscr{H}^{-1})$ is a Hilbert space, we can choose a subsequence $\{n_j\}_{j=1}^\infty$ for which $\partial_t u^{n_j} \rightharpoonup \partial_t u$ in $L_2(0,T;\mathscr{H}^{-1})$ and $\|\partial_t u\|_{L_2(0,T;\mathscr{H}^{-1})} \leqslant c_3$.

References

[BV] A. V. Babin and M. I. Vishik, *Attractors for evolution equations*, "Nauka", Moscow, 1989; English transl., North-Holland, Amsterdam, 1992.

[DL] G. Duvaut and J. L. Lions, *Les inéquations en mécanique et en physique*, Dunod, Paris, 1972.

[F] K. O. Friedrichs, *Spektraltheorie halbbeschränkter Operatoren und ihre Anwendung auf Spektralzerlegung von Differentialoperatoren*. I, II, Math. Ann. **109** (1934), 465–487, 685–713.

[L1] O. A. Ladyzhenskaya, *On nonlinear problems of continuum mechanics*, Proc. Internat. Congr. Math. (Moscow, 1966), "Nauka", Moscow, 1968, pp. 560–573; English transl. in Amer. Math. Soc. Transl. (2) **70** (1968).

[L2] ———, *New equations for the description of motion of viscous incompressible fluids and global solvability of boundary value problems for them*, Trudy Mat. Inst. Steklov. **102** (1967), 85–104; English transl. in Proc. Steklov Inst. Math. **102** (1967).

[L3] ———, *On some modifications of the Navier–Stokes equations for large gradients of velocity*, Zap. Nauchn. Sem. Leningrad. Otdel. Mat. Inst. Steklov (LOMI) **7** (1968), 126–154; English transl. in Sem. Math. V. A. Steklov Math. Inst. Leningrad **7** (1968).

[L4] ———, *Mathematical problems in the dynamics of a viscous incompressible fluid*, 2nd rev. aug. ed., "Nauka", Moscow, 1970; English transl. of 1st ed., *The mathematical theory of viscous incompressible flow*, Gordon and Breach, New York, 1963.

[L5] ———, *On dynamical system generated by the Navier–Stokes equations*, Zap. Nauchn. Sem. Leningrad. Otdel. Mat. Inst. Steklov (LOMI) **27** (1972), 91–114; English transl. in J. Soviet Math. **3** (1975), no. 4.

[L6] ———, *On finding the minimal global attractors for the Navier–Stokes equations and other partial differential equations*, Uspekhi Mat. Nauk **42** (1987), no. 6 (258), 25–60; English transl. in Russian Math. Surveys **42** (1987).

[L7] ———, *Attractors for semi-groups and evolution equations*, Lezioni Lincee, Roma 1988, Cambridge Univ. Press, Cambridge, 1991.

[L8] ———, *On the property of instant smoothing of solutions to the Navier–Stokes equations and their approximations in the domains with nonsmooth boundaries*, Proc. French-Japanese Sympos. (Luminy, October 1991) (to appear).

[L9] ———, *First boundary value problem for the Navier–Stokes equations in domain with nonsmooth boundaries*, C. R. Acad. Sci. Paris Sér. I Math. **314** (1992), 253–258.

[L10] ———, *Boundary value problems of mathematical physics*, "Nauka", Moscow, 1973; English transl., Springer-Verlag, Berlin, 1985.

[LM] J. L. Lions and E. Magenes, *Problèmes aux limites non homogènes et applications*. Vols. I, III, Dunod, Paris, 1968, 1970.

[MM] P. P. Mosolov and V. P. Myasnikov, *Mechanics of rigid plastic media*, "Nauka", Moscow, 1981. (Russian)

[S] G. A. Seregin, *On the dynamical system associated with two-dimensional equations of the motion of the Bingham fluid*, Zap. Nauchn. Sem. Leningrad. Otdel. Mat. Inst. Steklov (LOMI) **188** (1991), 128–142; English transl. in J. Math. Sci. **70** (1994), no. 3.

[T] R. T. Temam, *Infinite-dimensional dynamical system in mechanics and physics*, Springer-Verlag, Berlin, 1988.

Translated by the authors

Amer. Math. Soc. Transl.
(2) Vol. **164**, 1995

Elliptic Differential Inequalities, Embedding Theorems, and Variational Problems

V. A. Malyshev

Contents

§0. Introduction
§1. Embedding theorems
§2. Differential inequalities in the space $\mathscr{D}'(\Omega)$
§3. Differential inequalities in the space $L_2(\Omega)$
§4. Variational problems
REFERENCES

§0. Introduction

Every undergraduate who has taken calculus knows the following two statements. Let u be a sufficiently smooth function on an interval (a, b).

The function u is increasing on the interval (a, b) if and only if

$$\frac{d}{dx}u \geqslant 0 \tag{0.1}$$

for all $x \in (a, b)$.

The function u is convex on the interval (a, b) if and only if

$$\frac{d^2}{dx^2}u \geqslant 0 \tag{0.2}$$

for all $x \in (a, b)$.

Unfortunately, the differential inequalities (0.1) and (0.2) do not define the entire sets of increasing and convex functions because in classical analysis not all functions are differentiable. However, we can take the differential inequalities (0.1) and (0.2) to be differential inequalities in the sense of the theory of distributions. Then the following two statements **[H1]** improve the situation.

The set of all solutions of the differential inequality (0.1) in the space of distributions $\mathscr{D}'(a, b)$ is equal to the set of all increasing functions.

The set of all solutions of the differential inequality (0.2) in the space of distributions $\mathscr{D}'(a, b)$ is equal to the set of all convex functions.

1991 *Mathematics Subject Classification*. Primary 35J85, 46E35; Secondary 26D10, 35R99, 49J40.

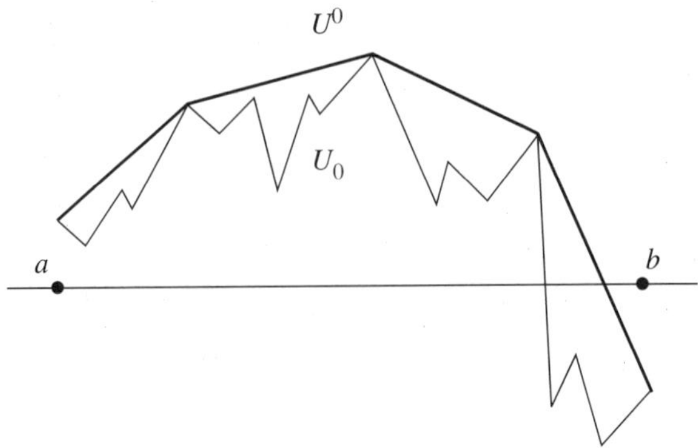

This paper is devoted to elliptic differential inequalities

$$\mathscr{L}u \geqslant 0$$

with an elliptic differential operator $\mathscr{L}$. These differential inequalities will be considered in the sense of the theory of distributions in the spaces $\mathscr{D}'(\Omega)$ and $\mathscr{L}(\Omega)$. But in this Introduction we would like to describe the main topics of the paper by means of the simplest differential inequalities (0.1) and (0.2) with $\mathscr{L} = d/dx$ and $\mathscr{L} = d^2/dx^2$.

0.1. Embedding theorems. Traditionally, embedding theorems deal with the continuous inclusion of one linear topological space into another. The following two statements look like embedding theorems, but deal with continuous inclusions of convex sets into linear spaces. The first statement is a version of the Dini theorem, and the second follows from **[LS]**.

Let a sequence of continuous increasing functions $u_1, u_2, \dots$ converge to a continuous increasing function u in the space $\mathscr{D}(a,b)$. Then the sequence $u_1, u_2, \dots$ converges to the function u in the space $C(a,b)$.

Let a sequence of convex functions $u_1, u_2, \dots$ converge to a convex function u in the space $\mathscr{D}(a,b)$. Then the sequence $u_1, u_2, \dots$ converges to u in the space $C(a,b)$.

These statements look like embedding theorems because the convergence in the space $\mathscr{D}(a,b)$ is weaker than the convergence in $C(a,b)$. But the sets of increasing and convex functions are not linear spaces. This is an importaint feature of these embedding theorems.

0.2. Variational problems with one-sided restrictions. Let there be given a function $U_0 \in W_2^1(a,b)$. Consider the variational problem

$$\left\| -\frac{d}{dx} U \right\|_{L_2(a,b)} \xrightarrow[U \in U_0 + \overset{\circ}{W}{}_2^1(a,b):\, U \geqslant U_0]{} \min \tag{0.3}$$

Here the barrier-function U_0 plays the role of the one-sided restriction.

It can be proved that the solution U_0 of the variational problem (0.3) has the following form. Let $x \in (a,b)$; then

$$U^0(x) = \inf_{\alpha,\beta} \{ \alpha x + \beta : \alpha t + \beta \geqslant U_0(t) \text{ for all } a \leqslant t \leqslant b \}.$$

The function U^0 is shown in the figure.

Therefore, to construct the function U^0 means to construct its convex hull.

Note that

$$-\frac{d^2}{dx^2}U^0 = \Big(\frac{d}{dx}\Big)\Big(-\frac{d}{dx}\Big)U^0 \geq 0.$$

This is the relationship of the variational problem (0.3) with the differential inequality (0.2).

Let $U_0 \in W_2^2(a,b)$. Consider the variational problem

$$\Big\|\Big(-\frac{d}{dx}U\Big)^2\Big\|_{L_2(a,b)} \xrightarrow[U \in U_0 + \overset{\circ}{W}{}_2^2(a,b): U \geqslant U_0]{} \min \tag{0.4}$$

Let $U^0 \in W_2^2(a,b)$ be the solution of the problem.

Note that $U_0 \in C^1(a,b)$ and $U^0 \in C^1(a,b)$. But it can be proved that the function U^0 satisfies the differential inequality

$$\frac{d^4}{dx^4}U^0 = \Big(\frac{d}{dx}\Big)^2\Big(-\frac{d}{dx}\Big)^2 U^0 \geq 0.$$

This differential inequality implies the inclusion $U^0 \in C^2(a,b)$. Therefore, the function U^0 has one additional continuous derivative.

0.3. Variational problems with differential restrictions. Let us consider the following problems of monotone and convex filtration.

Suppose that an "input signal" $u_* \in L_2(a,b)$ is a continuous increasing function or a convex function. Therefore, the input signal u_* satisfies the differential inequality (0.1) or (0.2).

Let $u_0 = u_* + \xi$ be an "output signal" where $\xi \in L_2(a,b)$ is an additive random error.

Thus we only know the form of the input signal u_* (it may be increasing or convex) and the output signal u_0. Our aim is to restore the input signal u_*.

To restore u_*, we propose to solve the variational problem

$$\|u - u_0\|_{L_2(a,b)} \xrightarrow[u \in L_2(a,b): \frac{d}{dx}u \geqslant 0]{} \min \tag{0.5}$$

if we know that u_* is increasing, and to solve the variational problem

$$\|u - u_0\|_{L_2(a,b)} \xrightarrow[u \in L_2(a,b): \frac{d^2}{dx^2}u \geqslant 0]{} \min \tag{0.6}$$

if we know that u_* is convex.

Let u^0 be the solution of problems (0.5) or (0.6). The inequality

$$\|u^0 - u_*\|_{L_2(a,b)} \leqslant \|\xi\|_{L_2(a,b)}$$

shows that u^0 can be used as an estimate for the input signal u_*.

Moreover, if the error ξ is "sufficiently random," then u^0 lies near u_* in $L_2(a,b)$. By the embedding theorems considered above, u^0 lies near u_* in $C(a,b)$.

0.4. Duality. The variational problems (0.3) and (0.4) are dual to the variational problems (0.5) and (0.6) in the following sense.

Let $U_0 \in W_2^1(a,b)$ and $u_0 = -dU_0/dx \in L_2(a,b)$. If U^0 and u^0 are the solutions of the variational problems (0.3) and (0.5), then $u^0 = -dU^0/dx$.

Let $U_0 \in W_2^2(a,b)$ and $u_0 = \big(-d/dx\big)^2 U_0 \in L_2(a,b)$. If U^0 and u^0 are the solutions of the variational problems (0.4) and (0.6), then $u^0 = \big(-d/dx\big)^2 U^0$.

Therefore, the solution u^0 of the variational problem (0.5) can be constructed by the following algorithm.

1. Define the function U_0 such that $U_0 = \int_x^b u_0(t)\,dt$.
2. Define a function U^0 as the solution of the variational problem (0.3).
3. Define the solution $u^0(x) = -du^0(x)/dx$.

This algorithm of the best monotone L_2-approximation was studied in the paper **[M1]**, and the problem was considered in **[SW1]**. A similar problem of the best monotone L_p-approximation was studied in **[SW2, SW3, SWY]** for $1 < p < \infty$ and in **[DH1, DH2, HL, HMS, LT, L]** for $p = 1$ and $p = \infty$.

1. Embedding theorems

1.1. Borel measures and Bessel potentials. Let Ω be an open set in the space $\mathbb{R}^n$. As usual, the double inclusion $\omega \Subset \Omega$ means that the open set $\omega \subset \Omega$ has a compact closure $\operatorname{Cl}\omega$ contained in Ω.

A distribution $\Lambda \in \mathscr{D}'(\Omega)$ has *zero order* if, given $\omega \Subset \Omega$, there exists $C > 0$ such that

$$|(\Lambda, \varphi)| \leqslant C\|\varphi\|_{C(\operatorname{Cl}\omega)}$$

for all $\varphi \in C_0^\infty(\omega)$. The space of zero order distributions will be denoted by $\mathscr{D}'(\Omega)^0$. Note that the space $\mathscr{D}'(\Omega)^0$ inherits its topology from the space $\mathscr{D}'(\Omega)$. In particular, $\lim_{k\to\infty}\Lambda_k = \Lambda$ in $\mathscr{D}'(\Omega)^0$ means that

$$\lim_{k\to\infty}(\Lambda_k, \varphi) = (\Lambda, \varphi)$$

for each function $\varphi \in C_0^\infty(\Omega)$.

It follows directly from the definition that any distribution $\Lambda \in \mathscr{D}'(\Omega)^0$ can be extended, as a linear continuous functional, from the space $C_0^\infty(\Omega)$ of test functions onto the space $C_0^0(\Omega)$ of continuous functions with compact support. Therefore, for each distribution $\Lambda \in \mathscr{D}'(\Omega)^0$ there exists a unique real Borel measure μ_Λ such that

$$(\Lambda, \varphi) = \int \varphi\, d\mu_\Lambda$$

for all $\varphi \in C_0^\infty(\Omega)$.

Let $\mathscr{M}(\Omega)$ be the space of all real Borel measures on the set Ω. This space is dual to the space $C_0^0(\Omega)$ and therefore it has the weak topology. In particular, the relation $\lim_{k\to\infty}\Lambda_k = \Lambda$ in $\mathscr{M}(\Omega)$ means that

$$\lim_{k\to\infty}(\Lambda_k, \varphi) = (\Lambda, \varphi)$$

for each function $\varphi \in C_0^0(\Omega)$.

The identification of the zero order distribution $\Lambda \in \mathscr{D}'(\Omega)^0$ and the Borel measure $\mu_\Lambda \in \mathscr{M}(\Omega)$ shows that $\mathscr{D}'(\Omega)^0$ and $\mathscr{M}(\Omega)$ as linear spaces are equal. Nevertheless, these spaces are not equal as linear topological spaces because $\mathscr{D}'(\Omega)^0$ has more convergent sequences than $\mathscr{M}(\Omega)$.

Let $s \in \mathbb{R}$. As usual, we denote by $H_2^s(\mathbb{R}^n)$ the space of distributions $u \in \mathfrak{S}'(\mathbb{R}^n)$ such that the Fourier transformation $\mathscr{F}u$ is in $L_2(\mathbb{R}^n)^{\text{loc}}$ and the value

$$\|u\|_{H_2^s(\mathbb{R}^n)} = \int |\mathscr{F}u(\xi)|^2 \left(1 + \|\xi\|^2\right)^s d\xi$$

is finite. Distributions u belonging to $H_2^s(\mathbb{R}^n)$ are called *Bessel potentials*. We define the Fourier transformation $\mathscr{F}$ of a function $\varphi \in \mathfrak{S}(\mathbb{R}^n)$ by the rule

$$\mathscr{F}\varphi(\xi) = (2\pi)^{-n/2} \int e^{-i(x,\xi)} \varphi(x)\, dx.$$

The space of Bessel potentials $H_2^s(\mathbb{R}^n)$ is a Hilbert space with the scalar product

$$(u, v)_{H_2^s(\mathbb{R}^n)} = \int \mathscr{F}u(\xi)\overline{\mathscr{F}v(\xi)} \left(1 + \|\xi\|^2\right)^s d\xi.$$

The local space of Bessel potentials will be denoted by $H_2^s(\Omega)^{\text{loc}}$. It consists of distributions $u \in \mathscr{D}'(\Omega)$ such that $\varphi u \in H_2^s(\mathbb{R}^n)$ for all $\varphi \in C_0^\infty(\Omega)$, where the distribution φu is equal to zero outside the set Ω. The topology of $H_2^s(\Omega)^{\text{loc}}$ is defined by the system of seminorms

$$p_\varphi(u) = \|\varphi u\|_{H_2^s(\mathbb{R}^n)},$$

where $\varphi \in C_0^\infty(\Omega)$.

1.2. Elementary properties of positive distributions. A test function $\varphi \in C_0^\infty(\Omega)$ is said to be *positive* if $\varphi(x) \geqslant 0$ for all $x \in \Omega$. A distribution $\Lambda \in \mathscr{D}'(\Omega)$ is said to be *positive* if $(\Lambda, \varphi) \geqslant 0$ for each positive test function $\varphi \in C_0^\infty(\Omega)$. To signify that the distribution Λ is positive, we shall use the inequality $\Lambda \geqslant 0$. The delta-function δ_x, where $x \in \Omega$, is a fundamental example of a positive distribution. We shall denote the set of all positive test functions by $\overset{+}{C}{}_0^\infty(\Omega)$, and the set of all positive distributions by $\overset{+}{\mathscr{D}}{}'(\Omega)$.

NOTE. Let H be a real linear space. A nonempty subset $V \subset H$ is said to be a *linear semigroup* if the inclusions $u \in V$ and $v \in V$ imply $u + v \in V$ and the inequality $\lambda \geqslant 0$ and the inclusion $u \in V$ imply $\lambda u \in V$. This definition shows that V is convex. If $H = V - V$, then the linear semigroup V is said to be *reproducing*. The reproducibility of the linear semigroup V means that, given $u \in H$, there exist $u_+, u_- \in V$ such that $u = u_+ - u_-$. A linear semigroup K is said to be a *cone* if the inclusions $u \in K$ and $-u \in K$ imply that $u = 0$.

Let $\omega \Subset \Omega$. By $\varkappa_\omega \in \overset{+}{C}{}_0^\infty(\Omega)$ we shall denote a test function such that $\varkappa_\omega(x) = 1$ for all x from a domain containing the set $\operatorname{Cl}\omega$ and $\varkappa_\omega(x) \leqslant 1$ for all $x \in \Omega$.

The following well-known theorem describes the set of positive distributions.

THEOREM 1.1. *The set* $\overset{+}{\mathscr{D}}{}'(\Omega)$ *is a closed cone in the space* $\mathscr{D}'(\Omega)^0$.

PROOF. Let $\Lambda \in \overset{+}{\mathscr{D}}'(\Omega)$ and $\omega \Subset \Omega$. If $\varphi \in C_0^\infty(\omega)$, then

$$-\|\varphi\|_{C(\operatorname{Cl}\omega)}\varkappa_\omega \leqslant \varphi \leqslant \|\varphi\|_{C(\operatorname{Cl}\omega)}\varkappa_\omega.$$

Therefore,

$$-\|\varphi\|_{C(\operatorname{Cl}\omega)}(\Lambda, \varkappa_\omega) \leqslant (\Lambda, \varphi) \leqslant \|\varphi\|_{C(\operatorname{Cl}\omega)}(\Lambda, \varkappa_\omega).$$

Hence

$$|(\Lambda, \varphi)| \leqslant (\Lambda, \varkappa_\omega)\|\varphi\|_{C(\operatorname{Cl}\omega)}.$$

Therefore, the distribution Λ has zero order.

Obviously, the set $\overset{+}{\mathscr{D}}'(\Omega)$ is a linear semigroup. Let $\Lambda \in \overset{+}{\mathscr{D}}'(\Omega)$ and $-\Lambda \in \overset{+}{\mathscr{D}}'(\Omega)$. We shall prove that $\Lambda = 0$. Let $\varphi \in C_0^\infty(\Omega)$, and let $\omega \Subset \Omega$ be a domain such that $\operatorname{supp}\varphi \subset \omega$. Then

$$|(\Lambda, \varphi)| \leqslant (\Lambda, \varkappa_\omega)\|\varphi\|_{C(\operatorname{Cl}\Omega)} \quad \text{and} \quad |(-\Lambda, \varphi)| \leqslant (-\Lambda, \varkappa_\omega)\|\varphi\|_{C(\operatorname{Cl}\Omega)}.$$

Therefore, $\Lambda = 0$ and the set $\overset{+}{\mathscr{D}}'(\Omega)$ is a cone in the space $\mathscr{D}'(\Omega)$.

The cone $\overset{+}{\mathscr{D}}'(\Omega) \subset \mathscr{D}'(\Omega)$ of positive distributions is dual to the cone $\overset{+}{C}{}_0^\infty(\Omega) \subset C_0^\infty(\Omega)$ of positive test functions. Therefore, $\overset{+}{\mathscr{D}}'(\Omega)$ is closed in $\mathscr{D}'(\Omega)$. □

COROLLARY 1.1. *For each positive distribution Λ there exists a unique positive Borel measure μ_Λ such that*

$$(\Lambda, \varphi) = \int \varphi \, d\mu_\Lambda$$

for all $\varphi \in C_0^\infty(\Omega)$.

COROLLARY 1.2. *Let $\Lambda \in \overset{+}{\mathscr{D}}'(\Omega)$. Then*

$$\Lambda \in \operatorname{Clco}\{\delta_x : x \in \operatorname{supp}\Lambda\},$$

where Clco *is the closure of the conical hull in the space $\mathscr{D}'(\Omega)$.*

1.3. Embedding theorems. By the definition, $\overset{+}{\mathscr{D}}'(\Omega) \subset \mathscr{D}'(\Omega)$. Therefore, the cone $\overset{+}{\mathscr{D}}'(\Omega)$ gets its topology from $\mathscr{D}'(\Omega)$.

THEOREM 1.2. *Let $s < -n/2$. Then the operator*

$$\mathrm{IN}\colon \overset{+}{\mathscr{D}}'(\Omega) \to H_2^s(\Omega)^{\mathrm{loc}}$$

is sequentially continuous.

PROOF. 1. We shall prove that the topology in the space $H_2^s(\Omega)^{\mathrm{loc}}$ can be defined by a countable system of seminorms $p_{\psi_1}, p_{\psi_2}, \dots$, where $\psi_1, \psi_2, \dots \in \overset{+}{C}{}_0^\infty(\Omega)$.

Note that in this part of proof the number s may be arbitrary. Later it will be essential that the functions $\psi_1, \psi_2, \dots$ be positive.

The construction of the system $\psi_1, \psi_2, \dots$ is based on the following statement. Let $\varphi \in C_0^\infty(\Omega)$. Then the mapping $\varphi^*\colon H_2^s(\mathbb{R}^n) \to H_2^s(\mathbb{R}^n)$ such that $\varphi^*\colon u \longmapsto \varphi u$ is continuous **[R]**.

Now let $S_1 \subset S_2 \cdots \subset \Omega$ be compact subsets such that $\Omega = S_1 \cup S_2 \cup \cdots$, and let $\psi_1, \psi_2, \ldots \in \overset{+}{C}{}_0^\infty(\Omega)$ be test functions such that ψ_k is equal to 1 in a domain containing S_k.

We shall prove that any open set

$$O(\varphi_1, \ldots, \varphi_m; \lambda_1, \ldots, \lambda_m) = \left\{ u \in \mathscr{D}'(\Omega) : \left\| \varphi_i u \right\|_{H_2^s(\mathbb{R}^n)} < \lambda_i,\ 1 \leqslant i \leqslant m \right\},$$

where $\varphi_1, \ldots, \varphi_m \in C_0^\infty(\Omega)$ and $\lambda_1, \ldots, \lambda_m > 0$, contains an open set

$$O(\psi_k, \tau) = \left\{ u \in \mathscr{D}'(\Omega) : \left\| \psi_k u \right\|_{H_2^s(\mathbb{R}^n)} < \tau \right\}.$$

Let k satisfy $\operatorname{supp} \varphi_1 \cup \cdots \cup \operatorname{supp} \varphi_m \subset S_k$, and let $C > 0$ satisfy

$$\left\| \varphi_i u \right\|_{H_2^s(\mathbb{R}^n)} \leqslant C \left\| u \right\|_{H_2^s(\mathbb{R}^n)}, \qquad 1 \leqslant i \leqslant m,$$

for all $u \in H_2^s(\mathbb{R}^n)$. Define $\tau = C^{-1} \min\{\lambda_1, \ldots, \lambda_m\}$. Then

$$\left\| \varphi_i u \right\|_{H_2^s(\mathbb{R}^n)} = \left\| \varphi_i \psi_k u \right\|_{H_2^s(\mathbb{R}^n)} \leqslant C \left\| \psi_k u \right\|_{H_2^s(\mathbb{R}^n)}, \qquad 1 \leqslant i \leqslant m.$$

Therefore, $O(\psi_k, \tau) \subset O(\varphi_1, \ldots, \varphi_m; \lambda_1, \ldots, \lambda_m)$. Thus the topology of the space $H_2^s(\Omega)^{\text{loc}}$ can be defined by the countable system of seminorms $p_{\psi_1}, p_{\psi_2}, \ldots$, where the test functions $\psi_1, \psi_2, \ldots \in C_0^\infty(\Omega)$ are positive.

2. Let $\Lambda \in \overset{+}{\mathscr{D}}{}'(\Omega)$. We claim that $\Lambda \in H_2^s(\Omega)^{\text{loc}}$. Let $\varphi \in C_0^\infty(\Omega)$. The distribution $\varphi\Lambda \in \mathscr{D}'(\mathbb{R}^n)^0$ has compact support. Therefore,

$$\mathscr{F}(\varphi\Lambda)(\xi) = (2\pi)^{-n/2} \int e^{-i(x,\xi)}\, d\mu_{\varphi\Lambda}(x).$$

Consequently

$$\left\| \mathscr{F}(\varphi\Lambda) \right\|_{C(\mathbb{R}^n)} \leqslant (2\pi)^{-n/2} |\mu_{\varphi\Lambda}|\{\operatorname{supp} \varphi\},$$

where $|\mu_{\varphi\Lambda}|$ is the variation of the measure $\mu_{\varphi\Lambda}$. By this inequality,

$$\begin{aligned} \left\| \varphi\Lambda \right\|^2_{H_2^s(\mathbb{R}^n)} &= \int |\mathscr{F}(\varphi\Lambda)(\xi)|^2 \left(1 + \|\xi\|^2\right)^s d\xi \\ &\leqslant (2\pi)^{-n} |\mu_{\varphi\Lambda}|\{\operatorname{supp} \varphi\}^2 \int \left(1 + \|\xi\|^2\right)^s d\xi < \infty. \end{aligned}$$

Therefore, $\varphi\Lambda \in H_2^s(\mathbb{R}^n)$, and so $\Lambda \in H_2^s(\Omega)^{\text{loc}}$, as claimed.

Let $\Lambda \in \overset{+}{\mathscr{D}}{}'(\Omega)$. By $\mathrm{IN}(\Lambda)$ we denote the distribution Λ in $H_2^s(\Omega)^{\text{loc}}$. Let us verify that the linear operator $\mathrm{IN} \colon \overset{+}{\mathscr{D}}{}'(\Omega) \to H_2^s(\Omega)^{\text{loc}}$ is a monomorphism. Let $\Lambda \neq 0$ and $\mathrm{IN}(\Lambda) = 0$. If $\varphi \in C_0^\infty(\Omega)$, then $\mathscr{F}(\varphi\Lambda) = 0$. Hence $\varphi\Lambda = 0$. Therefore, $\Lambda = 0$. This contradiction shows that IN is a monomorphism.

We shall prove that the operator IN is sequentially continuous. Let $\Lambda_0, \Lambda_1, \Lambda_2, \ldots \in \overset{+}{\mathscr{D}}{}'(\Omega)$ and $\lim_{k\to\infty} \Lambda_k = \Lambda_0$ in $\mathscr{D}'(\Omega)$. We must prove that $\lim_{k\to\infty} \Lambda_k = \Lambda_0$ in $H_2^s(\Omega)^{\text{loc}}$. For this it is sufficiently to prove that

$$\lim_{k\to\infty} \left\| \psi\Lambda_k - \psi\Lambda_0 \right\|_{H_2^s(\mathbb{R}^n)} = 0,$$

where $\psi \in \overset{+}{C}{}_0^\infty(\Omega)$.

Obviously

$$\begin{aligned}
&\lim_{k\to\infty}\left\{\mathscr{F}(\psi\Lambda_k)(\xi)-\mathscr{F}(\psi\Lambda_0)(\xi)\right\}\\
&\quad=(2\pi)^{-n/2}\lim_{k\to\infty}\left\{\left(\psi\Lambda_k,e^{-i(\cdot,\xi)}\right)-\left(\psi\Lambda_0,e^{-i(\cdot,\xi)}\right)\right\}\\
&\quad=(2\pi)^{-n/2}\lim_{k\to\infty}\left\{\left(\Lambda_k,\psi e^{-i(\cdot,\xi)}\right)-\left(\Lambda_0,\psi e^{-i(\cdot,\xi)}\right)\right\}=0
\end{aligned}$$

for all $\xi\in\mathbb{R}^n$. Similarly

$$\begin{aligned}
&\left|\mathscr{F}(\psi\Lambda_k)(\xi)-\mathscr{F}(\psi\Lambda_0)(\xi)\right|\\
&\quad=(2\pi)^{-n/2}\left|\left(\psi\Lambda_k,e^{-i(\cdot,\xi)}\right)-\left(\psi\Lambda_0,e^{-i(\cdot,\xi)}\right)\right|\\
&\quad=(2\pi)^{-n/2}\left|\int e^{-i(x,\xi)}\,d\mu_{\psi\Lambda_k}(x)-\int e^{-i(x,\xi)}\,d\mu_{\psi\Lambda_0}(x)\right|\\
&\quad\leqslant(2\pi)^{-n/2}\int 1\,d\mu_{\psi\Lambda_k}+\int 1\,d\mu_{\psi\Lambda_0}\\
&\quad\leqslant(2\pi)^{-n/2}\int \psi\,d\mu_{\Lambda_k}+\int \psi\,d\mu_{\Lambda_0}\\
&\quad\leqslant(2\pi)^{-n/2}\left\{(\Lambda_k,\psi)+(\Lambda_0,\psi)\right\}\leqslant C
\end{aligned}$$

where $C>0$ does not depend on $k=1,2,\ldots$. Indeed,

$$\lim_{k\to\infty}(\Lambda_k,\psi)=(\Lambda_0,\psi)$$

by assumption.

Now by the Lebesgue theorem it follows that

$$\left\|\psi\Lambda_k-\psi\Lambda_0\right\|^2_{H_2^s(\mathbb{R}^n)}=\int\left|\mathscr{F}(\psi\Lambda_k)(\xi)-\mathscr{F}(\psi\Lambda_0)(\xi)\right|^2\left(1+\|\xi\|^2\right)^s d\xi\xrightarrow[k\to\infty]{}0.$$

Thus the embedding operator IN is sequentially continuous. □

It is well known that the cone $\overset{+}{\mathscr{D}}{}'(\Omega)$ is reproducing in $\mathscr{D}'(\Omega)^0$. Therefore,

$$\mathscr{D}'(\Omega)=\overset{+}{\mathscr{D}}{}'(\Omega)-\overset{+}{\mathscr{D}}{}'(\Omega).$$

Consequently $\mathscr{D}'(\Omega)^0\subset H_2^s(\Omega)^{\mathrm{loc}}$, and we can define the linear operator

$$\mathrm{IN}\colon\mathscr{D}'(\Omega)^0\to H_2^s(\Omega)^{\mathrm{loc}}.$$

We shall prove that this operator is not sequentially continuous. In other words, the embedding operator $\mathrm{IN}\colon\overset{+}{\mathscr{D}}{}'(\Omega)\to H_2^s(\Omega)^{\mathrm{loc}}$ cannot be continued from the cone $\overset{+}{\mathscr{D}}{}'(\Omega)$ onto the space $\mathscr{D}'(\Omega)^0$ as a sequentially continuous operator.

THEOREM 1.3. *Let $s<-n/2$. Then the operator $\mathrm{IN}\colon\mathscr{D}'(\Omega)^0\to H_2^s(\Omega)^{\mathrm{loc}}$ is not sequentially continuous.*

PROOF. We can assume that $0\in\Omega$. Let $m>-s$. Define

$$\Lambda_h^m=h^{1-m}\sum_{k=0}^{m}(-1)^{m-k}C_m^k\delta_{khe_1},$$

where $e_1=(1,0,\ldots,0)$ and $h>0$ is a sufficiently small number.

If $\varphi \in C_0^\infty(\Omega)$, then

$$\left(\Lambda_h^m, \varphi\right) = h^{1-m} \sum_{k=0}^{m} (-1)^{m-k} C_m^k \varphi(khe_1) = \frac{\partial^m \varphi(\tau e_1)}{\partial x_1^m} h,$$

where $0 < \tau < mh$. Therefore, $\Lambda_h^m \to 0$ as $h \searrow 0$ in $\mathscr{D}'(\Omega)^0$. We show that

$$\left\|\Lambda_h^m\right\|_{H_2^s(\mathbb{R}^n)} \xrightarrow[h\searrow 0]{} \infty.$$

Note that

$$\mathscr{F}\left(\Lambda_h^m\right)(\xi) = (2\pi)^{-n/2} h^{1-m} \sum_{k=0}^{m} (-1)^{m-k} C_m^k e^{-ikh\xi_1}.$$

Obviously

$$\lim_{\varepsilon \searrow 0} \varepsilon^{-m} \left| \sum_{k=0}^{m} (-1)^{m-k} C_m^k e^{-ik\varepsilon} \right| = 1.$$

Therefore, let

$$\varepsilon^{-m} \left| \sum_{k=0}^{m} (-1)^{m-k} C_m^k e^{-ik\varepsilon} \right| \geqslant 1/2$$

for $0 \leqslant \varepsilon \leqslant \varepsilon_0$. Further, let $h^{(m-1)/(2m)} + h \leqslant \varepsilon_0$. Define the cube

$$Q_h = \big\{ \xi \in \mathbb{R}^n : h^{(m-1)/(2m)} \leqslant h\xi_1 \leqslant h^{(m-1)/(2m)} + h,$$
$$0 \leqslant \xi_2 \leqslant 1, \dots, 0 \leqslant \xi_n \leqslant 1 \big\}.$$

Let $\xi \in Q_h$. Then

$$\begin{aligned}
&\left|\mathscr{F}\left(\Lambda_h^m\right)(\xi)\right|^2 \left(1 + \|\xi\|^2\right)^s \\
&= (2\pi)^{-n} \left| h^{1-m} \sum_{k=0}^{m} (-1)^{m-k} C_m^k e^{-ikh\xi_1} \right|^2 \left(1 + \|\xi\|^2\right)^s \\
&= (2\pi)^{-n} \left| (h\xi_1)^{-m} \sum_{k=0}^{m} (-1)^{m-k} C_m^k e^{-ikh\xi_1} \right|^2 (h\xi_1)^{2m} h^{2-2m} \left(1 + \|\xi\|^2\right)^s \\
&\geqslant Ch^{1-m} \left(n + \left(1 + h^{-1+(m-1)/(2m)}\right)^2 \right)^s \geqslant Ch^{(1-m)(m+s)/m},
\end{aligned}$$

where $C > 0$ does not depend on h. By this inequality we have

$$\left\|\Lambda_h^m)\right\|_{H_2^s(\mathbb{R}^n)}^2 \geqslant Ch^{(m-1)(m+s)/m}.$$

Therefore, $\mathrm{IN}\colon \mathscr{D}'(\Omega)^0 \longrightarrow H_2^s(\Omega)^{\mathrm{loc}}$ is not sequentially continuous. □

Recall that $\mathscr{M}(\Omega)$ as a linear space is equal to the space $\mathscr{D}'(\Omega)$ but $\mathscr{M}(\Omega)$ contains fewer convergent sequences than $\mathscr{D}'(\Omega)$. We shall prove that the embedding operator $\mathrm{IN}\colon \overset{+}{\mathscr{D}}'(\Omega) \longrightarrow H_2^s(\Omega)^{\mathrm{loc}}$ can be continued from the cone $\overset{+}{\mathscr{D}}'(\Omega)$ onto the space $\mathscr{M}(\Omega)$ as a sequentially continuous operator.

THEOREM 1.4. *Let $s < -n/2$. Then the operator* $\mathrm{IN}\colon \mathscr{M}(\Omega) \longrightarrow H_2^s(\Omega)^{\mathrm{loc}}$ *is sequentially continuous.*

PROOF. Let $\Lambda_1, \Lambda_2, \dots \in \mathscr{M}(\Omega)$ and $\lim_{k\to\infty} \Lambda_k = 0$ in $\mathscr{M}(\Omega)$. We must prove that

$$\lim_{k\to\infty} \Lambda_k = 0 \quad \text{in } H_2^s(\Omega)^{\text{loc}}.$$

Let $\varphi \in C_0^\infty(\Omega)$. Pick a compact set $S \subset \Omega$ containing a neighborhood of the support $\operatorname{supp}\varphi$. Obviously, $\varphi\Lambda_1, \varphi\Lambda_2, \dots$ are in $C(S)^*$, the dual space to the Banach space $C(S)$. By the assumption,

$$\lim_{k\to\infty} (\varphi\Lambda_k, f) = 0$$

for all $f \in C(S)$. Therefore, there exists a constant C such that

$$|(\varphi\Lambda_k, f)| \leqslant C\|f\|_{C(S)}$$

for all $k = 1, 2, \dots$ and $f \in C(S)$.

Consequently, for each $\xi \in \mathbb{R}^n$ we have the relation

$$\lim_{k\to\infty} \mathscr{F}(\varphi\Lambda_k)(\xi) = (2\pi)^{-n/2} \lim_{k\to\infty} \left(\varphi\Lambda_k, e^{-i(\cdot,\xi)}\right) = 0$$

and the inequality

$$\left|\mathscr{F}(\varphi\Lambda_k)(\xi)\right| = (2\pi)^{-n/2}\left|\left(\varphi\Lambda_k, e^{-i(\cdot,\xi)}\right)\right| \leqslant (2\pi)^{-n/2}C,$$

where $k = 1, 2, \dots$. Therefore,

$$\left|\mathscr{F}(\varphi\Lambda_k)(\xi)\right|^2\left(1 + \|\xi\|^2\right)^s \leqslant (2\pi)^{-n}C^2\left(1 + \|\xi\|^2\right)^s \in L_1(\mathbb{R}^n).$$

By definition,

$$\left\|\varphi\Lambda_k\right\|^2_{H_2^s(\mathbb{R}^n)} = \int \left|\mathscr{F}(\varphi\Lambda_k)(\xi)\right|^2\left(1 + \|\xi\|^2\right)^s d\xi.$$

Hence by the Lebesgue theorem we have

$$\lim_{k\to\infty} \left\|\varphi\Lambda_k\right\|^s_{h_2^s(\mathbb{R}^n)} = 0.$$

This proves the statement. □

2. Differential inequalities in the space $\mathscr{D}'(\Omega)$

2.1. Notation. Let $\mathscr{L} = \sum_{|\alpha|\leqslant l} a_\alpha \partial^\alpha$ be an elliptic differential operator of degree l with constant real coefficients, where $\alpha = (\alpha_1, \dots, \alpha_n)$ is a multi-index, $|\alpha| = \alpha_1 + \dots + \alpha_n$ is an order of the multi-index, and

$$\partial^\alpha = \frac{\partial^{\alpha_1}}{\partial x_1^{\alpha_1}} \cdots \frac{\partial^{\alpha_n}}{\partial x_n^{\alpha_n}}$$

is a multidifferential. Note that in the multidimensional case $n \geqslant 2$ the elliptic differential operator $\mathscr{L}$ must have even order $l \geqslant 2$.

Let E be the real fundamental solution of the differential operator $\mathscr{L}$. Recall that $\mathscr{L}E = \delta_0$ by definition. Let

$$\check{\mathscr{L}} = \sum_{|\alpha|\leqslant l} (-1)^{|\alpha|} a_\alpha \partial^\alpha$$

be the dual differential operator and $\check{E}(x) = E(-x)$ the fundamental solution of the differential operator $\check{\mathscr{L}}$.

2.2. A local representation theorem. One says that a distribution $u \in \mathscr{D}'(\Omega)$ *satisfies the differential inequality* $\mathscr{L}u \geqslant 0$ if the distribution $\Lambda = \mathscr{L}u$ is positive. If a distribution $u \in \mathscr{D}'(\Omega)$ satisfies the differential inequality $\mathscr{L}u \geqslant 0$, we shall say that the distribution u is $\mathscr{L}$*-positive*. Let

$$\overset{+}{\mathscr{D}}{}'_{\mathscr{L}}(\Omega) = \{\, u \in \mathscr{D}'(\Omega) : \mathscr{L}u \geqslant 0 \,\}$$

be the set of all $\mathscr{L}$-positive distributions. Obviously, the set $\overset{+}{\mathscr{D}}{}'_{\mathscr{L}}(\Omega)$ is a closed linear semigroup in the space $\mathscr{D}'(\Omega)$.

Let $\omega \Subset \Omega$. By $\Lambda^{\omega} \in \overset{+}{\mathscr{D}}{}'(\mathbb{R}^n)$, where $\Lambda \in \overset{+}{\mathscr{D}}{}'(\Omega)$, we shall denote a distribution such that

$$(\Lambda^{\omega}, \varphi) = \int_{\omega} \varphi \, d\mu_{\Lambda}$$

for $\varphi \in C_0^{\infty}(\mathbb{R}^n)$. Thus we see that the distribution Λ^{ω} has compact support $\operatorname{supp}\Lambda^{\omega} \subset \operatorname{Cl}\omega$. By $u\big|_{\omega} \in \mathscr{D}'(\omega)$, where $u \in \mathscr{D}'(\Omega)$, we shall denote a distribution such that $(u\big|_{\omega}, \varphi) = (u, \varphi)$, where $\varphi \in C_0^{\infty}(\omega)$.

Theorem 2.1. *Let* $u \in \overset{+}{\mathscr{D}}{}'_{\mathscr{L}}(\Omega)$. *Then for each domain* $\omega \Subset \Omega$ *there exists a function* $v \in C^{\infty}(\omega)$ *such that* $\mathscr{L}v = 0$ *and the distribution* u *in the domain* ω *has the representation*

$$u(x) = v(x) + \int_{\omega} E(x-y)\, d\mu_{\Lambda}(y),$$

where $\Lambda = \mathscr{L}u$.

Proof. Obviously,

$$\mathscr{L}\big[u\big|_{\omega} - (E * \Lambda^{\omega})\big|_{\omega}\big] = (\mathscr{L}u)\big|_{\omega} - (\mathscr{L}E * \Lambda^{\omega})\big|_{\omega} = \Lambda\big|_{\omega} - \Lambda^{\omega}\big|_{\omega} = 0.$$

By the relation $\mathscr{L}\big[u\big|_{\omega}\big] = \mathscr{L}\big[(E * \Lambda^{\omega})\big|_{\omega}\big]$, there exists a function $v \in C^{\infty}(\omega)$ such that $u\big|_{\omega} = v + (E * \Lambda^{\omega})\big|_{\omega}$. Therefore, the proof of the theorem will follow if we verify that the convolution $E * \Lambda^{\omega}$ is represented by the function

$$\int_{\omega} E(\cdot - y)\, d\mu_{\Lambda}(y) \in L_1(\mathbb{R}^n)^{\mathrm{loc}}.$$

This verification is based on the inclusion $E \in L_1(\mathbb{R}^n)^{\mathrm{loc}}$ and a standard application of Fubini's theorem.

Let $\omega_0 \Subset \mathbb{R}^n$. Then by the inequality $\mu\{\omega\} < \infty$ we have

$$\begin{aligned}\int_{\omega_0 \times \omega} |E(x-y)|\, dx\, d\mu_{\Lambda}(y) &= \int_{\omega}\left\{\int_{\omega_0} |E(x-y)|\, dx\right\} d\mu_{\Lambda}(y) \\ &\leqslant \int_{\omega}\left\{\int_{\omega_0 - \omega} |E(x)|\, dx\right\} d\mu_{\Lambda}(y) = \mu_{\Lambda}\{\omega\} \int_{\omega_0 - \omega} |E(x)|\, dx < \infty.\end{aligned}$$

Let $\varphi \in C_0^\infty(\mathbb{R}^n)$. Then

$$\begin{aligned}\left(\int_\omega E(\cdot - y)\,d\mu_\Lambda(y), \varphi\right) &= \int \left\{\int_\omega E(x-y)\,d\mu_\Lambda(y)\right\}\varphi(x)\,dx \\ &= \int_\omega \left\{\int E(x-y)\varphi(x)\,dx\right\} d\mu_\Lambda(y) \\ &= \int_\omega \left\{\int E(-y-x)\varphi(-x)\,dx\right\} d\mu_\Lambda(y) \\ &= \int_\omega \left\{\int E(-y-x)\check{\varphi}(x)\,dx\right\} d\mu_\Lambda(y) \\ &= \int_\omega (E * \check{\varphi})(-y)\,d\mu_\Lambda(y) = (\Lambda^\omega, (E*\check{\varphi})(0-\cdot)) \\ &= [\Lambda^\omega * (E * \check{\varphi})](0) = [(E*\Lambda^\omega)*\check{\varphi}](0) = (E*\Lambda, \varphi),\end{aligned}$$

where $\check{\varphi}(x) = \varphi(-x)$. This proves the equality

$$E * \Lambda^\omega = \int_\omega E(\cdot - y)\,d\mu_\Lambda(y). \qquad \square$$

NOTE. When $\mathscr{L} = \Delta$ and $n = 2$ Theorem 2.1 is the classical Riesz theorem of the local representation of subharmonic functions **[H1]**.

COROLLARY 2.1. *If $\mu_\Lambda\{\Omega\} < \infty$, then there exists a function $v \in C^\infty(\Omega)$ such that $\mathscr{L}v = 0$ and*

$$u(x) = v(x) + \int_\Omega E(x-y)\,d\mu_\Lambda(y)$$

in the domain Ω.

COROLLARY 2.2. *By Theorem* 2.1 *it follows that $\overset{+}{\mathscr{D}}{}'_{\mathscr{L}}(\Omega) \subset L_1(\Omega)^{\text{loc}}$. In particular, all $\mathscr{L}$-positive distributions are locally integrable fuctions.*

2.3. Embedding theorems. In **[H2]** it was proved that the embedding operator $\text{IN}\colon \overset{+}{\mathscr{D}}{}'_\Delta(\Omega) \to L_p(\Omega)^{\text{loc}}$ is sequentially continuous for $p < n/(n-2)$. We shall now obtain similar results for an arbitrary elliptic differential operator $\mathscr{L}$.

THEOREM 2.2. *Let $s < l - n/2$. The embedding operator*

$$\text{IN}\colon \overset{+}{\mathscr{D}}{}'_{\mathscr{L}}(\Omega) \to H_2^s(\Omega)^{\text{loc}}$$

is sequentially continuous.

PROOF. 1. Let $u \in \overset{+}{\mathscr{D}}{}'_{\mathscr{L}}(\Omega)$. We claim that $u \in H_2^s(\Omega)^{\text{loc}}$. Let $\omega \Subset \Omega$ and $\Lambda = \varkappa_\omega \mathscr{L}u$. Then $\Lambda \in H_2^{s-l}(\mathbb{R}^n)$. Therefore, $E * \Lambda \in H_2^s(\mathbb{R}^n)^{\text{loc}}$ by **[R]**. Then

$$\mathscr{L}\left[u\big|_\omega - (E*\Lambda)\big|_\omega\right] = (\mathscr{L}u)\big|_\omega - (\mathscr{L}E*\Lambda)\big|_\omega = \Lambda\big|_\omega - \Lambda\big|_\omega = 0.$$

Therefore, there exists a function $v \in C^\infty(\omega)$ such that $\mathscr{L}v = 0$ and $u\big|_\omega = v + (E*\Lambda)\big|_\omega$. This proves the inclusion $u \in H_2^s(\Omega)$.

2. Let $u_0, u_1, u_2, \ldots \in \overset{+}{\mathscr{D}}{}'_{\mathscr{L}}(\Omega)$ and $\lim_{k\to\infty} u_k = u_0$ in $\mathscr{D}'(\Omega)$. We must prove that $\lim_{k\to\infty} u_k = u_0$ in $H_2^s(\Omega)^{\text{loc}}$. The statement follows from the relation $\lim_{k\to\infty} \mathscr{L}u_k = \mathscr{L}u_0$, which, by Theorem 1.2, is true in $H_2^{s-l}(\Omega)^{\text{loc}}$.

Let $\varphi \in C_0^\infty(\Omega)$. Define the domains $\omega_1 \Subset \omega_2 \Subset \omega \Subset \Omega$ such that $\operatorname{supp}\varphi \subset \omega_1$. Let $\varkappa_{\omega_1} \in C_0^\infty(\omega_2)$ and $\varkappa_{\omega_2} \in C_0^\infty(\omega)$.

Let us write down the following representation in the domain ω:

$$u_k\big|_\omega = v_k + (E * \Lambda_k)\big|_\omega, \qquad k = 0, 1, 2, \dots,$$

where $\Lambda_k = \varkappa_\omega \mathscr{L} u_k$ and a function $v_k \in C^\infty(\omega)$ satisfies the equality $\mathscr{L} v_k = 0$.

Let the function $\psi \in C_0^\infty(\mathbb{R}^n)$ be equal to 1 in a neighborhood of the set $\operatorname{supp}\varkappa_\omega - \operatorname{supp}\varkappa_\omega$ and $F = \psi E$. By the proof of Theorem 1.2, it follows that

$$\lim_{k\to\infty} \{\mathscr{F}(\Lambda_k)(\xi) - \mathscr{F}(\Lambda_0)(\xi)\} = 0$$

and

$$\left|\mathscr{F}(\Lambda_k)(\xi) - \mathscr{F}(\Lambda_0)(\xi)\right| \leqslant C$$

for all $\xi \in \mathbb{R}^n$, where $C > 0$ does not depend on $k = 1, 2, \dots$. Hence

$$\left|\mathscr{F}(\Lambda_k)(\xi) - \mathscr{F}(\Lambda_0)(\xi)\right|^2 \left|\mathscr{F}F(\xi)\right|^2 (1 + \|\xi\|^2)^s \leqslant C^2 \left|\mathscr{F}F(\xi)\right|^2 (1 + \|\xi\|^2)^s$$

for all $\xi \in \mathbb{R}^n$. By definition,

$$\begin{aligned} &\left\|F * \Lambda_k - F * \Lambda_0\right\|^2_{H_2^s(\mathbb{R}^n)} \\ &\quad = \int \left|\mathscr{F}(F * \Lambda_k)(\xi) - \mathscr{F}(F * \Lambda_0)(\xi)\right|^2 (1 + \|\xi\|^2)^s \, d\xi \\ &\quad = \int \left|\mathscr{F}(\Lambda_k)(\xi) - \mathscr{F}(\Lambda_0)(\xi)\right|^2 \left|\mathscr{F}F(\xi)\right|^2 (1 + \|\xi\|^2)^s \, d\xi. \end{aligned}$$

From $E \in H_2^s(\mathbb{R}^n)^{\mathrm{loc}}$ it follows that $F \in H_2^s(\mathbb{R}^n)$. Therefore,

$$\lim_{k\to\infty} \left\|F * \Lambda_k - F * \Lambda_0\right\|_{H_2^s(\mathbb{R}^n)} = 0.$$

Obviously $\varkappa_{\omega_2}(E * \Lambda_k) = \varkappa_{\omega_2}(F * \Lambda_k)$ for $k = 0, 1, 2, \dots$. Therefore,

$$\lim_{k\to\infty} \left\|\varkappa_{\omega_2}(E * \Lambda_k) - \varkappa_{\omega_2}(E * \Lambda_0)\right\|_{H_2^s(\mathbb{R}^n)} = 0.$$

Hence $\lim_{k\to\infty} v_k = v_0$ in $H_2^s(\mathbb{R}^n)$. Therefore,

$$\lim_{k\to\infty} \left\|\varkappa_{\omega_1} v_k - \varkappa_{\omega_1} v_0\right\|_{H_2^s(\mathbb{R}^n)} = 0.$$

Obviously $\varphi u_k = \varphi \varkappa_{\omega_1} v_k + \varphi \varkappa_{\omega_2}(E * \Lambda_k)$ for $k = 0, 1, 2, \dots$. Therefore,

$$\lim_{k\to\infty} \left\|\varphi u_k - \varphi u_0\right\|_{H_2^s(\mathbb{R}^n)} = 0.$$

Thus the operator IN is sequentially continuous. □

COROLLARY 2.3. *Let $l > n/2$ and $0 \leqslant s < l - n/2$. Then the embedding operator* IN: $\overset{+}{\mathscr{D}}{}'_{\mathscr{L}}(\Omega) \to W_2^s(\Omega)^{\mathrm{loc}}$ *is sequentially continuous.*

COROLLARY 2.4. *For $l > n$ the embedding operator* IN: $\overset{+}{\mathscr{D}}{}'_{\mathscr{L}}(\Omega) \to C^{l-n-1}(\Omega)$ *is sequentially continuous.*

PROOF. Let $l - n/2 - 1 < s < l - n/2$. Then $0 \leqslant l - n - 1 < s - n/2$. Therefore we have the embedding $H_2^s(\mathbb{R}^n) \hookrightarrow C^{l-n-1}(\mathbb{R}^n)$. The topology in $C^m(\Omega)$ is defined by the system of seminorms $p_\varphi(u) = \left\|\varphi u\right\|^m_C C(\mathbb{R}^n)$, where $\varphi \in C_0^\infty(\Omega)$. Hence the operator IN is sequentially continuous. □

2.4. Spline-approximation in the space $L_1(\Omega)^{\mathrm{loc}}$. We shall call a distribution $U \in \mathscr{D}'(\Omega)$ an $\mathscr{L}$*-spline* of rank k if the support $\operatorname{supp} \mathscr{L} U$ of the distribution $\Lambda = \mathscr{L} U$ is finite and the distribution Λ has order k.

Let $x_1, \ldots, x_k \in \Omega$. Then the function

$$u(x) = v(x) + \sum_{i=1}^{k} \lambda_i E(x - x_i),$$

where the function $v \in C^\infty(\Omega)$ satisfies the differential equation $\mathscr{L} v = 0$, is an $\mathscr{L}$-spline of rank 0. Obviously, the $\mathscr{L}$-spline u is an $\mathscr{L}$-positive function if and only if $\lambda_1 \geqslant 0, \ldots, \lambda_k \geqslant 0$.

NOTE. In the one-dimensional case the fundamental solution E of the operator $\mathscr{L} = d^l/dx^l$ has the form

$$E(x) = \begin{cases} x^{l-1}/(l-1)! & \text{for } x \geqslant 0, \\ 0 & \text{for } x < 0. \end{cases}$$

Therefore, a d^l/dx^l-spline u of rank 0 is a polynomial spline.

THEOREM 2.3. *Let an open set $\Omega \subset \mathbb{R}^n$ be bounded and $u \in \overset{+}{\mathscr{D}}{}'_{\mathscr{L}}(\Omega)$. Then there is a sequence of $\mathscr{L}$-positive $\mathscr{L}$-splines $u_1, u_2, \ldots$ in the linear semigroup $\overset{+}{\mathscr{D}}{}'_{\mathscr{L}}(\Omega)$ such that*

$$u = \lim_{k \to \infty} u_k \quad \textit{in } L_1(\Omega)^{\mathrm{loc}}$$

and

$$\operatorname{supp} \mathscr{L} u_k \subset \operatorname{supp} \mathscr{L} u, \quad \mu_{\mathscr{L} u_k}\{\Omega\} \leqslant \mu_{\mathscr{L} u}\{\Omega\} \quad \textit{for all } k = 1, 2, \ldots.$$

PROOF. Define $\omega(r) = \Omega \setminus \bigcup_{x \in \partial\Omega} \mathbb{D}^n(x, r)$, where $\mathbb{D}^n(x, r) = \{y \in \mathbb{R}^n : \|y - x\| \leqslant r\}$. Let $r > 0$ be so small that $\omega(r) \neq \varnothing$.

We shall prove that the set $\omega(r)$ is open. For this it is sufficient to prove that the set

$$\tau(r) = \bigcup_{x \in \partial\Omega} \mathbb{D}^n(x, r)$$

is closed. Let $x_1, x_2, \ldots \in \tau(r)$ and $x_0 = \lim_{k \to \infty} x_k$. There are $y_1, y_2, \cdots \in \partial\Omega$ such that $\|y_k - x_k\| \leqslant r$ for all $k = 1, 2, \ldots$. We can assume that $y_0 = \lim_{k \to \infty} y_k$, where $y_0 \in \partial\Omega$. Obviously $\|y_0 - x_0\| \leqslant r$. Hence the set $\tau(r)$ is closed.

We shall prove that the set $\Omega \setminus \omega(r)$ cannot be represented as $\Omega \setminus \omega(r) = K \cup F$, where $K \neq \varnothing$ is a compact set, $F \neq \varnothing$ is a closed set in Ω, and $K \cap F = \varnothing$.

Suppose such a representation exists. Pick points $x_0 \in \partial\Omega$ and $y_0 \in K$ such that

$$\|x_0 - y_0\| = \operatorname{dist}(\partial\Omega, K).$$

Then $x_0 \neq y_0$ and the open interval (x_0, y_0) is contained in F. The set F is closed. Therefore, $y_0 \in F \cap K \neq \varnothing$. This contradiction proves the statement.

By Theorem 2.1, there exists a fuction $v \in C^\infty(\omega(r/2))$ such that $\mathscr{L} v = 0$ and

$$u(x) = v(x) + \int_{\omega(r/2)} E(x - y)\, d\mu_\Lambda(y)$$

in the domain $\omega(r/2)$ where $\Lambda = \mathscr{L} u$.

Let $t = (t_1, \dots, t_n) \in \mathbb{Z}^n$ be an integer vector. Define

$$F_t(k) = \omega(r/2) \cap \operatorname{supp} \Lambda \cap \Delta_t(k)$$

where

$$\Delta_t(k) = \Big[\frac{t_1}{k}, \frac{t_1+1}{k}\Big) \times \dots \times \Big[\frac{t_n}{k}, \frac{t_n+1}{k}\Big).$$

Let $T(k) = \{ t \in \mathbb{Z}^n : F_t(k) \neq \varnothing \}$. Define

$$w_k(x) = \sum_{t \in T(k)} \mu_\Lambda\{F_t(k)\} E[x - x_t(k)], \qquad w(x) = \int_{\omega(r/2)} E(x-y)\, d\mu_\Lambda(y),$$

where $x_t(k) \in F_t(k)$ for each $t \in T(k)$.

We prove that $w = \lim_{k\to\infty} w_k$ in $L_1(\mathbb{R}^n)^{\mathrm{loc}}$. Let $\omega \Subset \mathbb{R}^n$. Then

$$\begin{aligned}
&\int_\omega |w(x) - w_k(x)|\, dx \\
&\quad = \int_\omega \Big| \sum_{t\in T(k)} \int_{F_t(k)} \{E(x-y) - E(x - x_t(k))\}\, d\mu_\Lambda(y) \Big|\, dx \\
&\quad \leqslant \int_\omega \sum_{t\in T(k)} \int_{F_t(k)} |E(x-y) - E(x - x_t(k))|\, d\mu_\Lambda(y)\, dx \\
&\quad = \sum_{t\in T(k)} \int_{F_t(k)} \int_\omega |E(x-y) - E(x - x_t(k))|\, dx\, d\mu_\Lambda(y) \\
&\quad \leqslant \mu_\Lambda\{\omega(r/2)\} \sup_{h \in (n/k)\mathbb{D}^n} \int_{\omega-\omega} |E(x+h) - E(x)|\, dx.
\end{aligned}$$

Hence the relation

$$\lim_{k\to\infty} \int_\omega |w(x) - w_k(x)|\, dx = 0$$

follows from $\mu_\Lambda\{\omega(r/2)\} < \infty$ and

$$\lim_{k\to\infty} \sup_{h\in(n/k)\mathbb{D}^n} \int_{\omega-\omega} |E(x+h) - E(x)|\, dx = 0.$$

Therefore, $w = \lim_{k\to\infty} w_k$ in $L_1(\mathbb{R}^n)^{\mathrm{loc}}$.

The set $\Omega \setminus \omega(r)$ cannot be represented in the form $\Omega \setminus \omega(r) = K \cup F$, where $K \neq \varnothing$ is a compact set, $F \neq \varnothing$ is a closed set in Ω and $K \cap F = \varnothing$. Hence the Runge theorem **[H1]** implies the existence of functions $v_1, v_2, \dots \in C^\infty(\mathbb{R}^n)$ such that $\mathscr{L} v_k = 0$ and $v = \lim_{k\to\infty} v_k$ in $C^\infty(\omega(r/2))$.

Define $u_k = v_k + w_k$. Note that $u_k \in L_1(\mathbb{R}^n)^{\mathrm{loc}}$, because $v_k \in C^\infty(\mathbb{R}^n)$ and $w_k \in L_1(\mathbb{R}^n)^{\mathrm{loc}}$. Obviously $u = \lim_{k\to\infty} u_k$ in $L_1(\omega(r))$. By definition, the function u_k is an $\mathscr{L}$-positive $\mathscr{L}$-spline, and $\operatorname{supp} \mathscr{L} u_k \subset \operatorname{supp} \mathscr{L} u$ and $\mu_{\mathscr{L} u_k}\{\Omega\} \leqslant \mu_{\mathscr{L} u}\{\Omega\}$ for all $k = 1, 2, \dots$.

To finish the proof we construct an $\mathscr{L}$-positive $\mathscr{L}$-spline u_k such that

$$\|u - u_k\|_{L_1(\omega(2^{-k}))} \leqslant 1/k, \qquad \operatorname{supp} \mathscr{L} u_k \subset \operatorname{supp} \mathscr{L} u, \qquad \mu_{\mathscr{L} u_k}\{\Omega\} \leqslant \mu_{\mathscr{L} u}\{\Omega\}$$

for all $k = 1, 2, \dots$. Obviously $u = \lim_{k\to\infty} u_k$ in the space $L_1(\Omega)^{\mathrm{loc}}$. $\square$

COROLLARY 2.5. *Let $l > n/2$ and $0 \leqslant s < l - n/2$. Then any $\mathscr{L}$-positive function u has an $\mathscr{L}$-positive $\mathscr{L}$-spline approximation in $W_2^s(\Omega)^{\mathrm{loc}}$. In particular, if $l > n$, then it has an $\mathscr{L}$-positive $\mathscr{L}$-spline approximation in the space $C^{l-n-1}(\Omega)$.*

3. Differential inequalities in the space $L_2(\Omega)$

3.1. Linear semigroups in Hilbert spaces. Let H be a real Hilbert space with scalar product $(\cdot,\cdot)$ and norm $\|\cdot\|$.

Let $V \subset H$ be a closed linear semigroup. One can prove that V has a direct sum representation $V = P \oplus K$, where $P = V \cap (-V)$ is a closed linear space and $K = V \cap P^{\perp}$ is a closed cone.

Let $V \subset H$ be a linear semigroup. We define the dual semigroup V^* by

$$V^* = \bigcap_{v \in V} \{\, u \in H : (u,v) \geqslant 0 \,\}.$$

It follows that V^* is closed in the Hilbert space H.

Let $V \subset H$ be a nonempty convex subset and $u^0 \in V$. The set

$$V\{u^0\} = \{\, \lambda(u - u^0) : \lambda \geqslant 0, u \in V \,\}$$

is a linear semigroup, called the *semigroup of possible directions* at the point u^0.

Let $K \subset H$ be a closed cone. A linearly independent family $S \subset K$ will be called a *generating family* of the cone K if $\operatorname{Clco} S = K$, where $\operatorname{Clco} S$ is the closure of the conical hull of the set S.

Let $S \subset K$ be a generating family of the cone K. Obviously, $v \in K^*$ if and only if $(u,v) \geqslant 0$ for all $u \in S$.

Let $V \subset H$ be a nonempty closed convex set in the Hilbert space H. It is well known that for each $u_0 \in H$ the minimization problem

$$(*) \qquad \|u - u_0\| \xrightarrow[u \in V]{} \min$$

has a unique solution $u^0 \in V$. Therefore, we can define a map $\mathfrak{P}_V : H \to V$ by the rule $\mathfrak{P}_V : u_0 \longmapsto u^0$. The map $\mathfrak{P}_V$ is called the *projection* onto V.

It can be proved that the point $u \in V$ is the solution of the minimization problem $(*)$ if and only if

$$u^0 \in u_0 + V\{u^0\}^*.$$

Let $V \subset H$ be a closed linear semigroup V. In this case a point $u^0 \in V$ is the solution of the problem of minimization $(*)$ if and only if $u^0 \in u_0 + V^*$ and $(u^0 - u_0, u^0) = 0$.

By this criterion, it follows that the point $u^0 \in V$ is the solution of the minimization problem $(*)$ if and only if this point is the solution of the minimization problem

$$(**) \qquad \|u\| \xrightarrow[u \in u_0 + V^*]{} \min.$$

Therefore, the problems $(*)$ and $(**)$ are dual.

It is clear from $(*)$ and $(**)$ that

$$\mathfrak{P}_V(u) = u + \mathfrak{P}_{V^*}(-u), \quad u \in H.$$

Note that the vectors $\mathfrak{P}_V(u)$ and $\mathfrak{P}_{V^*}(-u)$ are orthogonal.

It can be proved that

$$\left\|\mathfrak{P}_V(u_* + \xi) - u_*\right\| \leqslant \left(\mathfrak{P}_V(\xi) - \xi, u_*\right)^{1/2} + \left\|\mathfrak{P}_V(\xi)\right\|$$

for all $u_* \in V$ and $\xi \in H$.

3.2. Linear semigroups of differential inequalities. We shall suppose that $l > n/2$ and the domain $\Omega \subset \mathbb{R}^n$ is bounded. Consider the set

$$V_{\mathscr{L}}(\Omega) = \left\{ u \in L_2(\Omega) : \mathscr{L}u \geqslant 0 \right\}$$

of solutions of the differential inequality $\mathscr{L}u \geqslant 0$ in the space $L_2(\Omega)$. The inequality $l > n/2$ implies the inclusion $\mathscr{D}'_{\mathscr{L}}(\Omega) \subset L_2(\Omega)^{\mathrm{loc}}$. Therefore, the only solutions $u \in \mathscr{D}'(\Omega)$ of the differential inequality $\mathscr{L}u \geqslant 0$ that are not included in the sets $V_{\mathscr{L}}(\Omega)$ are those that increase too quickly near the boundary $\partial\Omega$.

THEOREM 3.1. *The set $V_{\mathscr{L}}(\Omega)$ is a closed linear semigroup such that*

$$V_{\mathscr{L}}(\Omega) = P_{\mathscr{L}}(\Omega) \oplus K_{\mathscr{L}}(\Omega),$$

where $P_{\mathscr{L}}(\Omega) = \left\{ u \in L_2(\Omega) : \mathscr{L}u = 0 \right\}$ is a closed linear subspace and $K_{\mathscr{L}}(\Omega) = \left\{ u \in P_{\mathscr{L}}(\Omega)^{\perp} : \mathscr{L}u \geqslant 0 \right\}$ is a closed cone.

PROOF. Obviously the set $V_{\mathscr{L}}(\Omega)$ is a linear semigroup in $L_2(\Omega)$. We show that it is closed. Let $u_1, u_2, \ldots \in V_{\mathscr{L}}(\Omega)$ and $\lim_{k\to\infty} u_k = u$ in $L_2(\Omega)$. If $\varphi \in \overset{+}{C}{}_0^{\infty}(\Omega)$, then

$$\left(\mathscr{L}u_k, \varphi\right) = \left(u_k, \check{\mathscr{L}}\varphi\right)_{L_2(\Omega)} \xrightarrow[k\to\infty]{} \left(u, \check{\mathscr{L}}\varphi\right)_{L_2(\Omega)} = \left(\mathscr{L}u, \varphi\right).$$

By assumption, $(\mathscr{L}u_k, \varphi) \geqslant 0$ for all $1 \leqslant k < \infty$. Hence $(\mathscr{L}u, \varphi) \geqslant 0$. Thus $u \in V_{\mathscr{L}}(\Omega)$ and the set $V_{\mathscr{L}}(\Omega)$ is closed.

Similarly, the space $P_{\mathscr{L}}(\Omega)$ and the linear semigroup are closed in $L_2(\Omega)$.

Let us prove that

$$P_{\mathscr{L}}(\Omega) = V_{\mathscr{L}}(\Omega) \cap \left(-V_{\mathscr{L}}(\Omega)\right).$$

The inclusion $P_{\mathscr{L}}(\Omega) \subset V_{\mathscr{L}}(\Omega) \cap \left(-V_{\mathscr{L}}(\Omega)\right)$ is obvious. Let $u \in V_{\mathscr{L}}(\Omega)$ and $u \in -V_{\mathscr{L}}(\Omega)$. Then $\mathscr{L}u \in \overset{+}{\mathscr{D}}{}'(\Omega)$ and $\mathscr{L}u \in \overset{+}{\mathscr{D}}{}'(\Omega)$. Therefore, by Theorem 1.1 it follows that $\mathscr{L}u = 0$. Hence $u \in P_{\mathscr{L}}(\Omega)$. Thus the linear semigroup $K_{\mathscr{L}}(\Omega)$ is a closed cone and

$$V_{\mathscr{L}}(\Omega) = P_{\mathscr{L}}(\Omega) \oplus K_{\mathscr{L}}(\Omega),$$

by the definition of $K_{\mathscr{L}}(\Omega)$. □

3.3. Main integral equality.

THEOREM 3.2. *Let $u \in V_{\mathscr{L}}(\Omega)$. If a function $V \in \overset{\circ}{W}{}_2^l(\Omega)$ satisfies the inequality $V\big|_{\operatorname{supp}\mathscr{L}u} \geqslant 0$, then V is $\mu_{\mathscr{L}u}$-integrable and*

$$\int_{\Omega} V \, d\mu_{\mathscr{L}u} = \left(u, \check{\mathscr{L}}V\right)_{L_2(\Omega)}. \tag{3.1}$$

PROOF. Since $l > n/2$, we have the embedding $W_2^l(\Omega) \hookrightarrow C(\operatorname{Cl}\Omega)$. Hence the function V is continuous.

To prove (3.1), consider the cases $\mu_{\mathscr{L}U}\{\Omega\} < \infty$ and $\mu_{\mathscr{L}u}\{\Omega\} = \infty$. In the first case the function V may have any sign on $\operatorname{supp}\mathscr{L}u$.

Let $\mu_{\mathscr{L}u}\{\Omega\} < \infty$. In the space $C_0^\infty(\Omega)$ select a sequence $\varphi_1, \varphi_2, \dots$ such that

$$V = \lim_{k\to\infty} \varphi_k \quad \text{in } W_2^l(\Omega). \tag{3.2}$$

By the inequality $l > n/2$, the relation (3.2) is true in the space $C(\operatorname{Cl}\Omega)$. But

$$\int_\Omega \varphi_k \, d\mu_{\mathscr{L}u} = (\mathscr{L}u, \varphi_k) = \left(u, \check{\mathscr{L}}\varphi_k\right)_{L_2(\Omega)}.$$

Hence the representation (3.1) follows from the inequality $\mu_{\mathscr{L}u}\{\Omega\} < \infty$.

Let $\mu_{\mathscr{L}u}\{\Omega\} = \infty$. By the Hedberg theorem **[He]** there is a sequence $w_1, w_2, \dots \in C_0^\infty(\Omega)$ such that

$$V = \lim_{k\to\infty} w_k V \tag{3.3}$$

in $W_2^l(\Omega)$ and $0 \leqslant w_k \leqslant 1$ for all $k = 1, 2, \dots$. Define $V_k = w_k V$. Since $\mu_{\mathscr{L}u}\{\operatorname{supp} w_k\} < \infty$, we have

$$\int_\Omega V_k \, d\mu_{\mathscr{L}u} = \left(u, \check{\mathscr{L}} V_k\right)_{L_2(\Omega)}.$$

Obviously

$$\left(u, \check{\mathscr{L}} V\right)_{L_2(\Omega)} = \lim_{k\to\infty} \left(u, \check{\mathscr{L}} V_k\right)_{L_2(\Omega)}$$

and

$$\int_\Omega V_k \, d\mu_{\mathscr{L}u} = \int_{\operatorname{supp}\mathscr{L}u} V_k \, d\mu_{\mathscr{L}u}.$$

By the inequality $l > n/2$, the relation (3.3) is true in the space $C(\operatorname{Cl}\Omega)$. Therefore, by Fatou's theorem,

$$\int_{\operatorname{supp}\mathscr{L}u} V \, d\mu_{\mathscr{L}u} \leqslant \left(u, \check{\mathscr{L}} V\right)_{L_2(\Omega)}.$$

By the inequality $0 \leqslant V_k(x) \leqslant V(x)$, where $x \in \operatorname{supp}\mathscr{L}u$, we obtain

$$\int_{\operatorname{supp}\mathscr{L}u} V \, d\mu_{\mathscr{L}u} \geqslant \int_{\operatorname{supp}\mathscr{L}u} V_k \, d\mu_{\mathscr{L}u}.$$

Hence

$$\int_{\operatorname{supp}\mathscr{L}u} V \, d\mu_{\mathscr{L}u} \geqslant \left(u, \check{\mathscr{L}} V\right)_{L_2(\Omega)}.$$

Thus

$$\int_{\operatorname{supp}\mathscr{L}u} V \, d\mu_{\mathscr{L}u} = \left(u, \check{\mathscr{L}} V\right)_{L_2(\Omega)}.$$

Therefore, the equality

$$\int_\Omega V \, d\mu_{\mathscr{L}u} = \int_{\operatorname{supp}\mathscr{L}u} V \, d\mu_{\mathscr{L}u}$$

proves the representation (3.1). □

REMARK. A similar formula was proved in **[BB]**.

3.4. Isomorphisms. Recall that the closure of the space $C_0^\infty(\Omega)$ in the Sobolev space $W_2^l(\Omega)$ is denoted by $\overset{\circ}{W}{}_2^l(\Omega)$.

In **[M2]** it was proved that the spaces $\overset{\circ}{W}{}_2^l(\Omega)$ and $P_{\mathscr{L}}(\Omega)^\perp$ are canonically continuously isomorphic:

$$\check{E}*\colon P_{\mathscr{L}}(\Omega)^\perp \to \overset{\circ}{W}{}_2^l(\Omega), \qquad \check{\mathscr{L}}\colon \overset{\circ}{W}{}_2^l(\Omega) \to P_{\mathscr{L}}(\Omega)^\perp,$$

where

$$\big(\check{E} * u\big)(x) = \int_\Omega E(y-x)u(y)\,dy$$

for $u \in L_2(\Omega)$. This statement leads to the following analogous results. Let

$$\overset{+}{W}{}_2^l(\Omega) = \big\{\, V \in \overset{\circ}{W}{}_2^l(\Omega) : V \geqslant 0 \,\big\}$$

be the cone of positive functions in the space $\overset{\circ}{W}{}_2^l(\Omega)$.

Theorem 3.3. *The mappings*

$$\check{E}*\colon V_{\mathscr{L}}(\Omega) \to \overset{+}{W}{}_2^l(\Omega), \qquad \check{\mathscr{L}}\colon \overset{+}{W}{}_2^l(\Omega) \to V_{\mathscr{L}}(\Omega)^*$$

are continuous isomorphisms.

Proof. We shall prove that

$$\check{E} * u \in \overset{+}{W}{}_2^l(\Omega) \quad \text{for } u \in V_{\mathscr{L}}(\Omega)^*, \qquad \check{\mathscr{L}} V \in V_{\mathscr{L}}(\Omega)^* \quad \text{for } V \in \overset{+}{W}{}_2^l(\Omega).$$

Let $u \in V_{\mathscr{L}}(\Omega)^*$. Then $u \in P_{\mathscr{L}}(\Omega)^\perp$. Hence $\check{E} * u \in \overset{\circ}{W}{}_2^l(\Omega)$. But

$$\big(\check{E} * u\big)(x) = \big(E(\cdot - x), u\big)_{L_2(\Omega)} \quad \text{and} \quad E(\cdot - x)\big|_\Omega \in V_{\mathscr{L}}(\Omega)$$

for all $x \in \Omega$. Therefore $\big(\check{E} * u\big)(x) \geqslant 0$ and $\check{E} * u \in \overset{+}{W}{}_2^l(\Omega)$.

Let $V \in \overset{+}{W}{}_2^l(\Omega)$. Then by Theorem 3.2 we have

$$\big(u, \check{\mathscr{L}} V\big)_{L_2(\Omega)} = \int_\Omega V\, d\mu_{\mathscr{L}u} \geqslant 0$$

for all $u \in V_{\mathscr{L}}(\Omega)$. Hence $\check{\mathscr{L}} V \in V_{\mathscr{L}}(\Omega)^*$. □

Let $\overset{+}{W}{}_2^l(\Omega)^*_{\mathscr{L}}$ be the dual linear semigroup for the cone $\overset{+}{W}{}_2^l(\Omega)$ in the Hilbert space $\overset{\circ}{W}{}_2^l(\Omega)$ with scalar product $(U, V)_{\mathscr{L}} = (\check{\mathscr{L}} U, \check{\mathscr{L}} V)_{L_2(\Omega)}$.

Theorem 3.4. *The mappings*

$$\check{E}*\colon K_{\mathscr{L}}(\Omega) \to \overset{+}{W}{}_2^l(\Omega)^*_{\mathscr{L}}, \qquad \check{\mathscr{L}}\colon \overset{+}{W}{}_2^l(\Omega)^*_{\mathscr{L}} \to K_{\mathscr{L}}(\Omega)$$

are isometric isomorphisms.

Proof. We shall prove that

$$\check{E} * u \in \overset{+}{W}{}_2^l(\Omega)^*_{\mathscr{L}} \quad \text{if } u \in K_{\mathscr{L}}(\Omega), \qquad \check{\mathscr{L}} V \in K_{\mathscr{L}}(\Omega) \quad \text{if } V \in \overset{+}{W}{}_2^l(\Omega)^*_{\mathscr{L}}.$$

Let $u \in K_{\mathscr{L}}(\Omega)$. Consider $V \in \overset{+}{W}{}_2^l(\Omega)^*_{\mathscr{L}}$. Then

$$\bigl(\check{E} * u, V\bigr)_{\mathscr{L}} = \bigl(\check{\mathscr{L}}(\check{E} * u), \check{\mathscr{L}} V\bigr)_{L_2(\Omega)} = \bigl(u, \check{\mathscr{L}} V\bigr)_{L_2(\Omega)} = \int_\Omega V\, d\mu_{\mathscr{L}u} \geqslant 0.$$

Hence $\check{E} * u \in \overset{+}{W}{}_2^l(\Omega)^*_{\mathscr{L}}$.

Let $V \in \overset{+}{W}{}_2^l(\Omega)^*_{\mathscr{L}}$. Consider a positive function $\varphi \in C_0^\infty(\Omega)$. Then

$$\bigl(\mathscr{L}\check{\mathscr{L}} V, \varphi\bigr) = \bigl(\check{\mathscr{L}} V, \check{\mathscr{L}} \varphi\bigr)_{L_2(\Omega)} = \bigl(V, \varphi\bigr)_{\mathscr{L}} \geqslant 0.$$

Hence $\check{\mathscr{L}} V \in V_{\mathscr{L}}(\Omega)$. But $\check{\mathscr{L}} V \in P_{\mathscr{L}}(\Omega)^\perp$. Therefore, $\check{\mathscr{L}} V \in K_{\mathscr{L}}(\Omega)$. □

Let $S \subset \Omega$. Consider the cone

$$\overset{+}{W}{}_2^l(\Omega, S) = \bigl\{\, V \in \overset{+}{W}{}_2^l(\Omega) : V\big|_S = 0 \,\bigr\}$$

of positive functions in the space $\overset{\circ}{W}{}_2^l(\Omega)$ that vanishes on S.

Theorem 3.5. *Let $u \in V_{\mathscr{L}}(\Omega)$. The mappings*

$$\check{E} *\colon V_{\mathscr{L}}(\Omega)\{u\}^* \to \overset{+}{W}{}_2^l(\Omega, \operatorname{supp} \mathscr{L}u),$$

$$\check{\mathscr{L}}\colon \overset{+}{W}{}_2^l(\Omega, \operatorname{supp} \mathscr{L}u) \to V_{\mathscr{L}}(\Omega)\{u\}^*$$

are continuous isomorphisms.

Proof. We shall prove that

$$\check{E} * v \in \overset{+}{W}{}_2^l(\Omega, \operatorname{supp} \mathscr{L}u) \quad \text{if } v \in V_{\mathscr{L}}(\Omega)\{u\}^*,$$

$$\check{\mathscr{L}} V \in V_{\mathscr{L}}(\Omega)\{u\}^* \quad \text{if } V \in \overset{+}{W}{}_2^l(\Omega, \operatorname{supp} \mathscr{L}u).$$

1. Let $v \in V_{\mathscr{L}}(\Omega)\{u\}^*$. Then $v \in V_{\mathscr{L}}(\Omega)^*$. Therefore, $\check{E} * v \in \overset{+}{W}{}_2^l(\Omega)$ by Theorem 3.3. Let $V = \check{E} * v$. We prove that $V\big|_{\operatorname{supp} \mathscr{L}u} = 0$. Suppose that $x_0 \in \operatorname{supp} \mathscr{L}u$ and $V(x_0) > 0$. Denote by $\omega \Subset \Omega$ the domain such that $x_0 \in \omega$ and $V(x) \geqslant \varepsilon$ for all $x \in \omega$ where $\varepsilon > 0$. Let $\Lambda = \mathscr{L}u$. Then obviously $\Lambda - \Lambda^\omega\big|_\Omega \in \overset{+}{\mathscr{D}}{}'(\Omega)$. Let $w = (\check{E} * \Lambda^\omega)\big|_\Omega$. Obviously $w \in V_{\mathscr{L}}(\Omega)$ and

$$\mathscr{L}(u - w) = \Lambda - \Lambda^\omega\big|_\Omega.$$

Therefore, $u - w \in V_{\mathscr{L}}(\Omega)$ and $-w \in V_{\mathscr{L}}(\Omega)\{u\}$. Hence $-(w, v)_{L_2(\Omega)} \geqslant 0$.

On the other hand,

$$(w,v)_{L_2(\Omega)} = (w, \mathscr{L}\check{V})_{L_2(\Omega)} = \int_\Omega V\, d\mu_{\mathscr{L}w} \geqslant \int_\omega V\, d\mu_{\mathscr{L}u} \geqslant \varepsilon\mu_{\mathscr{L}u}\{\omega\} > 0.$$

This contradiction proves that $V\big|_{\operatorname{supp}\mathscr{L}u} = 0$. Hence

$$\check{E} * v \in \overset{+}{W}{}_2^l(\Omega, \operatorname{supp}\mathscr{L}u).$$

2. Let $V \in \overset{+}{W}{}_2^l(\Omega, \operatorname{supp}\mathscr{L}u)$. Consider $w \in V_{\mathscr{L}}(\Omega)$. We prove that

$$\big(w - u, \check{\mathscr{L}}V\big)_{L_2(\Omega)} \geqslant 0.$$

By Theorem 3.2 it follows that

$$\big(u, \check{\mathscr{L}}V\big)_{L_2(\Omega)} = \int_\Omega V\, d\mu_{\mathscr{L}u} = 0.$$

Therefore,

$$\big(w - u, \check{\mathscr{L}}V\big)_{L_2(\Omega)} = \int_\Omega V\, d\mu_{\mathscr{L}w} - \int_\Omega V\, d\mu_{\mathscr{L}u} = \int_\Omega V d\,\mu_{\mathscr{L}w} \geqslant 0.$$

Thus $\check{\mathscr{L}}V \in V_{\mathscr{L}}(\Omega)\{u\}^*$. □

THEOREM 3.6. *Let* $u \in K_{\mathscr{L}}(\Omega)$. *The mappings*

$$\check{E}*\colon \operatorname{Cl}\big(K_{\mathscr{L}}(\Omega)\{u\}\big) \to \overset{+}{W}{}_2^l\big(\Omega, \operatorname{supp}\mathscr{L}u\big)^*_{\mathscr{L}},$$

$$\check{\mathscr{L}}\colon \overset{+}{W}{}_2^l\big(\Omega, \operatorname{supp}\mathscr{L}u\big)^*_{\mathscr{L}} \to \operatorname{Cl}\big(K_{\mathscr{L}}(\Omega)\{u\}\big)$$

are isometric isomorphisms.

PROOF. We shall prove that

$$\check{E} * v \in \overset{+}{W}{}_2^l\big(\Omega, \operatorname{supp}\mathscr{L}u\big) \quad \text{if } u \in \operatorname{Cl}\big(K_{\mathscr{L}}(\Omega)\{u\}\big),$$

$$\check{\mathscr{L}}V \in \operatorname{Cl}\big(K_{\mathscr{L}}(\Omega)\{u\}\big) \quad \text{if } V \in \overset{+}{W}{}_2^l\big(\Omega, \operatorname{supp}\mathscr{L}u\big)^*_{\mathscr{L}}.$$

1. Let $v \in K_{\mathscr{L}}(\Omega)$ and $V \in \overset{+}{W}{}_2^l(\Omega, \operatorname{supp}\mathscr{L}u)$. Then

$$\begin{aligned}\big(\check{E} * (v-u), V\big)_{\mathscr{L}} &= \big(v - u, \check{\mathscr{L}}V\big)_{L_2(\Omega)}\\ &= \int_\Omega V\, d\mu_{\mathscr{L}v} - \int_\Omega V\, d\mu_{\mathscr{L}u} = \int_\Omega V\, d\mu_{\mathscr{L}v} \geqslant 0.\end{aligned}$$

Therefore $\check{E} * (v-u) \in \overset{+}{W}{}_2^l(\Omega, \operatorname{supp}\mathscr{L}u)^*_{\mathscr{L}}$. Thus $\check{E} * w \in \overset{+}{W}{}_2^l(\Omega, \operatorname{supp}\mathscr{L}u)^*_{\mathscr{L}}$ for all $w \in K_{\mathscr{L}}(\Omega)\{u\}$. Hence $\check{E} * w \in \overset{+}{W}{}_2^l(\Omega, \operatorname{supp}\mathscr{L}u)^*_{\mathscr{L}}$ for all $w \in \operatorname{Cl} K_{\mathscr{L}}(\Omega)\{u\}$.

2. Let $V \in \overset{+}{W}{}_2^l(\Omega, \operatorname{supp}\mathscr{L}u)^*_{\mathscr{L}}$. If $U \in \overset{+}{W}{}_2^l(\Omega, \operatorname{supp}\mathscr{L}u)$, then

$$\big(V, U\big)_{\mathscr{L}} = \big(\check{\mathscr{L}}V, \check{\mathscr{L}}U\big)_{L_2(\Omega)} \geqslant 0.$$

Therefore,

$$\check{\mathscr{L}}V \in V_{\mathscr{L}}(\Omega)\{u\}^{**} = \operatorname{Cl}\big(V_{\mathscr{L}}(\Omega)\{u\}\big).$$

Note that $V \in \overset{\circ}{W}{}_2^l(\Omega)$. Hence $\check{\mathscr{L}}V \in \operatorname{Cl}(K_{\mathscr{L}}(\Omega)\{u\})$. □

3.5. A generating family of the cone $K_{\mathscr{L}}(\Omega)$. Let $y \in \mathbb{R}^n$. Define the function $\gamma_y = \mathfrak{P}_{P_{\mathscr{L}}(\Omega)^\perp}\{E(\cdot - y)\big|_\Omega\}$, where $\mathfrak{P}_{P_{\mathscr{L}}(\Omega)^\perp} : L_2(\Omega) \longrightarrow P_{\mathscr{L}}(\Omega)^\perp$ is the projection.

Obviously $\gamma_y \neq 0$ if and only if $y \in \Omega$. Moreover, the inclusion $E(\cdot - y)\big|_\Omega \in V_{\mathscr{L}}(\Omega)$ implies the inclusion $\gamma_y \in K_{\mathscr{L}}(\Omega)$.

Let $S \subset \Omega$. Define the cone $K_{\mathscr{L}}(\Omega, S) = \{ u \in K_{\mathscr{L}}(\Omega) : \operatorname{supp} \mathscr{L}u \subset S \}$.

THEOREM 3.7. $K_{\mathscr{L}}(\Omega, S) = \operatorname{Clco}\{ \gamma_y : y \in S \}$.

PROOF. We shall prove the dual relation

$$K_{\mathscr{L}}(\Omega, S)^* = \operatorname{Clco}\{ \gamma_y : y \in S \}^*.$$

Let $v \in K_{\mathscr{L}}(\Omega, S)^*$ and $y \in S$. Then $(v, \gamma_y)_{L_2(\Omega)} \geqslant 0$ because $\gamma_y \in K_{\mathscr{L}}(\Omega, S)$. Hence $v \in \operatorname{Clco}\{ \gamma_y : y \in S \}^*$.

Let $v \in \operatorname{Clco}\{ \gamma_y : y \in S \}^*$. Then

$$v = \mathfrak{P}_{P_{\mathscr{L}}(\Omega)}(v) + \mathfrak{P}_{P_{\mathscr{L}}(\Omega)^\perp}(v).$$

The inclusion $\mathfrak{P}_{P_{\mathscr{L}}(\Omega)}(v) \in K_{\mathscr{L}}(\Omega, S)^*$ is obvious. Therefore, we can suppose that $v \in P_{\mathscr{L}}(\Omega)^\perp$.

Let $V = \check{E} * v$ and $y \in S$. Then

$$V(y) = (\check{E} * v)(y) = \big(E(\cdot - y), v\big)_{L_2(\Omega)} = \big(\gamma_y, v\big)_{L_2(\Omega)} \geqslant 0.$$

Let $u \in K_{\mathscr{L}}(\Omega, S)$. Then $V(y) \geqslant 0$ for all $y \in \operatorname{supp} \mathscr{L}u$. Consequently

$$(u, v) = (u, \check{\mathscr{L}} V)_{L_2(\Omega)} = \int_\Omega V \, d\mu_{\mathscr{L}u} \geqslant 0.$$

Therefore, $v \in K_{\mathscr{L}}(\Omega, S)^*$. □

COROLLARY 3.1. *Let $u \in K_{\mathscr{L}}(\Omega)$. Then*

$$u \in \operatorname{Clco}\{ \gamma_y : y \in \operatorname{supp} \mathscr{L}u \}.$$

Therefore, the family $\{ \gamma_y : y \in \Omega \}$ is the generating family of the cone $K_{\mathscr{L}}(\Omega)$.

THEOREM 3.8. *The set $\{ \gamma_y : y \in \operatorname{Cl}\Omega \}$ is compact.*

PROOF. Define the mapping $\gamma \colon \operatorname{Cl}\Omega \longrightarrow L_2(\Omega)$ by $\gamma \colon y \longmapsto \gamma_y$. The mapping γ is continuous, injective on Ω, and equal to 0 on $\partial\Omega$. Therefore its image $\{ \gamma_y : y \in \operatorname{Cl}\Omega \}$ in $L_2(\Omega)$ is homeomorphic to the topological quotient space $\operatorname{Cl}\Omega/\partial\Omega$. The elements of this quotient space are all separate points of the set Ω and the whole set $\partial\Omega$. (For example, if Ω is an n-dimensional ball, then the quotient space is homeomorphic to the n-dimensional sphere). The quotient space is compact. Therefore the set $\{ \gamma_y : y \in \operatorname{Cl}\Omega \}$ is compact. □

3.6. The semigroup of possible directions. Let $u \in V_{\mathscr{L}}(\Omega)$. Then the closed linear semigroup $\operatorname{Cl}\big(V_{\mathscr{L}}(\Omega)\{u\}\big)$ has the following representation:

$$\operatorname{Cl}\big(V_{\mathscr{L}}(\Omega)\{u\}\big) = P^u_{\mathscr{L}}(\Omega) \oplus K^u_{\mathscr{L}}(\Omega),$$

where

$$P^u_{\mathscr{L}}(\Omega) = \operatorname{Cl}\big(V_{\mathscr{L}}(\Omega)\{u\}\big) \cap -\operatorname{Cl}\big(V_{\mathscr{L}}(\Omega)\{u\}\big)$$

is a closed space and

$$K^u_{\mathscr{L}}(\Omega) = \operatorname{Cl}\big(V_{\mathscr{L}}(\Omega)\{u\}\big) \cap P^u_{\mathscr{L}}(\Omega)^{\perp}$$

is a closed cone.

THEOREM 3.9. *The equality* $P^u_{\mathscr{L}}(\Omega) = L_2(\Omega)$ *is true if and only if* $\operatorname{supp} \mathscr{L}u = \Omega$.

PROOF. The relation

$$L_2(\Omega) = \operatorname{Cl}\big(V_{\mathscr{L}}(\Omega)\{u\}\big) \cap -\operatorname{Cl}\big(V_{\mathscr{L}}(\Omega)\{u\}\big)$$

is equivalent to

$$P_{\mathscr{L}}(\Omega)^{\perp} = \operatorname{Cl}\big(K_{\mathscr{L}}(\Omega)\{u\}\big) \cap -\operatorname{Cl}\big(K_{\mathscr{L}}(\Omega)\{u\}\big).$$

By Theorem 3.6, the last relation is equivalent to

$$\overset{\circ}{W}{}^l_2(\Omega) = \overset{+}{W}{}^l_2\big(\Omega, \operatorname{supp} \mathscr{L}u\big)^*_{\mathscr{L}} \cap -\overset{+}{W}{}^l_2\big(\Omega, \operatorname{supp} \mathscr{L}u\big)^*_{\mathscr{L}}.$$

Let $\operatorname{supp} \mathscr{L}u = \Omega$. Then $\overset{+}{W}{}^l_2(\Omega, \operatorname{supp} \mathscr{L}u) = 0$. Therefore,

$$\overset{\circ}{W}{}^l_2(\Omega) = \overset{+}{W}{}^l_2\big(\Omega, \operatorname{supp} \mathscr{L}u\big)^*_{\mathscr{L}}.$$

Thus $P^u_{\mathscr{L}}(\Omega) = L_2(\Omega)$.

Let $P^u_{\mathscr{L}}(\Omega) = L_2(\Omega)$. Suppose that the set $\omega = \Omega \setminus \operatorname{supp} \mathscr{L}u$ is not empty. Define $v \in C_0^\infty(\omega)$ so that $v = 0$. Then $v \in \overset{\circ}{W}{}^l_2(\Omega)$. Hence $(v, \varphi)_{\mathscr{L}} \geqslant 0$ and $(v, -\varphi)_{\mathscr{L}} \geqslant 0$ for all functions $\varphi \in \overset{+}{C}{}_0^\infty(\omega) \subset \overset{+}{W}{}^l_2(\Omega, \operatorname{supp} \mathscr{L}u)$. Therefore

$$(v, \varphi)_{\mathscr{L}} = \big(\check{\mathscr{L}}v, \check{\mathscr{L}}\varphi\big)_{L_2(\Omega)} = \big(\mathscr{L}\check{\mathscr{L}}v, \varphi\big) = 0$$

for all $\varphi \in \overset{+}{C}{}_0^\infty(\omega)$. Hence $\mathscr{L}\check{\mathscr{L}}v = 0$ and $v = 0$. The contradiction proves that $\operatorname{supp} \mathscr{L}u = \Omega$. $\square$

COROLLARY 3.2. *Let* $u \in V_{\mathscr{L}}(\Omega)$. *The equality* $\operatorname{Cl} V_{\mathscr{L}}(\Omega)\{u\} = L_2(\Omega)$ *is true if and only if* $\operatorname{supp} \mathscr{L}u = \Omega$.

THEOREM 3.10. *Let $l > n/2+1$. If $u = E(\cdot - y)\big|_{\Omega}$, where $y \in \Omega$, then*

$$P^u_{\mathscr{L}}(\Omega) = P_{\mathscr{L}}(\Omega) \oplus \operatorname{Lin}\{\gamma_y, \gamma_y^1, \dots, \gamma_y^n\},$$

where

$$\gamma_y^i = \mathfrak{P}_{P_{\mathscr{L}}(\Omega)^\perp}\Big\{\frac{\partial}{\partial x_i} E(\cdot - y)\big|_{\Omega}\Big\}, \quad 1 \leqslant i \leqslant n.$$

PROOF. Let $v \in P^u_{\mathscr{L}}(\Omega)$. Then

$$v \in \operatorname{Cl}\big(V_{\mathscr{L}}(\Omega)\{u\}\big) \cap -\operatorname{Cl}\big(V_{\mathscr{L}}(\Omega)\{u\}\big),$$

and so

$$w \in \operatorname{Cl}\big(K_{\mathscr{L}}(\Omega)\{u\}\big) \cap -\operatorname{Cl}\big(K_{\mathscr{L}}(\Omega)\{u\}\big),$$

where $w = \mathfrak{P}_{P_{\mathscr{L}}(\Omega)^\perp}(v)$. Hence

$$W \in \overset{+}{W}{}_2^l(\Omega, \operatorname{supp}\mathscr{L}u)^*_{\mathscr{L}} \cap -\overset{+}{W}{}_2^l(\Omega, \operatorname{supp}\mathscr{L}u)^*_{\mathscr{L}},$$

where $W = \check{E} * w$. Thus

$$(W, \varphi)_{\mathscr{L}} = (\check{\mathscr{L}} W, \check{\mathscr{L}}\varphi)_{L_1(\Omega)} = \big(w, \check{\mathscr{L}}\varphi\big)_{L_2(\Omega)} = \big(\mathscr{L}w, \varphi\big) = \big(\mathscr{L}v, \varphi\big) = 0$$

for all $\varphi \in \overset{+}{W}{}_2^l(\Omega, \operatorname{supp}\mathscr{L}u)$. Since

$$\overset{+}{C}{}_0^\infty(\Omega \setminus \operatorname{supp}\mathscr{L}u) \subset \overset{+}{W}{}_2^l(\Omega, \operatorname{supp}\mathscr{L}u),$$

the last equation shows that $\operatorname{supp}\mathscr{L}v \subset \operatorname{supp}\mathscr{L}u = \{y\}$. Therefore,

$$\mathscr{L}v = \sum_{|\alpha|<p} a_\alpha \partial^\alpha \delta_y.$$

We shall prove that $a_\alpha = 0$ for all $|\alpha| \geqslant 2$.

We may suppose that $y = 0$. Suppose a positive test function ψ has a support in a sufficiently small neighborhood of 0 and is equal to 1 in another neighborhood of 0. Define $\varphi(x) = x_k^2\psi(x),$, $1 \leqslant k \leqslant n$. Then $\varphi \in \overset{+}{W}{}_2^l(\Omega, \operatorname{supp}\mathscr{L}u)$. Therefore,

$$(\mathscr{L}v, \varphi) = \Big(\sum_{|\alpha|<p} a_\alpha \partial^\alpha \delta_y, \varphi\Big) = 2a_{2A^k} = 0, \quad 1 \leqslant k \leqslant n,$$

where $A^k = (0, \dots, 0, 1, 0, \dots, 0)$.

Let $|\beta| \geqslant 2$. Define $\lambda > 0$ so that the function $\|x\|^2 + \lambda x^\beta$ is positive on the support $\operatorname{supp} \psi$. Let $\varphi(x) = (\|x\|^2 + \lambda x^\beta)\psi(x)$. Then $\varphi \in \overset{+}{W}{}_2^l(\Omega, \operatorname{supp} \mathscr{L} u)$. Therefore,

$$(\mathscr{L} v, \varphi) = \Big(\sum_{|\alpha|<p} a_\alpha \partial^\alpha \delta_y, \varphi \Big) = \lambda \beta! a_\beta = 0.$$

Thus we have proved that

$$\mathscr{L} v = \sum_{|\alpha|<2} a_\alpha \partial^\alpha \delta_y.$$

Therefore,

$$v = v_0 + a_0 E(\cdot - y)\big|_\Omega + \sum_{k=1}^n a_k \frac{\partial}{\partial x_k} E(\cdot - y)\big|_\Omega,$$

where $\mathscr{L} v_0 = 0$. Hence we obtain the inclusion

$$P^u_{\mathscr{L}}(\Omega) \subset P_{\mathscr{L}}(\Omega) \oplus \operatorname{Lin}\{\gamma_y, \gamma_y^1, \dots, \gamma_y^n\}.$$

Obviously $P^u_{\mathscr{L}}(\Omega) \supset P_{\mathscr{L}}(\Omega)$ and $P^u_{\mathscr{L}}(\Omega) \ni E(\cdot - y)\big|_\Omega$. Therefore, to prove the inclusion

$$P^u_{\mathscr{L}}(\Omega) \supset P_{\mathscr{L}}(\Omega) \oplus \operatorname{Lin}\{\gamma_y, \gamma_y^1, \dots, \gamma_y^n\}$$

we must verify that

$$P^u_{\mathscr{L}}(\Omega) \ni \frac{\partial}{\partial x_k} E(\cdot - y)\big|_\Omega, \quad 1 \leqslant k \leqslant n.$$

We shall continue to assume that $y = 0$.

Let us prove that

$$\frac{\partial}{\partial x_k} E\Big|_\Omega = \lim_{\lambda \to 0} \frac{E(\cdot + \lambda A^k) - E}{\lambda}\Big|_\Omega \tag{3.4}$$

in $L_2(\Omega)$. Let a function $\varphi \in C_0^\infty(\Omega)$ be equal to 1 on a neighborhood of $\operatorname{Cl}\Omega$. Define $H = \varphi E$. Then

$$H(\cdot + \lambda A^k)\big|_\Omega = E(\cdot + \lambda A^k)\big|_\Omega$$

if λ is sufficiently small. Thereofore,

$$\begin{aligned} \left\| \frac{E(\cdot + \lambda A^k) - E}{\lambda} - \frac{\partial}{\partial x_k} E \right\|_{L_2(\Omega)} &= \left\| \frac{H(\cdot + \lambda A^k) - H}{\lambda} - \frac{\partial}{\partial x_k} H \right\|_{L_2(\Omega)} \\ &\leqslant \left\| \frac{H(\cdot + \lambda A^k) - H}{\lambda} - \frac{\partial}{\partial x_k} H \right\|_{L_2(\mathbb{R}^n)} \\ &= \left\| \Big(\frac{e^{i\lambda\xi_k - 1}}{\lambda} - i\xi \Big) \mathscr{F} H \right\|_{L_2(\mathbb{R}^n)} = \left\| f(i\lambda\xi_k)\xi_k \mathscr{F} H \right\|_{L_2(\mathbb{R}^n)}, \end{aligned}$$

where

$$f(t) = \sum_{m=2}^\infty \frac{(it)^{m-1}}{m!} = \frac{\sin t}{t} - 1 - i\frac{1 - \cos t}{t}.$$

Obviously $|f(t)| \leqslant C$, $t \in \mathbb{R}$, and $\lim_{t \to 0} f(t) = 0$.

The inequality $l > n/2 + 1$ and the inclusion $E \in H_2^s(\mathbb{R}^n)^{\text{loc}}$, where $s < l - n/2$, imply that $E \in W_2^l(\mathbb{R}^n)^{\text{loc}}$. Therefore, $H \in W_2^1(\mathbb{R}^n)$. Hence $\xi_k \mathscr{F} H \in L_2(\mathbb{R}^n)$.

Let $\varepsilon > 0$. Define $r > 0$ so that

$$\left\|f(i\lambda\xi_k)\xi_k\mathscr{F}H\right\|_{L_2(\mathbb{R}^n\setminus r\mathbb{D}^n)} \leqslant \varepsilon^2/2,$$

and $\delta > 0$ so that

$$\left\|f(i\lambda\xi_k)\xi_k\mathscr{F}H\right\|^2_{L_2(r\mathbb{D}^n)} \leqslant \varepsilon^2/2$$

for all $|\lambda| \leqslant \delta$. Thus

$$\left\|f(i\lambda\xi_k)\xi_k\mathscr{F}H\right\|_{L_2(\mathbb{R}^n)} \leqslant \varepsilon$$

for all $|\lambda| \leqslant \delta$. Therefore

$$\lim_{\lambda\to 0}\left\|\frac{E(\cdot+\lambda A^k)-E}{\lambda} - \frac{\partial}{\partial x_k}E\right\|_{L_2(\Omega)} = 0.$$

This proves the limit (3.4).

If $\lambda > 0$, then

$$\frac{E(\cdot+\lambda A^k)-E}{\lambda}\bigg|_\Omega \in V_{\mathscr{L}}(\Omega)\{u\}$$

and

$$-\frac{E(\cdot-\lambda A^k)-E}{-\lambda}\bigg|_\Omega = \frac{E(\cdot-\lambda A^k)-E}{\lambda}\bigg|_\Omega \in V_{\mathscr{L}}(\Omega)\{u\}.$$

Therefore,

$$\frac{\partial}{\partial x_k}E\bigg|_\Omega \in \operatorname{Cl} V_{\mathscr{L}}(\Omega)\{u\} \quad\text{and}\quad \frac{\partial}{\partial x_k}E\bigg|_\Omega \in -\operatorname{Cl} V_{\mathscr{L}}(\Omega)\{u\}.$$

These inclusions prove the theorem. □

COROLLARY 3.3. *Let $l > n/2+1$ and consider the $\mathscr{L}$-spline $u \in V_{\mathscr{L}}(\Omega)$. Then the space $P^u_{\mathscr{L}}(\Omega)$ is equal to the space of $\mathscr{L}$-splines of rank* 1 *with nodes in the set* $\operatorname{supp}\mathscr{L}u$.

3.7. Spline approximation in the space $L_2(\Omega)$.

THEOREM 3.11. *Let $u \in V_{\mathscr{L}}(\Omega)$. Then in the linear semigroup $V_{\mathscr{L}}(\Omega)$, there is a sequence of $\mathscr{L}$-splines $u_1, u_2, \ldots$ such that*

$$u = \lim_{k\to\infty} u_k$$

in the space $L_2(\Omega)$ and

$$\operatorname{supp}\mathscr{L}u_k \subset \operatorname{supp}\mathscr{L}u, \qquad \mu_{\mathscr{L}u_k}\{\Omega\} \leqslant \mu_{\mathscr{L}u}\{\Omega\}$$

for all $k = 1, 2, \ldots$.

PROOF. 1. Let $\mu_{\mathscr{L}u}\{\Omega\} = \infty$. Note that $u = v + w$, where $v \in P_{\mathscr{L}}(\Omega)$ and $w \in K_{\mathscr{L}}(\Omega)$. By Corollary 3.1, $w \in \operatorname{Clco}\{\gamma_y : y \in \operatorname{supp}\mathscr{L}u\}$. This proves the theorem in this case.

2. Let $\mu_{\mathscr{L}u}\{u\} < \infty$. In this case by Corollary 2.1 there is a function $v \in C^\infty(\Omega)$ such that $\mathscr{L}v = 0$ and $u = v + (E * \Lambda^\Omega)\big|_\Omega$, where $\Lambda = \mathscr{L}u$. The inclusion $E * \Lambda^\Omega \in L_2(\mathbb{R}^n)^{\mathrm{loc}}$ clearly implies that $(E * \Lambda^\Omega)\big|_\Omega \in L_2(\Omega)$. Therefore, $v \in P_{\mathscr{L}}(\Omega)$.

Let $\Omega \Subset \Omega_0 \Subset \mathbb{R}^n$. Then $(E * \Lambda^\Omega)\big|_{\Omega_0} \in \mathscr{D}'_{\mathscr{L}}(\Omega_0)$. Hence by Theorem 2.3, there is a sequence of $\mathscr{L}$-splines $w_1, w_2, \ldots \in \mathscr{D}'_{\mathscr{L}}(\Omega_0)$ such that

$$(E * \Lambda^\Omega)\big|_{\Omega_0} = \lim_{k\to\infty} w_k$$

in $L_1(\Omega_0)^{\mathrm{loc}}$ and

$$\operatorname{supp} \mathscr{L} w_k \subset \operatorname{supp} \mathscr{L} u, \qquad \mu_{\mathscr{L} w_k}\{\Omega\} \leqslant \mu_{\mathscr{L} u}\{\Omega\}$$

for all $k = 1, 2, \ldots$. By Theorem 2.2, this is true in $L_2(\Omega_0)^{\mathrm{loc}}$. Hence

$$(E * \Lambda^\Omega)\big|_{\Omega} = \lim_{k\to\infty} w_k\big|_{\Omega}$$

in $L_2(\Omega)$.

Obviously the sequence of $\mathscr{L}$-positive $\mathscr{L}$-splines can be taken in the form

$$u_k = v + w_k\big|_{\Omega}, \qquad k = 1, 2, \ldots. \qquad \square$$

3.8. Differential inequalities with bounded measure. Let $0 \leqslant \lambda \leqslant \infty$. Consider the convex sets

$$V^\lambda_{\mathscr{L}}(\Omega) = \{u \in V_{\mathscr{L}}(\Omega) : \mu_{\mathscr{L} u}\{\Omega\} \leqslant \lambda\}, \qquad K^\lambda_{\mathscr{L}}(\Omega) = V^\lambda_{\mathscr{L}}(\Omega) \cap P_{\mathscr{L}}(\Omega)^\perp.$$

Note that $V^0(\Omega) = P_{\mathscr{L}}(\Omega)$ and $V^\infty_{\mathscr{L}}(\Omega) = V_{\mathscr{L}}(\Omega)$. Obviously

$$V^\lambda_{\mathscr{L}}(\Omega) = P_{\mathscr{L}}(\Omega) \oplus K^\lambda_{\mathscr{L}}(\Omega).$$

THEOREM 3.12. *If $0 < \lambda < \infty$, then the set $V^\lambda_{\mathscr{L}}(\Omega)$ is closed, the set $K^\lambda_{\mathscr{L}}(\Omega)$ is compact, and $K^\lambda_{\mathscr{L}}(\Omega) = \lambda \operatorname{Cl} \operatorname{conv}\{\gamma_y : y \in \Omega\}$.*

PROOF. 1. Let $u_1, u_2, \ldots \in V^\lambda_{\mathscr{L}}(\Omega)$ and $u = \lim_{k\to\infty} u_k$ in $L_2(\Omega)$. By the inclusion $V^\lambda_{\mathscr{L}}(\Omega) \subset V_{\mathscr{L}}(\Omega)$, it follows that $u \in V_{\mathscr{L}}(\Omega)$. We shall prove that $\mu_{\mathscr{L} u}\{\Omega\} \leqslant \lambda$. Note that

$$\mathscr{L} u = \lim_{k\to\infty} \mathscr{L} u_k \quad \text{in } \mathscr{D}'(\Omega).$$

Let $\omega \Subset \Omega$ and $\varphi = \varkappa_\omega$. Then

$$\begin{aligned} \mu_{\mathscr{L} u}\{\omega\} &= \int_\omega \varphi \, d\mu_{\mathscr{L} u} \leqslant \int_\Omega \varphi \, d\mu_{\mathscr{L} u} = (\mathscr{L} u, \varphi) \\ &= \lim_{k\to\infty} (\mathscr{L} u_k, \varphi) = \lim_{k\to\infty} \int_\Omega \varphi \, d\mu_{\mathscr{L} u_k} \leqslant \lambda. \end{aligned}$$

Obviously $\Omega = \bigcup_{\omega \Subset \Omega} \omega$. Therefore, $\mu_{\mathscr{L} u}\{\Omega\} \leqslant \lambda$ and $u \in V^\lambda_{\mathscr{L}}(\Omega)$. Thus the set $V^\lambda_{\mathscr{L}}(\Omega)$ is closed.

2. We prove that the set $K^\lambda_{\mathscr{L}}(\Omega)$ is compact. Note that $K^\lambda_{\mathscr{L}}(\Omega) = \lambda K^1_{\mathscr{L}}(\Omega)$. Therefore, we can assume that $\lambda = 1$.

Let $u \in K^1_{\mathscr{L}}(\Omega)$. By Theorem 3.11, there is a sequence of $\mathscr{L}$-splines $u_1, u_2, \dots \in V_{\mathscr{L}}(\Omega)$ such that $u = \lim_{k\to\infty} u_k$ in $L_2(\Omega)$ and

$$\operatorname{supp} \mathscr{L} u_k \subset \operatorname{supp} \mathscr{L} u, \qquad \mu_{\mathscr{L}u_k}\{\Omega\} \leqslant \mu_{\mathscr{L}u}\{\Omega\}$$

for all $k = 1, 2, \dots$. The $\mathscr{L}$-spline u_k has the representation $u_k = v_k + w_k$, where $v_k = \mathfrak{P}_{P_{\mathscr{L}}(\Omega)}(u_k)$ and $w_k = \mathfrak{P}_{P_{\mathscr{L}}(\Omega)^\perp}(u_k)$. Note that

$$w_k = \sum_{i=1}^{m_k} \lambda_{ki} \gamma_{y_{ki}},$$

where $\lambda_{ki} \geqslant 0$ and $y_{ki} \in \operatorname{supp} \mathscr{L} u$. Obviously

$$\|v_k\|^2_{l_2(\Omega)} = (u_k, v_k)_{L_2(\Omega)} = (u_k - u, v_k)_{L_2(\Omega)} \leqslant \|u_k - u\|_{L_2(\Omega)} \|v_k\|_{L_2(\Omega)}.$$

Therefore, $u_k = \lim_{k\to\infty} w_k$ in $L_2(\Omega)$. Moreover, $\mu_{\mathscr{L}w_k}\{\Omega\} = \sum_{i=1}^{m_k} \lambda_{ki}$, and hence $\sum_{i=1}^{m_k} \lambda_{ki} \leqslant 1$. Thus we have the inclusion

$$w_k = \left(1 - \sum_{i=1}^{m_k} \lambda_{ki}\right) 0 + \sum_{i=1}^{m_k} \lambda_{ki} \gamma_{y_{ki}} \in \operatorname{conv}\{\, \gamma_y : y \in \Omega \,\}.$$

Therefore, $u \in \operatorname{Cl}\operatorname{conv}\{\, \gamma_y : y \in \Omega \,\}$, and so

$$K^1_{\mathscr{L}}(\Omega) \subset \operatorname{Cl}\operatorname{conv}\{\, \gamma_y : y \in \Omega \,\}.$$

Obviously $\gamma_y \in K^1_{\mathscr{L}}(\Omega)$ for all $y \in \Omega$. Therefore,

$$K^1_{\mathscr{L}}(\Omega) \supset \operatorname{Cl}\operatorname{conv}\{\, \gamma_y : y \in \Omega \,\}.$$

This proves the relation $K^1_{\mathscr{L}}(\Omega) = \operatorname{Cl}\operatorname{conv}\{\, \gamma_y : y \in \Omega \,\}$. Obviously

$$\operatorname{Cl}\operatorname{conv}\{\, \gamma_y : y \in \Omega \,\} = \operatorname{Cl}\operatorname{conv}\{\, \gamma_y : y \in \operatorname{Cl}\Omega \,\}.$$

By Theorem 3.8, the set $\{\, \gamma_y : y \in \operatorname{Cl}\Omega \,\}$ is compact. Therefore, the set $K^1_{\mathscr{L}}(\Omega)$ is compact as the closure of the convex hull of the compact set **[R]**. □

4. Variational problems

4.1. Preliminary remarks. Let $U_0 \in W^l_2(\Omega)$ and $u_0 \in L_2(\Omega)$. We shall consider the variational problems

$$\|\check{\mathscr{L}} U\|_{L_2(\Omega)} \xrightarrow[U \in U_0 + \overset{\circ}{W}{}^l_2(\Omega) : U \geqslant U_0]{} \min$$

and

$$\|u - u_0\|_{L_2(\Omega)} \xrightarrow[u \in L_2(\Omega) : \mathscr{L} u \geqslant 0]{} \min .$$

The first is the variational problem with barrier $U \geqslant U_0$, and the second is the problem of L_2-approximation with differential restriction $\mathscr{L} u \geqslant 0$. These variational problems are dual in the sense that if the functions U_0 and u_0 are related by $U_0 = \check{E} * u_0$, then the solutions U^0 and u^0 of the variational problems are connected by the equality

$u^0 = \check{\mathscr{L}} U^0$. Later we shall prove that the solution of the first variational problem is an $\mathscr{L}\check{\mathscr{L}}$-positive function $U^0 \in U_0 + \overset{+}{W}{}_2^l(\Omega)$ such that

$$\operatorname{supp} \mathscr{L}\check{\mathscr{L}} U^0 \subset \{ x \in \Omega : U^0(x) = U_0(x) \}.$$

Therefore, U^0 can be constructed by solving the interpolational problem with the one-sided restriction $U^0 \geqslant U_0$ and the differential restriction $\mathscr{L}\check{\mathscr{L}} U^0 \geqslant 0$.

4.2. Variational problems without a restriction on the measure. Let $S \subset \Omega$ be a closed subset. Define a linear space

$$P_{\mathscr{L}}(\Omega, S) = \{ u \in L_2(\Omega) : \operatorname{supp} \mathscr{L} u \subset S \}$$

and consider the sets

$$V_{\mathscr{L}}(\Omega, S) = V_{\mathscr{L}}(\Omega) \cap P_{\mathscr{L}}(\Omega, S), \qquad K_{\mathscr{L}}(\Omega, S) = K_{\mathscr{L}}(\Omega) \cap P_{\mathscr{L}}(\Omega, S).$$

The set S is closed in Ω. Therefore, the space $P_{\mathscr{L}}(\Omega, S)$ is closed in $L_2(\Omega)$. Hence the sets $V_{\mathscr{L}}(\Omega, S)$ and $K_{\mathscr{L}}(\Omega, S)$ are closed in $L_2(\Omega)$. Obviously

$$V_{\mathscr{L}}(\Omega, S) = P_{\mathscr{L}}(\Omega) \oplus K_{\mathscr{L}}(\Omega, S).$$

THEOREM 4.1. *Let $u \in V_{\mathscr{L}}(\Omega, S)$. The inclusion $v \in V_{\mathscr{L}}(\Omega, S)\{u\}^*$ holds if and only if v can be represented in the form $v = \check{\mathscr{L}} V$, where the function $V \in \overset{\circ}{W}{}_2^l(\Omega)$ satisfies $V|_S \geqslant 0$ and $V|_{\operatorname{supp} \mathscr{L} u} = 0$.*

PROOF. *Necessity.* Let $v \in V_{\mathscr{L}}(\Omega, S)\{u\}^*$. By the inclusions

$$P_{\mathscr{L}}(\Omega) \subset V_{\mathscr{L}}(\Omega, S) \subset V_{\mathscr{L}}(\Omega, S)\{u\}$$

it follows that

$$P_{\mathscr{L}}(\Omega)^{\perp} \supset V_{\mathscr{L}}(\Omega, S)^* \supset V_{\mathscr{L}}(\Omega, S)\{u\}^*.$$

Therefore, $v \in P_{\mathscr{L}}(\Omega)^{\perp}$. Hence $v = \check{\mathscr{L}} V$, where $V = \check{E} * v \in \overset{\circ}{W}{}_2^l(\Omega)$.

Let $x \in S$. Then the relation $V(x) = \big(E(\cdot - x), v\big)_{L_2(\Omega)}$ and the inclusions $E(\cdot - x)|_{\Omega} \in V_{\mathscr{L}}(\Omega, S)$ and $v \in V_{\mathscr{L}}(\Omega, S)^*$ imply that $V(x) \geqslant 0$. Therefore, $V|_S \geqslant 0$.

Let $\operatorname{supp} \mathscr{L} u \neq 0$ and $x \in \operatorname{supp} \mathscr{L} u$. By the proof of Theorem 3.6, it follows that $V(x) = 0$. Therefore, $V|_{\operatorname{supp} \mathscr{L} u} = 0$.

Sufficiency. Let $v = \check{\mathscr{L}} V$, and let the function $V \in \overset{\circ}{W}{}_2^l(\Omega)$ satisfy $V|_S \geqslant 0$ and $V|_{\operatorname{supp} \mathscr{L} u} = 0$. From the inclusion $w \in V_{\mathscr{L}}(\Omega, S)$, it follows that $V(x) \geqslant 0$ for all $x \in \operatorname{supp} \mathscr{L} u \cup \operatorname{supp} \mathscr{L} w$. Therefore, by Theorem 3.2,

$$\begin{aligned}\big(w - u, v\big)_{L_2(\Omega)} &= \big(w - u, \check{\mathscr{L}} V\big)_{l_2(\Omega)} \\ &= \int_{\Omega} V \, d\mu_{\mathscr{L} w} - \int_{\Omega} V \, d\mu_{\mathscr{L} u} = \int_{\Omega} V \, d\mu_{\mathscr{L} w} \geqslant 0.\end{aligned}$$

Hence $v \in V_{\mathscr{L}}(\Omega, S)\{u\}^*$. □

THEOREM 4.2. *Given $U_0 \in W_2^l(\Omega)$, there exists a unique $\mathscr{L}\check{\mathscr{L}}$-positive function $U^0 \in U_0 + \overset{\circ}{W}{}_2^l(\Omega)$ such that $U^0|_S \geqslant U_0|_S$ and $\operatorname{supp} \mathscr{L}\check{\mathscr{L}} U^0 \subset \{ x \in S : U^0(x) = U_0(x) \}$.*

The function U^0 is the solution of the variational problem

$$\|\check{\mathscr{L}} U\|_{l_2(\Omega)} \xrightarrow[U \in U_0 + \overset{\circ}{W}{}^l_2(\Omega):\, U|_S \geqslant U_0|_S]{} \min . \tag{4.1}$$

The mapping $\mathfrak{F}\colon W^l_2(\Omega) \longrightarrow W^l_2(\Omega)$, $\mathfrak{F}_{\mathscr{L}}\colon U_0 \longmapsto U^0$, is continuous.

Let $u_0 \in L_2(\Omega)$. Then the solution $u^0 \in V_{\mathscr{L}}(\Omega, S)$ of the variational problem

$$\|u - u_0\|_{L_2(\Omega)} \xrightarrow[u \in V_{\mathscr{L}}(\Omega,S)]{} \min \tag{4.2}$$

can be represented in the form $u^0 = \check{\mathscr{L}} \mathfrak{F}_{\mathscr{L}}(\check{E} * u_0)$.

PROOF. 1. *Existence.* We shall prove that the function

$$U^0 = U_0 + \check{E} * \mathfrak{P}_{V_{\mathscr{L}}(\Omega,S)^*}(-\check{\mathscr{L}} U_0) \tag{4.3}$$

satisfies all the conditions.

Indeed, by (4.3) it follows that $\check{\mathscr{L}} U^0 = \check{\mathscr{L}} U_0 + \mathfrak{P}_{V_{\mathscr{L}}(\Omega,S)^*}(-\check{\mathscr{L}} U_0)$. Hence

$$\check{\mathscr{L}} U^0 = \mathfrak{P}_{V_{\mathscr{L}}(\Omega,S)}(-\check{\mathscr{L}} U_0). \tag{4.4}$$

Therefore, $\check{\mathscr{L}} U^0 \in V_{\mathscr{L}}(\Omega, S)$ and the function U^0 is $\mathscr{L}\check{\mathscr{L}}$-positive.

Since $P_{\mathscr{L}}(\Omega) \subset V_{\mathscr{L}}(\Omega, S)$, we have $P_{\mathscr{L}}(\Omega)^\perp \subset V_{\mathscr{L}}(\Omega, S)^*$. Therefore,

$$\mathfrak{P}_{V_{\mathscr{L}}(\Omega,S)^*}(-\check{\mathscr{L}} U_0) \in P_{\mathscr{L}}(\Omega)^\perp.$$

Thus $U^0 \in U_0 + \overset{\circ}{W}{}^l_2(\Omega)$.

From (4.4) we get

$$\check{\mathscr{L}} U^0 \in \check{\mathscr{L}} U_0 + V_{\mathscr{L}}(\Omega, S)\{\check{\mathscr{L}} U^0\}^*.$$

From Theorem 4.1 it follows that $\check{\mathscr{L}} V = \check{\mathscr{L}} U^0 - \check{\mathscr{L}} U_0$, where the function $V \in \overset{\circ}{W}{}^l_2(\Omega)$ satisfies the relations $V|_S \geqslant 0$ and $V|_{\operatorname{supp} \mathscr{L}\check{\mathscr{L}} U^0} = 0$. Therefore, $U^0|_S \geqslant U_0|_S$ and

$$\operatorname{supp} \mathscr{L}\check{\mathscr{L}} U^0 \subset \{\, x \in \Omega : U^0(x) = U_0(x) \,\}.$$

The inclusion $\operatorname{supp} \mathscr{L}\check{\mathscr{L}} U^0 \subset S$ follows from the inclusion $\check{\mathscr{L}} U^0 \in V_{\mathscr{L}}(\Omega, S)$.

2. *Uniqueness.* Let the $\mathscr{L}\check{\mathscr{L}}$-positive function $U^1 \in U_0 + \overset{\circ}{W}{}^l_2(\Omega)$ satisfy the inequality $U^1|_S \geqslant U_0|_S$ and the inclusion

$$\operatorname{supp} \mathscr{L}\check{\mathscr{L}} U^1 \subset \{\, x \in S : U^1(x) = U_0(x) \,\}.$$

We shall prove that $U^1 = U^0$, where the function U^0 is defined by (4.3). To this end we prove that

$$\check{\mathscr{L}} U^1 \in \check{\mathscr{L}} U_0 + V_{\mathscr{L}}(\Omega, S)\{\check{\mathscr{L}} U^1\}^*. \tag{4.5}$$

Let $V = U^1 - U_0$. Then the function $V \in \overset{\circ}{W}{}^l_2(\Omega)$ satisfies $V|_S \geqslant 0$ and $V|_{\operatorname{supp} \mathscr{L}\check{\mathscr{L}} U^1} = 0$. By Theorem 4.1, we have $V \in V_{\mathscr{L}}(\Omega, S)\{\check{\mathscr{L}} U^1\}^*$. Thus (4.5) follows from the relation $\check{\mathscr{L}} U^1 - \check{\mathscr{L}} U_0 = \check{\mathscr{L}} V$.

By (4.5) it follows that $\check{\mathscr{L}} U^1 = \mathfrak{P}_{V_{\mathscr{L}}(\Omega,S)}(\check{\mathscr{L}} U_0)$. Therefore, $\check{\mathscr{L}} U^1 = \check{\mathscr{L}} U^0$. Thus, $U^1 - U_0 \in \overset{\circ}{W}{}^l_2(\Omega)$, $U^0 - U_0 \in \overset{\circ}{W}{}^l_2(\Omega)$, and $\check{\mathscr{L}}(U^1 - U_0) = \check{\mathscr{L}}(U^0 - U_0)$. Therefore, $U^1 - U_0 = U^0 - U_0$, and so $U^1 = U^0$.

3. *Variational problems.* By (4.4) it follows that the function $\check{\mathscr{L}} U^0$ is the solution of the variational problem

$$\|u\|_{L_2(\Omega)} \xrightarrow[u\in\check{\mathscr{L}} U_0+V_{\mathscr{L}}(\Omega,S)^*]{} \min .$$

Let the function $U \in U_0 + \overset{\circ}{W}{}^l_2(\Omega)$ satisfy the inequality $U\big|_S \geqslant U_0\big|_S$. We shall prove that

$$\check{\mathscr{L}} U \in \check{\mathscr{L}} U_0 + V_{\mathscr{L}}(\Omega, S).$$

Indeed, if $u \in V_{\mathscr{L}}(\Omega, S)$, then $\operatorname{supp} \mathscr{L} u \subset S$. Therefore,

$$\big(\check{\mathscr{L}}(U - U_0), u\big)_{L_2(\Omega)} = \int_\Omega (U - U_0)\, d\mu_{\mathscr{L} u} \geqslant 0.$$

By the inclusion $\check{\mathscr{L}} U \in \check{\mathscr{L}} U_0 + V_{\mathscr{L}}(\Omega, S)^*$,

$$\|\check{\mathscr{L}} U^0\|_{L_2(\Omega)} \leqslant \|\check{\mathscr{L}} U\|_{L_2(\Omega)}.$$

Therefore, U^0 is the solution of problem (4.1).

By (4.3),

$$\mathfrak{F}_{\mathscr{L}}(U) = U + \check{E} * \mathfrak{P}_{V_{\mathscr{L}}(\Omega,S)^*}(-\check{\mathscr{L}} U).$$

Therefore, the mapping $\mathfrak{F}_{\mathscr{L}}\colon W^l_2(\Omega) \to W^l_2(\Omega)$ is continuous.

Let $u_0 \in L_2(\Omega)$ and $U_0 = \check{E} * u_0$. Then by (4.4) we have $u^0 = \check{\mathscr{L}} U^0$, where $U^0 = \mathfrak{F}_{\mathscr{L}}(U_0)$ is the solution of the variational problem (4.2). □

COROLLARY 4.1. *If $0 \leqslant s < 2l - n/2$, then the mapping*

$$\mathfrak{F}_{\mathscr{L}}\colon W^l_2(\Omega) \to W^l_2(\Omega) \cap W^s_2(\Omega)^{\mathrm{loc}}$$

is continuous.

PROOF. Indeed, by Theorem 4.2, the mapping $\mathfrak{F}_{\mathscr{L}}\colon W^l_2(\Omega) \to W^l_2(\Omega)$ is continuous and $\mathfrak{F}_{\mathscr{L}}(U) \in \overset{+}{\mathscr{D}}{}'_{\mathscr{L}\check{\mathscr{L}}}(\Omega)$ for all $U \in W^l_2(\Omega)$. Therefore, the mapping

$$\mathfrak{F}_{\mathscr{L}}\colon W^l_2(\Omega) \to W^l_2(\Omega) \cap \overset{+}{\mathscr{D}}{}'_{\mathscr{L}\check{\mathscr{L}}}(\Omega)$$

is sequentially continuous. By Theorem 2.2 the mapping in: $\overset{+}{\mathscr{D}}{}'_{\mathscr{L}\check{\mathscr{L}}}(\Omega) \to W^s_2(\Omega)^{\mathrm{loc}}$ is sequentially continuous for all $0 \leqslant s < 2l - n/2$. Therefore, the mapping

$$\mathfrak{F}_{\mathscr{L}}\colon W^l_2(\Omega) \to W^l_2(\Omega) \cap W^s_2(\Omega)^{\mathrm{loc}}$$

is continuous. □

MINIMIZATION ALGORITHM. Let $u_0 \in L_2(\Omega)$. By Theorem 4.2, the solution u^0 of the variational problem (4.2) can be represented in the form

$$u^0 = \check{\mathscr{L}} \mathfrak{F}_{\mathscr{L}}(\check{E} * u_0).$$

Therefore, the algorithm of calculation the function u^0 by the function u_0 consists in three steps.

1. At the first step we define $U_0 = \check{E} * u_0$.

2. At the second step we define the $\mathscr{L}\check{\mathscr{L}}$-positive function $U^0 \in U_0 + \overset{\circ}{W}{}^l_2(\Omega)$ such that $U^0\big|_S \geqslant U_0\big|_S$ and $\operatorname{supp} \mathscr{L}\check{\mathscr{L}} U^0 \subset \{\, x \in S : U^0(x) = U_0(x) \,\}$.

3. At the third step we define the solution $u^0 = \check{\mathscr{L}} U^0$.

The operators $\check{E}*$ and $\check{\mathscr{L}}$ in the first and at the third steps of the algorithm are linear, and their superposition is equal to the unit operator in $L_2(\Omega)$. The operator $\mathfrak{F}_{\mathscr{L}}$ in the second step is nonlinear.

4.3. Variational problems with a restriction on the measure. Let $0 \leqslant \lambda \leqslant \infty$. Recall that

$$V_{\mathscr{L}}^{\lambda}(\Omega) = \{ u \in V_{\mathscr{L}}(\Omega) : \mu_{\mathscr{L}u}\{\Omega\} \leqslant \lambda \},$$
$$K_{\mathscr{L}}^{\lambda}(\Omega) = \{ u \in K_{\mathscr{L}}(\Omega) : \mu_{\mathscr{L}u}\{\Omega\} \leqslant \lambda \}.$$

Define

$$V_{\mathscr{L}}^{\lambda}(\Omega, S) = V_{\mathscr{L}}^{\lambda}(\Omega) \cap P_{\mathscr{L}}(\Omega, S), \qquad K_{\mathscr{L}}^{\lambda}(\Omega, S) = K_{\mathscr{L}}^{\lambda}(\Omega) \cap P_{\mathscr{L}}(\Omega, S).$$

Note that the set $V_{\mathscr{L}}^{\lambda}(\Omega, S)$ is closed and for $\lambda < \infty$ the set $K_{\mathscr{L}}^{\lambda}(\Omega, S)$ is compact in the space $L_2(\Omega)$. Obviously,

$$V_{\mathscr{L}}^{\lambda}(\Omega, S) = P_{\mathscr{L}}(\Omega) \oplus K_{\mathscr{L}}^{\lambda}(\Omega, S).$$

Besides,

$$V_{\mathscr{L}}^{0}(\Omega, S) = P_{\mathscr{L}}(\Omega), \qquad V_{\mathscr{L}}^{\infty}(\Omega, S) = V_{\mathscr{L}}(\Omega, S).$$

THEOREM 4.3. *Let $0 < \lambda < \infty$ and $u \in V_{\mathscr{L}}^{\lambda}(\Omega, S)$. If $\mu_{\mathscr{L}u}\{\Omega\} < \lambda$, then $V_{\mathscr{L}}^{\lambda}(\Omega, S)\{u\}^* = V_{\mathscr{L}}(\Omega, S)\{u\}^*$. If $\mu_{\mathscr{L}u}\{\Omega\} = \lambda$, then $v \in V_{\mathscr{L}}^{\lambda}(\Omega, S)\{u\}^*$ if and only if the function v can be represented in the form $v = \check{\mathscr{L}} V$, where the function $V \in \overset{\circ}{W}{}_2^l(\Omega)$ is constant on $\operatorname{supp} \mathscr{L}u$ and*

$$V\big|_{\operatorname{supp} \mathscr{L}u} = \inf_{x \in S} V(x) \leqslant 0$$

PROOF. 1. Let $\mu_{\mathscr{L}u}\{\Omega\} < \lambda$. We prove the equality

$$\operatorname{Cl} V_{\mathscr{L}}^{\lambda}(\Omega, S)\{u\} = \operatorname{Cl} V_{\mathscr{L}}(\Omega, S)\{u\}.$$

Obviously, $V_{\mathscr{L}}^{\lambda}(\Omega, S)\{u\} \subset V_{\mathscr{L}}(\Omega, S)\{u\}$. Therefore,

$$\operatorname{Cl} V_{\mathscr{L}}^{\lambda}(\Omega, S)\{u\} \subset \operatorname{Cl} V_{\mathscr{L}}(\Omega, S)\{u\}.$$

Let $v \in V_{\mathscr{L}}(\Omega, S)\{u\}$. Then $v = \gamma(w - u)$, where $\gamma > 0$ and $w \in V_{\mathscr{L}}(\Omega, S)$. Theorem 3.11 implies that there exists a sequence $\mathscr{L}$-positive $\mathscr{L}$-splines $w_1, w_2, \dots$ such that $w = \lim_{k \to \infty} w_k$ in $L_2(\Omega)$ and $\operatorname{supp} \mathscr{L}w_k \subset \operatorname{supp} \mathscr{L}w$ for all $k = 1, 2, \dots$.

We shall prove that $w_k - u \in V_{\mathscr{L}}^{\lambda}(\Omega, S)\{u\}$. Indeed, let $0 < \varepsilon_k < 1$ and $\varepsilon_k \mu_{\mathscr{L}w_k}\{\Omega\} \leqslant \lambda - \mu_{\mathscr{L}u}\{\Omega\}$. Then $\varepsilon_k w_k + (1 - \varepsilon_k)u \in V_{\mathscr{L}}^{\lambda}(\Omega, S)$, because

$$\mu_{\mathscr{L}\{\varepsilon_k w_k + (1-\varepsilon_k)u\}}\{\Omega\} = \varepsilon_k \mu_{\mathscr{L}w_k}\{\Omega\} + (1 - \varepsilon_k)\mu_{\mathscr{L}u}\{\Omega\} \leqslant \lambda.$$

From the relation $\{\varepsilon_k w_k + (1 - \varepsilon_k)u\} - u = \varepsilon_k(w_k - u)$ it follows that $w_k - u \in V_{\mathscr{L}}^{\lambda}(\Omega, S)\{u\}$. Therefore, $w - u \in \operatorname{Cl} V_{\mathscr{L}}^{\lambda}(\Omega, S)\{u\}$. Hence, $v \in \operatorname{Cl} V_{\mathscr{L}}^{\lambda}(\Omega, S)\{u\}$. Thus,

$$\operatorname{Cl} V_{\mathscr{L}}^{\lambda}(\Omega, S)\{u\} \supset \operatorname{Cl} V_{\mathscr{L}}(\Omega, S)\{u\},$$

and so

$$\operatorname{Cl} V_{\mathscr{L}}^{\lambda}(\Omega, S)\{u\} = \operatorname{Cl} V_{\mathscr{L}}(\Omega, S)\{u\}.$$

2. Let $\mu_{\mathscr{L}u}\{\Omega\} = \lambda$. Because $\lambda > 0$, the set $\operatorname{supp} \mathscr{L}u$ is not empty.

Necessity. Let $v \in V^{\lambda}_{\mathscr{L}}(\Omega, s)\{u\}^*$. Note that $P_{\mathscr{L}}(\Omega) \subset V^{\lambda}_{\mathscr{L}}(\Omega, S)\{u\}$. Therefore, $P_{\mathscr{L}}(\Omega)^{\perp} \supset V^{\lambda}_{\mathscr{L}}(\Omega, S)\{u\}^*$. Hence, $v = \check{\mathscr{L}} V$, where $V = \check{E} * v \in \overset{\circ}{W}{}^l_2(\Omega)$.

Let $w \in V^{\lambda}_{\mathscr{L}}(\Omega, S)$. Then

$$(w-u, v)_{L_2(\Omega)} = (w-u, \check{\mathscr{L}} u)_{L_2(\Omega)} = \int_\Omega V \, d\mu_{\mathscr{L}w} - \int_\Omega V \, d\mu_{\mathscr{L}u} \geqslant 0.$$

Thus,

$$\int_\Omega V \, d\mu_{\mathscr{L}w} \geqslant \int_\Omega V \, d\mu_{\mathscr{L}u}. \tag{4.6}$$

Let $\upsilon = \inf_{x \in S} V(x)$. Then by (4.6) we obtain

$$\int_\Omega V \, d\mu_{\mathscr{L}w} \geqslant \upsilon \mu_{\mathscr{L}u}\{\Omega\} = \upsilon\lambda.$$

In particular, if $w = 0$ then $\upsilon \leqslant 0$.

Let $x_0 \in \operatorname{supp} \mathscr{L} u$. We shall prove that $V(x_0) = \upsilon$.

Let $V(x_0) > \upsilon$. The function V is continuous. Therefore, there exist $\delta > 0$ and a neighborhood $\omega \Subset \Omega$ of the point x_0 such that $V(x) \geqslant \upsilon + \delta$ for all $x \in \omega$. Therefore,

$$\begin{aligned}
\int_\Omega V \, d\mu_{\mathscr{L}u} &= \int_{\Omega\setminus\omega} V \, d\mu_{\mathscr{L}u} + \int_\omega V \, d\mu_{\mathscr{L}u} \\
&\geqslant \upsilon \mu_{\mathscr{L}u}\{\Omega \setminus \omega\} + (\upsilon + \delta)\mu_{\mathscr{L}u}\{\omega\} \\
&= \upsilon \mu_{\mathscr{L}u}\{\Omega\} + \delta \mu_{\mathscr{L}u}\{\omega\} = \upsilon\lambda + \delta\mu_{\mathscr{L}u}\{\omega\}.
\end{aligned}$$

Since $x_0 \in \operatorname{supp} \mathscr{L} u$, it follows that $\mu_{\mathscr{L}u}\{\omega\} > 0$. Let $x_1 \in S$ be a point such that $V(x_1) - \upsilon < \delta\mu_{\mathscr{L}u}\{\omega\}/\lambda$. Define $w = \lambda E(\cdot - x_1)\big|_\Omega$. Then $w \in V^{\lambda}_{\mathscr{L}}(\Omega, S)$ and

$$\int_\Omega V \, d\mu_{\mathscr{L}w} = \lambda V(x_1) < \upsilon\lambda + \delta\mu_{\mathscr{L}u}\{\omega\}.$$

Therefore,

$$\int_\Omega V \, d\mu_{\mathscr{L}u} \geqslant \upsilon\lambda + \delta\mu_{\mathscr{L}u}\{\omega\}.$$

These two inequalities contradict (4.6). Hence, the assumption $V(x_0) > \upsilon$ is wrong, and we have the equality $V(x_0) = \upsilon$.

Sufficiency. Let $v = \check{\mathscr{L}} V$, where the function V satisfies all the conditions. Let $w \in V^{\lambda}_{\mathscr{L}}(\Omega, S)$. We prove that $(w-u, v)_{L_2(\Omega)} \geqslant 0$. Indeed,

$$\begin{aligned}
(w-u, v)_{L_2(\Omega)} &= (w-u, \check{\mathscr{L}} V)_{L_2(\Omega)} = \int_\Omega V \, d\mu_{\mathscr{L}w} - \int_\Omega V \, d\mu_{\mathscr{L}u} \\
&= \int_\Omega V \, d\mu_{\mathscr{L}w} - \upsilon\lambda \geqslant \upsilon\mu_{\mathscr{L}w}\{\Omega\} - \upsilon\lambda = -\upsilon(\lambda - \mu_{\mathscr{L}w}\{\Omega\}) \geqslant 0,
\end{aligned}$$

where $\upsilon = \inf_{x \in S} V(x)$. Therefore, $v \in V^{\lambda}_{\mathscr{L}}(\Omega, S)\{u\}^*$. □

THEOREM 4.4. *Let $0 \leqslant \tau \leqslant \infty$. Given $U_0 \in \overset{\circ}{W}{}^l_2(\Omega)$, there exists a unique $\mathscr{L}\check{\mathscr{L}}$-positive function $U^0_\tau \in U_0 + \overset{\circ}{W}{}^l_2(\Omega)$ such that $U^0_\tau\big|_S \geqslant U_0\big|_S - \tau$ and* $\operatorname{supp} \mathscr{L}\check{\mathscr{L}} U^0_\tau \subset$

$\{ x \in S : U_\tau^0(x) = U_0(x) - \tau \}$. *The function* U_τ^0 *is the solution of the variational problem*

$$(4.7) \qquad \|\check{\mathscr{L}} U\|_{L_2(\Omega)} \xrightarrow[U \in U_0 + \overset{\circ}{W}{}_2^l(\Omega) : U|_S \geqslant U_0|_S - \tau]{} \min .$$

The continuous function $\Lambda(\tau) = \mu_{\mathscr{L}\check{\mathscr{L}} U_\tau^0}\{\Omega\}$ *strictly decreases to* 0 *on the interval* $[0, \max\{0, \tau_\infty\}]$, *where*

$$\tau_\infty = - \inf_{x \in S} \{ U_\infty^0(x) - U_0(x) \}.$$

The mapping

$$\mathfrak{F}_{\mathscr{L}}^\tau : W_2^l(\Omega) \longrightarrow W_2^l(\Omega), \quad \mathfrak{F}_{\mathscr{L}} : U_0 \longmapsto U_\tau^0,$$

is continuous.

Let $u_0 \in L_2(\Omega)$ *and* $\lambda = \Lambda(\tau)$. *The solution* $u_\tau^0 \in V_{\mathscr{L}}^\lambda(\Omega, S)$ *of the variational problem*

$$(4.8) \qquad \|u - u_0\|_{L_2(\Omega)} \xrightarrow[u \in V_{\mathscr{L}}^\lambda(\Omega, S)]{} \min$$

can be represented in the form $u^0 = \check{\mathscr{L}} \mathfrak{F}_{\mathscr{L}}^\tau (\check{E} * u_0)$.

PROOF. 1. Let $0 \leqslant \lambda \leqslant \infty$ and $v_0 \in P_{\mathscr{L}}(\Omega)^\perp$. Define

$$v(\lambda)^0 = \mathfrak{P}_{V_{\mathscr{L}}^\lambda(\Omega,S)}(v_0) = \mathfrak{P}_{K_{\mathscr{L}}^\lambda(\Omega,S)}(v_0).$$

Let $V_0 = \check{E} * v_0$, $V(\lambda)^0 = \check{E} * v(\lambda)^0$. Then $V_0 \in \overset{\circ}{W}{}_2^l(\Omega)$ and $V(\lambda)^0 \in \overset{\circ}{W}{}_2^l(\Omega)$.

Obviously, $v(0)^0 = 0$ and $v(\infty)^0 = \mathfrak{P}_{V_{\mathscr{L}}(\Omega,S)}(v_0) = \mathfrak{P}_{K_{\mathscr{L}}(\Omega,S)}(v_0)$. Define

$$T(\lambda) = - \inf_{x \in S} \{ V(\lambda)^0(x) - V_0(x) \}.$$

Then

$$(4.9) \qquad V(\lambda)^0|_S \geqslant V_0|_S - T(\lambda).$$

We prove that

$$(4.10) \qquad \operatorname{supp} \mathscr{L} \check{\mathscr{L}} V(\lambda)^0 \subset \{ x \in S : V(\lambda)^0(x) = V_0(x) - T(\lambda) \}.$$

Let $\lambda_\infty = \mu_{\mathscr{L} v(\lambda)^0}\{\Omega\}$ and $\lambda \geqslant \lambda_\infty$. Then

$$v(\lambda)^0 = v(\infty)^0 \in V_{\mathscr{L}}^\lambda(\Omega, S).$$

Therefore, Theorem 4.2 implies the equality $T(\lambda) = 0$ and the inclusion (4.10). Suppose that $0 \leqslant \lambda < \lambda_\infty$. Then Theorem 4.3 and the inclusion

$$v(\lambda)^0 - v_0 \in V_{\mathscr{L}}^\lambda(\Omega, S)\{v(\lambda)^0\}^*$$

imply that $\mu_{\mathscr{L} v(\lambda)^0}\{\Omega\} = \lambda$. Indeed, if $\mu_{\mathscr{L} v(\lambda)^0}\{\Omega\} < \lambda$, then $v(\lambda)^0 = v(\infty)^0$. Therefore, we have a contradiction. Thus $\mu_{\mathscr{L} v(\lambda)^0}\{\Omega\} = \lambda$. Hence, Theorem 4.3 implies the inequality $T(\lambda) \geqslant 0$ and the inclusion (4.10).

2. Let us prove that the function $V(\lambda)^0$ is the solution of the variational problem

$$\|\check{\mathscr{L}} V\|_{L_2(\Omega)} \xrightarrow[V\in \overset{\circ}{W}{}_2^l(\Omega):V|_S \geqslant V_0|_S - T(\lambda)]{} \min .$$

Indeed, if $\lambda \geqslant \lambda_\infty$, then $T(\lambda) = 0$ and the statement follows from Theorem 4.2. Let $0 \leqslant \lambda < \lambda_\infty$. It is sufficient to prove that

$$(\check{\mathscr{L}} V, \check{\mathscr{L}} V(\lambda)^0)_{L_2(\Omega)} \geqslant (\check{\mathscr{L}} V(\lambda)^0, \check{\mathscr{L}} V(\lambda)^0)_{L_2(\Omega)}$$

for all functions $V \in \overset{\circ}{W}{}_2^l(\Omega)$ such that $V|_S \geqslant V_0|_S - T(\lambda)$. For $0 \leqslant \lambda < \lambda_\infty$ we have proved that $\lambda = \mu_{\mathscr{L} V(\lambda)}\{\Omega\}$. Hence the required inequality follows from the inequality

$$(\check{\mathscr{L}} V, \check{\mathscr{L}} V(\lambda)^0)_{L_2(\Omega)} = \int_\Omega V \, d\mu_{\mathscr{L}\check{\mathscr{L}} V(\lambda)^0} \geqslant \int_\Omega V_0 \, d\mu_{\mathscr{L}\check{\mathscr{L}} V(\lambda)^0} - \lambda T(\lambda)$$

and the equality

$$(\check{\mathscr{L}} V, \check{\mathscr{L}} V(\lambda)^0)_{L_2(\Omega)} = \int_\Omega V \, d\mu_{\mathscr{L}\check{\mathscr{L}} V(\lambda)^0} = \int_\Omega V_0 \, d\mu_{\mathscr{L}\check{\mathscr{L}} V(\lambda)^0} - \lambda T(\lambda).$$

3. Let us prove that the mapping $f : \lambda \longmapsto \mathfrak{P}_{K^\lambda_{\mathscr{L}}(\Omega,S)}(v_0)$ from $[0,\infty]$ into $P_{\mathscr{L}}(\Omega)^\perp$ is continuous. Obviously, $K^\lambda_{\mathscr{L}}(\Omega,S) = \lambda K^1_{\mathscr{L}}(\Omega,S)$ for all $0 < \lambda < \infty$. Then

$$\mathfrak{P}_{K^\lambda_{\mathscr{L}}(\Omega,S)}(v_0) = \mathfrak{P}_{\lambda K^1_{\mathscr{L}}(\Omega,S)}(\lambda v_0/\lambda) = \lambda \mathfrak{P}_{K^1_{\mathscr{L}}(\Omega,S)}(v_0/\lambda).$$

The set $K^1_{\mathscr{L}}(\Omega,S)$ is compact. Therefore, there is a constant C such that $\|v\|_{L_2(\Omega)} \leqslant C$ for all $v \in K^1_{\mathscr{L}}(\Omega,S)$. From the inequality

$$|f(\lambda) - f(0)| = \|\mathfrak{P}_{K^\lambda_{\mathscr{L}}(\Omega,S)}(v_0)\|_{L_2(\Omega)} = \lambda \|\mathfrak{P}_{\lambda K^1_{\mathscr{L}}(\Omega,S)}(v_0)\|_{L_2(\Omega)} \leqslant C\lambda$$

it follows that the mapping is continuous at the point $\lambda_0 = 0$. From the inequality

$$\begin{aligned}
|f(\lambda) - f(0)| &= \|\mathfrak{P}_{K^\lambda_{\mathscr{L}}(\Omega,S)}(v_0) - \mathfrak{P}_{K^{\lambda_0}_{\mathscr{L}}(\Omega,S)}(v_0)\|_{L_2(\Omega)} \\
&= \|\lambda \mathfrak{P}_{K^1_{\mathscr{L}}(\Omega,S)}(v_0/\lambda) - \lambda_0 \mathfrak{P}_{K^1_{\mathscr{L}}(\Omega,S)}(v_0/\lambda_0)\|_{L_2(\Omega)} \\
&\leqslant \|\lambda \mathfrak{P}_{K^1_{\mathscr{L}}(\Omega,S)}(v_0/\lambda) - \lambda_0 \mathfrak{P}_{K^1_{\mathscr{L}}(\Omega,S)}(v_0/\lambda)\|_{L_2(\Omega)} \\
&\quad + \|\lambda_0 \mathfrak{P}_{K^1_{\mathscr{L}}(\Omega,S)}(v_0/\lambda) - \lambda_0 \mathfrak{P}_{K^1_{\mathscr{L}}(\Omega,S)}(v_0/\lambda)\|_{L_2(\Omega)} \\
&= |1 - \lambda_0/\lambda| \|v_0\|_{L_2(\Omega)} + |\lambda_0/\lambda - 1| \|v_0\|_{L_2(\Omega)} \\
&= 2|1 - \lambda_0/\lambda| \|v_0\|_{L_2(\Omega)}
\end{aligned}$$

it follows that the mapping is continuous at any point $\lambda_0 > 0$.

Let us prove that the mapping is continuous at the point $\lambda_0 = \infty$. Let $\{\gamma_{x_i} : x_1, x_2, \ldots \in S\}$ be the generating family of the cone $K_{\mathscr{L}}(\Omega, S)$. Then

$$v(\infty)^0 = \lim_{k\to\infty} \mathfrak{P}_{\mathrm{co}\{\gamma_{x_1},\ldots,\gamma_{x_k}\}}(v_0)$$

in $L_2(\Omega)$. Let

$$\delta_k = \left\|\mathfrak{P}_{\mathrm{co}\{\gamma_{x_1},\ldots,\gamma_{x_k}\}}(v_0) - v_0\right\|_{L_2(\Omega)} - \left\|v(\infty)^0 - v_0\right\|_{L_2(\Omega)}.$$

Given k, there exists λ_k such that $\mathfrak{P}_{\mathrm{co}\{\gamma_{x_1},\ldots,\gamma_{x_k}\}}(v_0) \in V^{\lambda}_{\mathscr{L}}(\Omega, S)$ for all $\lambda \geqslant \lambda_k$. Therefore,

$$\left\|v(\lambda)^0 - v_0\right\|_{L_2(\Omega)} \leqslant \left\|v(\infty)^0 - v_0\right\|_{L_2(\Omega)} + \delta_k$$

for all $\lambda \geqslant \lambda_k$. Consequently,

$$\left\|v(\lambda)^0 - v_0\right\|^2_{L_2(\Omega)} \leqslant 2\delta_k \left\|v(\lambda)^0 - v_0\right\|_{L_2(\Omega)} + \delta_k^2$$

for all $\lambda \geqslant \lambda_k$. Let $\varepsilon > 0$, and let k satisfy the inequality

$$2\delta_k \left\|v(\infty)^0 - v_0\right\|_{L_2(\Omega)} + \delta_k^2 \leqslant \varepsilon^2.$$

Then $\left\|v(\lambda)^0 - v_0\right\|_{L_2(\Omega)} \leqslant \varepsilon$ for all $\lambda \geqslant \lambda_k$. Therefore, f is continuous at $\lambda_0 = \infty$.

4. *The function* T. Let us prove that the function T is continuous on $[0, \infty)$. Indeed, the mapping $\lambda \longmapsto v(\lambda)^0 - v_0$ is continuous from $[0, \infty)$ into $P_{\mathscr{L}}(\Omega)^{\perp}$. Therefore, the mapping $\lambda \longmapsto V(\lambda)^0 - V_0$ is continuous from $[0, \infty)$ into $\overset{\circ}{W}{}^l_2(\Omega)$. Hence, by the embedding $\overset{\circ}{W}{}^l_2(\Omega) \hookrightarrow C^0(\mathrm{Cl}\,\Omega)$ it follows that T is continuous.

Note that $T(\lambda) = 0$ for $\lambda_\infty \leqslant \lambda \leqslant \infty$. We shall prove that the function T strictly decreases on $[0, \lambda_\infty]$. The case $\lambda_\infty = 0$ is trivial. Let $\lambda_\infty > 0$.

Let $0 \leqslant \lambda_1 < \lambda_2 < \lambda_\infty$. Suppose that $T(\lambda_1) = T(\lambda_2) = \tau$. Then both $V(\lambda_1)^0$ and $V(\lambda_2)^0$ are solutions of the variational problem

$$\left\|\check{\mathscr{L}} V\right\|_{L_2(\Omega)} \xrightarrow[V \in \overset{\circ}{W}{}^l_2(\Omega):\, V|_S \geqslant V_0|_S - \tau]{} \min.$$

Therefore, $V(\lambda_1)^0 = V(\lambda_2)^0$. Consequently, $v(\lambda_1)^0 = v(\lambda_2)^0$. For $0 \leqslant \lambda < \lambda_\infty$ we have proved that $\mu_{\check{\mathscr{L}}v}\{\Omega\} = \lambda$. Therefore, $\lambda_1 = \lambda_2$. This contradiction shows that $T(\lambda_1) \neq T(\lambda_2)$. The function T is positive, and $\lim_{\lambda \nearrow \lambda_\infty} T(\lambda) = 0$. Therefore, T strictly decreases to 0 on $[0, \lambda_\infty]$.

5. *Existence*. Let $\tau = \infty$. Then a unique function U^0_∞ exists and has the form

$$U^0_\infty = U_0 + \check{E} * \mathfrak{P}_{P_{\mathscr{L}}(\Omega)^{\perp}}(-\mathscr{L} U_0).$$

Therefore, for $\tau \geqslant \max\{0, \tau_\infty\}$ the function $U^0_\tau = U^0_\infty$ satisfies all the conditions of the theorem. Later we shall prove that U^0_τ is unique for $\max\{0, \tau_\infty\} \leqslant \tau < \infty$.

Let $\tau = 0$. Then, by Theorem 4.2, it follows that a unique function U^0 exists and has the form

$$U^0 = U_0 + \check{E} * \mathfrak{P}_{V_{\mathscr{L}}(\Omega,S)}(-\mathscr{L} U_0).$$

Thus the existence of the function U^0_τ needs to be proved only in the case $0 < \tau_\infty$ for $0 < \tau < \tau_\infty$.

Let $V_0 = U_0 - U^0_\infty$ and $v_0 = \mathscr{L} V_0$. By the inclusion $V_0 \in \overset{\circ}{W}{}^l_2(\Omega)$, it follows that $v_0 \in P_{\mathscr{L}}(\Omega)^{\perp}$ and $V_0 = \check{E} * v_0$.

Let $0 \leqslant \lambda \leqslant \infty$. Define $W(\lambda)^0 = V(\lambda)^0 + U^0_\infty$. Then

$$T(\lambda) = \inf_{x \in S} \{ W(\lambda)^0(x) - U_0(x) \}.$$

Therefore,

$$W(\lambda)^0\big|_S \geqslant U_0\big|_S - T(\lambda) \tag{4.11}$$

and

$$\operatorname{supp} \mathscr{L}\check{\mathscr{L}} W(\lambda)^0 \subset \{ x \in S : W(\lambda)^0(x) = U_0(x) - T(\lambda) \}. \tag{4.12}$$

Note that the function $W(\lambda)^0 \in U_0 + \overset{\circ}{W}{}^l_2(\Omega)$ is $\mathscr{L}\check{\mathscr{L}}$-positive.

By using the definition of τ_∞ and the equation $W(\lambda)^0 = U^0_\infty$ we get the equation $T(0) = \tau_\infty$. Therefore, T is continuous and strictly decreases on $[0, \lambda_\infty]$ from τ_∞ to 0. Hence, given $\tau \in (0, \tau_\infty)$, there exists a unique point $\lambda \in (0, \lambda_\infty)$ such that $\tau = T(\lambda)$. By (4.11) and (4.12) it follows that the function $U_\tau = W(\lambda)$ satisfies all the requirements of the theorem.

6. *Variational problems.* Let us prove that the function $W(\lambda)^0$ is the solution of the variational problem

$$\|\check{\mathscr{L}} U\|_{L_2(\Omega)} \xrightarrow[U \in U_0 + \overset{\circ}{W}{}^l_2(\Omega) : U|_S \geqslant U_0|_S - T(\lambda)]{} \min.$$

Let the function $W_1(\lambda)^0$ be the solution of this problem. Then the function $V_1(\lambda)^0 = W_1(\lambda)^0 - U^0_\infty$ is the solution of the problem

$$\|\check{\mathscr{L}} V\|_{L_2(\Omega)} \xrightarrow[V \in \overset{\circ}{W}{}^l_2(\Omega) : V|_S \geqslant V_0|_S - T(\lambda)]{} \min.$$

Therefore, $V_1(\lambda)^0 = V(\lambda)^0$. Thus, $W_1(\lambda)^0 = W(\lambda)^0$.

Let us prove the uniqueness of the function U^0_τ. For $\tau > \tau_\infty$ this is obvious, and for $\tau = 0$ it follows from Theorem 4.2. Consider the case $0 > \tau \geqslant \tau_\infty$. In this case $U^0_\tau|_S \geqslant U_0|_S - T(\lambda)$ and $\operatorname{supp} \mathscr{L}\check{\mathscr{L}} U^0_\tau \subset \{ x \in S : U^0_\tau(x) = U(x) - T(\lambda) \}$, where the number $\lambda \in [0, \lambda_\infty)$ is such that $\tau = T(\lambda)$. Therefore, U^0_τ is the solution of the variational problem

$$\|\check{\mathscr{L}} U\|_{L_2(\Omega)} \xrightarrow[U \in U_0 + \overset{\circ}{W}{}^l_2(\Omega) : U|_S \geqslant U_0|_S - \tau]{} \min.$$

Hence the function U^0_τ is unique.

On the interval $[0, \infty]$ the function T is continuous and strictly decreases to 0. Moreover, $T(\lambda) = \tau$, where $\lambda \in [0, \lambda_\infty]$, implies $\lambda = \mu_{\mathscr{L}\check{\mathscr{L}} U^0_\tau}\{\Omega\}$, and $\lambda_\infty > 0$ implies the inequality $\tau_\infty = T(0) > 0$. Let Λ be the function inverse to T. Then on $[0, \tau_\infty]$ the function Λ is continuous and strictly decreases to 0. From $\Lambda(\tau) = \lambda$, where $\tau \in [0, \tau_\infty]$, it follows that $\tau = T(\lambda)$, where $\lambda \in [0, \lambda_\infty]$. Hence, $\lambda = \mu_{\mathscr{L}\check{\mathscr{L}} U^0_\tau}\{\Omega\}$. Thus, $\Lambda(\tau) = \mu_{\mathscr{L}\check{\mathscr{L}} U^0_\tau}\{\Omega\}$ for $\tau \in [0, \tau_\infty]$.

Let $u_0 \in L_2(\Omega)$ and $U_0 = \check{E} * u_0$. Define $V_0 = U_0 - U_\infty^0$ and $v_0 = \mathscr{L} V_0$. Then $v(\lambda)^0 = \mathfrak{P}_{K_{\mathscr{L}}^\lambda(\Omega,S)}(v_0)$ and $v(\lambda)^0 = \mathscr{L} V(\lambda)^0$. If $0 \leqslant \tau \leqslant \tau_\infty$ and $\tau = T(\lambda)$, where $0 \leqslant \lambda \leqslant \lambda_\infty$, then

$$\begin{aligned} u_\tau^0 &= \mathscr{L} U_\tau^0 = \mathscr{L} W(\lambda)^0 = \mathscr{L} U_\infty^0 + \mathscr{L} V(\lambda)^0 \\ &= \mathfrak{P}_{P_{\mathscr{L}}(\Omega)}(u_0) + \mathfrak{P}_{K_{\mathscr{L}}^\lambda(\Omega,S)}(u_0) = \mathfrak{P}_{V_{\mathscr{L}}^\lambda(\Omega,S)}(u_0). \end{aligned}$$

Moreover, $\lambda = \mu_{\mathscr{L}\check{\mathscr{L}} U_\tau^0}\{\Omega\}$. If $\tau > \tau_\infty$, then, obviously,

$$u_\tau^0 = \mathfrak{P}_{V_{\mathscr{L}}^0(\Omega,S)}(u_0) = \mathfrak{P}_{P_{\mathscr{L}}(\Omega)}(u_0). \qquad \square$$

COROLLARY 4.2. *Let $0 \leqslant \tau \leqslant \infty$ and $0 \leqslant s < 2l - n/2$. The mapping*

$$\mathfrak{F}\colon W_2^l(\Omega) \to W_2^l(\Omega) \cap W_2^s(\Omega)^{\mathrm{loc}}$$

is continuous.

References

[BB] H. Brézis and F. E. Browder, *Some proporties of higher order Sobolev spaces*, J. Math. Pures Appl. (9) **61** (1982), 245–259.

[DH1] R. Darst and R. Huotari, *Best L_1-approximation of bounded, approximately continuous functions on* $[0, 1]$ *by nondecreasing functions*, J. Approx. Theory **43** (1985), 178–189.

[DH2] ———, *Monotone approximation on an interval*, Classical Real Analysis (D. Waterman, editor), Contemp. Math., vol. 42, Amer. Math. Soc., Providence, RI, 1985, pp. 43–44.

[H1] L. Hörmander, *The analysis of linear partial differential operators.* I, Springer-Verlag, Berlin, 1983.

[H2] ———, *The analysis of linear partial differential operators.* II, Springer-Verlag, Berlin, 1983.

[He] L. I. Hedberg, *Approximation in Sobolev spaces and nonlinear potential theory*, Nonlinear Functional Analysis and Its Applications, Proc. Sympos. Pure Math., vol. 45, Part 1, Amer. Math. Soc., Providence, RI, 1986, pp. 473–480.

[HL] R. Huotari and D. Legg, *Best monotone approximation in* $L_1[0, 1]$, Proc. Amer. Math. Soc. **94** (1985), 279–282.

[HMS] R. Huotari, A. D. Meyerowitz, and M. Sheard, *Best monotone approximation in* $L_1[0, 1]$, J. Approx. Theory **47** (1986), 85–91.

[L] D. Legg, *Best monotone approximation in* $L^\infty[0, 1]$, Classical Real Analysis (D. Waterman, editor), Contemp. Math., vol. 42, Amer. Math. Soc., Providence, RI, 1985, pp. 105–111.

[LS] J. T. Lewis and O. Shisha, *L_p convergence of monotone functions and their uniform convergence*, J. Approx. Theory **14** (1975), 281–284.

[LT] D. Legg and D. Townsend, *Best monotone approximation in* $L_\infty[0, 1]$, J. Approx. Theory **42** (1984), 30–35.

[M1] V. A. Malyshev, *The solution of the problem of monotone approximation and the convex hull of a piecewise linear function*, Numerical Analysis: Methods, Algorithms, Applications (N. S. Bakhvalov et al., editors), Izdat. Moskov. Gos. Univ., Moscow, 1985, pp. 143–147. (Russian)

[M2] ———, *Hadamard conjecture and estimation of Green's functions*, Algebra i Analiz **4** (1992), no. 4, 1–44; English transl. in St. Petersburg Math. J. **4** (1993).

[R] W. Rudin, *Functional analysis*, McGraw-Hill, New York, 1973.

[SW1] J. J. Swetits and S. E. Weinstein, *Best monotone approximation and the Hardy-Littlewood maximal function*, Approx. Theory Appl. **5** (1989), no. 3, 35–39.

[SW2] ———, *Construction of best monotone approximation in l_p for* $1 < p < \infty$, Approx. Theory Appl. **5** (1989), no. 3, 69–77.

[SW3] ———, *Construction of best monotone approximation on* $L_p[0, 1]$, J. Approx. Theory **61** (1991), 118–130.

[SWY] J. J. Swetits, S. E. Weinstein, and Yuesheng Xu, *On the characterization and computation of best monotone approximation in* $L_p[0, 1]$ *for* $1 \leqslant p \leqslant \infty$, J. Approx. Theory **60** (1990), 58–69.

DEPARTMENT OF MATHEMATICS AND MECHANICS, ST. PETERSBURG STATE UNIVERSITY

Translated by the author

Amer. Math. Soc. Transl.
(2) Vol. **164**, 1995

Long Time Behavior of Flows Moving by Mean Curvature

V. I. Oliker and N. N. Uraltseva

Introduction and main results

In this paper we consider long time behavior of a flow of nonparametric surfaces in $\mathbb{R}^{n+1}$ such that the normal speed of propagation is equal to the mean curvature of the current surface. The flow that we consider starts as a bounded perturbation of a minimal surface, and it is assumed that the boundaries of the surfaces remain fixed for all time. This is a nonlinear degenerate parabolic flow, and we have shown earlier **[OU1, OU2]** that, in general, such flows will develop singularities in finite time. After a sufficiently long time these singularities will disappear and the flow will converge to the minimal surface spanning the boundary manifold. In this paper, we give the asymptotic representation of the flow as the time tends to ∞. In the special case of homogeneous boundary conditions, the corresponding representation was established in **[OU1]**.

In order to formulate our results, we need to introduce some notation. Let Ω be a bounded domain in $\mathbb{R}^n$, $n \geqslant 2$, with C^∞ boundary $\partial\Omega$, and φ a function in $C^\infty(\overline{\Omega})$ whose graph is a minimal surface, that is, such that $H(\varphi) = 0$ in Ω, where H is the mean curvature operator

$$H(\varphi) = \operatorname{div} \frac{D\varphi}{\sqrt{1+|D\varphi|^2}}, \qquad D\varphi = \operatorname{grad}\varphi, \quad |D\varphi|^2 = \sum_{i=1}^n \varphi_{x_i}^2.$$

Let $u(x,t)$ be a solution of the mean curvature evolution problem:

$$u_t = \sqrt{1+|Du|^2}H(u) \quad \text{in } \Omega \times [T,\infty), \tag{0.1}$$

$$u(x,t) = \varphi(x) \quad \text{in } \partial\Omega \times [T,\infty), \tag{0.2}$$

where T is such that $u(x,t) \in C^\infty(\overline{\Omega} \times [T,\infty))$. It was shown in **[OU2]** that such a T exists for the evolution (0.1), (0.2) considered on $\overline{\Omega} \times [0,\infty)$ provided that $u(x,0) = \varphi(x) + u_0(x)$ with $u_0(x) \in C_0^\infty(\overline{\Omega})$. Thus, T is the time moment when the possibly singular solution of the evolution problem in $\overline{\Omega} \times [0,\infty)$ becomes a classical solution.

1991 *Mathematics Subject Classification*. Primary Primary 35K60.
This research was partially supported by NSF Grant DMS-92-04490.

Let Σ be the minimal surface $(x, \varphi(x))$, $x \in \overline{\Omega}$, and let $g_{ij}(D\varphi) = \delta_{ij} + \varphi_{x_i}\varphi_{x_j}$, $i, j = 1, \dots, n$, be the coefficients of its metric induced from $\mathbb{R}^{n+1}$. Let

$$g^{ij}(D\varphi) = \delta_{ij} - \frac{\varphi_{x_i}\varphi_{x_j}}{\sqrt{1+|D\varphi|^2}}$$

be the elements of the inverse matrix $(g_{ij})^{-1}$. Further, let

$$L = \sqrt{1+|D\varphi|^2}\frac{\partial}{\partial x_i}\left(\frac{g^{ij}(D\varphi)}{\sqrt{1+|D\varphi|^2}}\frac{\partial}{\partial x_j}\right), \tag{0.3}$$

and let λ_1 and $\psi_1(x)$ be, respectively, the first eigenvalue and eigenfunction of $-L$ in Ω with Dirichlet boundary conditions, that is,

$$-\frac{\partial}{\partial x_i}\left(\frac{g^{ij}(D\varphi)}{\sqrt{1+|D\varphi|^2}}\frac{\partial \psi_1}{\partial x_j}\right) = \frac{\lambda_1\psi_1(x)}{\sqrt{1+|D\varphi|^2}} \quad \text{in } \Omega, \tag{0.4}$$

$$\psi = 0 \quad \text{on } \partial\Omega, \qquad \psi > 0 \quad \text{in } \Omega. \tag{0.5}$$

Note that $\lambda_1 > 0$.

THEOREM 1. *Let $u \in C^\infty(\overline{\Omega} \times [T, \infty))$ satisfy* (0.1) *and* (0.2). *Then for $t \geqslant T$*

$$u(x,t) = \varphi(x) + Ce^{-\lambda_1(t-T)}\psi_1(x) + O(e^{-\mu(t-T)}), \qquad x \in \Omega, \tag{0.6}$$

where C is a constant depending on $u(x,T)$, Ω, and $\mu = \mathrm{const} > \lambda_1$. In addition, for any $l \geqslant 1$ we have

$$\sup_{\Omega} |D^l(u - \varphi)| \leqslant C(l)e^{-\lambda_1(t-T)}, \tag{0.7}$$

where $C(l)$ is a constant depending on $u(x,T)$, Ω, and l.

A more geometric form of Theorem 1 can be obtained if we consider the asymptotics of the projection of $(0, \dots, 0, u(x,t) - \varphi(x))$ onto the unit normal $N(x,t)$ to the surface $(x, u(x,t))$, that is,

$$\frac{u(x,t) - \varphi(x)}{\sqrt{1+|Du|^2}}.$$

Let Δ_φ denote the Laplace operator on the surface Σ, and $\|b\|$ the norm of the second fundamental form of Σ.

THEOREM 2. *For $t \geqslant T$*

$$\frac{u(x,t) - \varphi(x)}{\sqrt{1+|Du|^2}} = Ce^{-\lambda_1(t-T)}\widehat{\varphi}_1(x) + O(e^{-\mu(t-T)}), \qquad x \in \Omega, \tag{0.8}$$

where $\widehat{\varphi}_1$ is the first eigenfunction of the operator $-L_\varphi = -\Delta_\varphi - \|b\|^2$ and λ_1 is its first eigenvalue (it is also the first eigenvalue of $-L$), $\mu > \lambda_1$.

The operator $L_\varphi \equiv \Delta_\varphi + \|b\|^2$ is the Jacobi operator and $-L_\varphi$ defines the second variation of the area functional; see **[S,** §3]. In our case, Σ is a stable minimal graph and the second variation is positive. Hence, $\lambda_1 > 0$. (The latter follows also from the form of (0.4) and the last part of Theorem 2.) Thus, geometrically, for large t, the function

$$\frac{u(x,t)-\varphi(x)}{\sqrt{1+|Du(x,t)|^2}}$$

mimics the behavior of self-similar solutions $Ce^{-\lambda_1 t}\widehat{\varphi}_1(x)$ to the problem

$$\begin{aligned} w_t &= L_\varphi w \quad \text{in } \Omega\times[0,\infty), \\ w(x,t) &= 0 \quad \text{on } \partial\Omega\times[0,\infty). \end{aligned}$$

The asymptotic behavior for mean curvature evolution was considered previously in several other cases. In **[H]** G. Huisken has shown that a flow of closed convex surfaces contracts smoothly to a point in finite time and becomes asymptotically spherical. In **[EH0, EH1]** K. Ecker and G. Huisken study solutions to (0.1) in the case when $\Omega = \mathbb{R}^n$ and $u(x,0)$ is a given locally Lipschitz function on $\mathbb{R}^n$. They show that after a suitable rescaling the flow smoothly converges to a solution of a stationary equation corresponding to expanding self-similar solutions.

1. Preliminaries

For $p=(p_1,\dots,p_n)$ put

$$F(p)=\sqrt{1+|p|^2}, \qquad F_i=\frac{\partial F}{\partial p_i}=\frac{p_i}{F(p)}, \qquad F_{ij}=\frac{\partial F_i}{\partial p_j}=\frac{1}{F}\left(\delta_{ij}-\frac{p_ip_j}{F^2}\right),$$

$$g^{ij}(p)=\delta_{i_j}-\frac{p_ip_j}{F^2(p)}, \qquad g_k^{ij}(p)=\frac{\partial g^{ij}(p)}{\partial p_k}.$$

In this notation the right-hand side of equation (0.1) assumes the form

$$\sqrt{1+|Du|^2}H(u)=F(Du)H(u)=g^{ij}(Du)u_{x_ix_j}.$$

Here and everywhere else in the paper summation over repeated indices is assumed.

Put $\widetilde{u}(x,t)=u(x,t)-\varphi(x)$. Then $\widetilde{u}$ is a solution of the initial-boundary value problem

$$\widetilde{u}_t-g^{ij}(D\varphi+D\widetilde{u})\widetilde{u}_{x_ix_j}-g^{ij}(D\varphi+D\widetilde{u})\varphi_{x_ix_j}=0 \quad \text{in } \Omega\times[T,\infty), \tag{1.1}$$

$$\widetilde{u}=0 \quad \text{on } \partial\Omega\times[T,\infty), \tag{1.2}$$

$$\widetilde{u}(x,T)=u(x,T)-\varphi(x) \quad \text{in } \overline{\Omega}\times\{T\}. \tag{1.3}$$

It was proved in **[OU2]** that for some constant $\nu>0$ depending on Ω, the following estimates hold:

$$\sup_{\Omega}|\widetilde{u}(x,t)|\leqslant Ce^{-\nu t} \quad \text{for } t\geqslant T, \tag{1.4}$$

$$\sup_{\partial\Omega}|D\widetilde{u}(x,t)|\leqslant Ce^{-\nu t} \quad \text{for } t\geqslant T. \tag{1.5}$$

These estimates and the usual maximum principle imply that

$$\sup_{\Omega}|D\widetilde{u}(x,t)|\leqslant C \quad \text{for } t\geqslant T. \tag{1.6}$$

Here and in the rest of the paper, unless stated otherwise, we denote by the same letter C various constants depending on the domain Ω, the function u, its derivatives at $t=T$, and the function φ and its derivatives.

Estimate (1.6) implies that equation (1.1) is uniformly parabolic and all the derivatives of u are uniformly bounded in $\overline{\Omega} \times [T, \infty)$.

Now we use (1.4) and results of M. Wiegner **[W]** on long time behavior of solutions to quasilinear parabolic equations to obtain the optimal rate of decay of $\widetilde{u}$.

PROPOSITION 1. *Let* $\widetilde{u}(x,t)$ *satisfy* (1.1)–(1.3). *Then*

$$\sup_{\Omega} |\widetilde{u}(x,t)| \leqslant Ce^{-\lambda_1(t-T)}, \qquad t \in [T, \infty), \tag{1.7}$$

with λ_1 *defined by* (0.4) *and* (0.5).

PROOF. Put

$$a_{ij}(x,p) = g^{ij}(D\varphi(x)+p), \qquad f(x,p) = g^{ij}(D\varphi(x)+p)\varphi_{x_i x_j}(x), \tag{1.8}$$

$$f_i(x,p) = \frac{\partial f(x,p)}{\partial p_i} = g_i^{kj}(D\varphi(x)+p)\varphi_{x_k x_j}(x) \tag{1.9}$$

The equation (1.1) now takes the form

$$\widetilde{u}_t - a_{ij}(x, D\widetilde{u})\widetilde{u}_{x_i x_j} = f(x, D\widetilde{u}). \tag{1.10}$$

For $x \in \overline{\Omega}$ and $p, \xi \in \mathbb{R}^n$ we have

$$\frac{|\xi|^2}{1+|D\varphi(x)+p|^2} \leqslant a_{ij}(x,p)\xi_i\xi_j \leqslant |\xi|^2. \tag{1.11}$$

Using the equation $H(\varphi) = 0$, we see that

$$f(x,0) = 0. \tag{1.12}$$

Now, since (1.4), (1.11), and (1.12) hold, we can apply Theorem 3.2 in **[W]** to the solution $\widetilde{u}$ of (1.10), (1.2), (1.3). According to this theorem, we have estimate (1.7) with λ_1 being the first eigenvalue of the operator

$$-\widetilde{L} = -a_{ij}(x,0)\frac{\partial^2}{\partial x_i \partial x_j} - f_i(x,0)\frac{\partial}{\partial x_j}$$

in Ω with homogeneous Dirichlet conditions on $\partial\Omega$. Thus, it only remains to show that $\widetilde{L} = L$. This will be done by direct computation.

From (0.3) we get

$$L = g^{ij}\frac{\partial^2}{\partial x_i \partial x_j} + b^i \frac{\partial}{\partial x_i}, \tag{1.13}$$

where

$$b^i(x) = \left[g_k^{ij} - \frac{g^{ij}}{F}F_k\right]\varphi_{x_k x_j}(x) \tag{1.14}$$

and g^{ij}, g_k^{ij}, F, and F_k are evaluated on $D\varphi(x)$. It follows from (1.8) that

$$a_{ij}(x,0) = g^{ij}(D\varphi(x)). \tag{1.15}$$

Next, we prove that the functions $b^i(x)$ coincide with $f_i(x,0)$, $i = 1, \ldots, n$. Indeed,

$$g_i^{kj} = -F^{-2}(\delta_{ki}\varphi_{x_j} + \delta_{ij}\varphi_{x_k}) + 2F^{-4}\varphi_{x_i}\varphi_{x_j}\varphi_{x_k}, \tag{1.16}$$

and by (1.14)

$$b^i(x) = F^{-2}(-\delta_{ki}\varphi_{x_j} - \delta_{kj}\varphi_{x_i} - \delta_{ij}\varphi_{x_k} + 3F^{-2}\varphi_{x_i}\varphi_{x_k}\varphi_{x_j})\varphi_{x_k x_j}. \tag{1.16′}$$

On the other hand, it follows from (1.9) and (1.16) that

$$f_i(x,0) = g_i^{kj}(D\varphi)\varphi_{x_k x_j} = F^{-2}(-\delta_{ki}\varphi_{x_j} - \delta_{ij}\varphi_{x_k} + 2F^{-2}\varphi_{x_i}\varphi_{x_j}\varphi_{x_k})\varphi_{x_k x_j}.$$

Thus, we obtain

$$\begin{aligned} b^i(x) &= f_i(x,0) + F^{-2}\varphi_{x_i}(-\delta_{kj} + F^{-2}\varphi_{x_k}\varphi_{x_j})\varphi_{x_k x_j} \\ &= f_i(x,0) - F^{-1}\varphi_{x_i}H(\varphi) = f_i(x,0) \end{aligned} \tag{1.17}$$

This and (1.15) imply $\widetilde{L} = L$. The proposition is proved. □

REMARK 1. Since $H(\varphi) = 0$ implies $F^2\varphi_{x_j x_j} = \varphi_{x_i}\varphi_{x_j}\varphi_{x_i x_j}$, from (1.16) we obtain

$$b^i(x) = \frac{2}{F^2}(\varphi_{x_j}\varphi_{x_i x_j} - \varphi_{x_i}\varphi_{x_j x_j}). \tag{1.17′}$$

We shall need this formula in the last section.

COROLLARY 1. *For each $l \geqslant 1$, there exists a constant $C(l)$ such that*

$$\sup_{\Omega} |D^l \widetilde{u}(x,t)| \leqslant C(l)e^{-\lambda_1(t-T)}, \qquad t \in [T,\infty). \tag{1.18}$$

To prove (1.18), we write (1.1) in the form

$$\widetilde{u}_t - A_{ij}(x,t)\widetilde{u}_{x_i x_j} - A_i(x,t)\widetilde{u}_{x_i} = 0 \quad \text{in } \overline{\Omega} \times [T,\infty), \tag{1.19}$$

where

$$A_{ij}(x,t) = g^{ij}(u(x,t)),$$

$$A_i(x,t) = \varphi_{x_k x_j}(x) \int_0^1 g_i^{jk}(\tau Du(x,t) + (1-\tau)D\varphi(x))\, d\tau.$$

Because of (1.6), equation (1.19) is uniformly parabolic and the $C^k(\overline{\Omega})$-norms of the coefficients are bounded uniformly in t for any $k \geqslant 0$. Using local Schauder-type estimates in $Q(t,t+1) = \overline{\Omega} \times [t,t+1]$ with t arbitrary (see **[LSU,** Chapter IV, Theorem 10.1]), we get

$$\|\widetilde{u}\|_{Q(t,t+1)}^{l+\alpha} \leqslant C(l) \max_{Q(t-1,t+1)} |\widetilde{u}|, \qquad \alpha \in (0,1), \tag{1.20}$$

where the norm on the left is the $C^{l+\alpha,(l+\alpha)/2}(Q(t,t+1))$-norm. Estimate (1.18) now follows from (1.20) and Proposition 1.

2. Proof of Theorem 1

Let $0 < \lambda_1 < \lambda_2 \leqslant \cdots$ be the spectrum of the operator $-L$ in Ω with homogeneous Dirichlet boundary condition and $\psi_1, \psi_2, \ldots$ the corresponding eigenfunctions. It follows from (1.13), (1.17) and (1.9) that

$$L = g^{ij}(D\varphi)\frac{\partial^2}{\partial x_i \partial x_j} + g_k^{ij}(D\varphi)\varphi_{x_i x_j}\frac{\partial}{\partial x_k}. \tag{2.1}$$

The functions ψ_i, $i = 1, 2, \ldots$, can be normalized so that they form a complete orthonormal system in the space $L^2(\Omega; 1/F(D\varphi))$ with the inner product

$$[v, h] = \int_\Omega \frac{v(x)h(x)}{F(D\varphi(x))}\,dx$$

Rewrite equation (0.1) in the form

$$\begin{aligned}\widetilde{u}_t - g^{ij}(D\varphi)\widetilde{u}_{x_i x_j} &- g_k^{ij}(D\varphi)\varphi_{x_i x_j}\widetilde{u}_{x_k} \\ &= [g^{ij}(Du) - g^{ij}(D\varphi)]u_{x_i x_j} - g_k^{ij}(D\varphi)\varphi_{x_i x_j}\widetilde{u}_{x_k}\end{aligned}$$

or

$$\widetilde{u}_t - L\widetilde{u} = f, \tag{2.2}$$

where $f = g_k^{ij}(D\varphi)\widetilde{u}_{x_k}\widetilde{u}_{x_i x_j} + O(|D\widetilde{u}|^2)$. By (1.18),

$$\sup_\Omega |f(x,t)| \leqslant Ce^{-2\lambda_1(t-T)}, \qquad t \geqslant T. \tag{2.3}$$

Now we are in the same circumstances as in **[OU1]** (see the end of §5), where the equation (2.2) with L replaced by the standard Laplace operator $\Delta = \partial^2/\partial x_1^2 + \cdots + \partial^2/\partial x_n^2$ was considered. Using the same arguments as in **[OU1]**, we complete the proof of our theorem.

3. Proof of Theorem 2

Everywhere in this section g_{ij} means $g_{ij}(D\varphi)$. Before we proceed with the proof of Theorem 2, we establish

LEMMA 1. *Suppose L is the operator defined by (0.3) and $\psi \in C^2(\Omega)$ satisfies $-L\psi = \lambda\psi$ in Ω for some $\lambda \in \mathbb{R}$. Then $\widehat{\psi} = \psi/F(D\varphi)$ satisfies the equation*

$$-\Delta_\varphi\widehat{\psi} - \|b\|^2\widehat{\psi} = \lambda\widehat{\psi},$$

where Δ_φ is the Laplace operator on the surface Σ and $\|b\|$ is the norm of its second fundamental form.

PROOF. The unit normal vector field on Σ is given by

$$N = \frac{1}{F(D\varphi)}(-D\varphi, 1), \tag{3.1}$$

and the coefficients of the second fundamental form are

$$b_{js} = \varphi_{x_j x_s}/F(D\varphi). \tag{3.1$'$}$$

Let $r(x) = (x, \varphi(x))$ be the position vector of the surface Σ. Using the Weingarten formulas, we get

$$N_{x_s} = -b_{si}g^{ik}r_{x_k}, \qquad \nabla_{sj}N = -(\nabla_j b_{si})g^{ik}r_{x_k} - b_{si}g^{ik}b_{kj}N, \tag{3.2}$$

where ∇_j and ∇_{sj} denote, respectively, the first and second covariant differentiation in the metric g_{ij}. By the Codazzi equations, $\nabla_j b_{si} = \nabla_i b_{sj}$. Thus, taking into account the fact that $H(\varphi) = g^{sj}b_{sj} = 0$, we arrive at the following well-known formula:

$$\Delta_\varphi N = g^{sj}\nabla_{sj}N = -\nabla_i(g^{sj}b_{s_j})g^{ik}r_{x_k} - \|b\|^2 N = -\|b\|^2 N. \tag{3.3}$$

We now compute $\Delta_\varphi(\psi N)$. We have

$$\Delta_\varphi(\psi N) = (\Delta_\varphi \psi)N + 2g^{ij}\psi_{x_i}N_{x_j} + \psi\Delta_\varphi N, \tag{3.4}$$

and it follows from (3.2) that

$$g^{ij}\psi_{x_i}N_{x_j} = -g^{ij}\psi_{x_i}b_{js}g^{sk}r_{x_k}.$$

Then, taking (3.1$'$) into account, we obtain

$$\begin{aligned} g^{ij}\psi_{x_i}N_{x_j} &= -\psi_{x_i}\left(\delta_{ij} - \frac{\varphi_{x_i}\varphi_{x_j}}{F^2}\right)\frac{\varphi_{x_j x_s}}{F}\left(\delta_{sk} - \frac{\varphi_{x_s}\varphi_{x_k}}{F^2}\right)r_{x_k} \\ &= -\frac{\psi_{x_i}}{F}(\varphi_{x_i x_k} - F^{-2}\varphi_{x_i}\varphi_{x_j}\varphi_{x_j x_k} \\ &\quad - F^{-2}\varphi_{x_s}\varphi_{x_k}\varphi_{x_i x_s} + F^{-4}\varphi_{x_i}\varphi_{x_j}\varphi_{x_j x_s}\varphi_{x_s}\varphi_{x_k})r_{x_k}. \end{aligned}$$

Let N^{n+1} and $N^{n+1}_{x_i}$ denote the $(n+1)$st components of the corresponding vectors. Then, using the equation $F^2\Delta\varphi = \varphi_{x_i}\varphi_{x_j}\varphi_{x_i x_j}$, we get

$$g^{ij}\psi_{x_i}N^{n+1}_{x_j} = -\frac{\psi_{x_i}}{F^3}(\varphi_{x_i x_s}\varphi_{x_s} - \varphi_{x_i}\Delta\varphi).$$

Projecting (3.4) onto the direction x_{n+1} and using (3.1) and (3.3), we get

$$\Delta_\varphi\left(\frac{\psi}{F}\right) = \frac{1}{F}\Delta_\varphi\psi - 2\frac{\psi_{x_i}}{F^3}(\varphi_{x_i x_s}\varphi_{x_s} - \varphi_{x_i}\Delta\varphi) - \frac{\psi}{F}\|b\|^2. \tag{3.5}$$

On the other hand, since Σ is minimal, we have

$$\Delta_\varphi\psi = g^{ij}\nabla_{ij}\psi = g^{ij}\psi_{x_i x_j}.$$

Hence, using (2.1) and (1.17$'$), we obtain

$$-\Delta_\varphi\left(\frac{\psi}{F}\right) - \|b\|^2\frac{\psi}{F} = -\frac{1}{F}L\psi = \lambda\frac{\psi}{F}.$$

The lemma is proved. □

To complete the proof of Theorem 2, we note that by (0.6)

$$\frac{\widetilde{u}(x,t)}{F(Du)} = Ce^{-\lambda_1(t-T)}\frac{\psi_1(x)}{F(D\varphi)}(1 + O(|D\widetilde{u}|) + O(e^{-\mu(t-T)}), \qquad \mu > \lambda_1.$$

Because of (0.7) we may write this as

$$\frac{\widetilde{u}(x,t)}{F(Du)} = Ce^{-\lambda_1(t-T)}\frac{\psi_1(x)}{F(D\varphi)} + O(e^{-\mu(t-T)}).$$

Theorem 2 is proved.

References

[EH0] K. Ecker and G. Huisken, *Mean curvature evolution of entire graphs*, Ann. of Math. (2) **130** (1989), 453–471.

[EH1] ———, *Interior estimates for hypersurfaces moving by mean curvature*, Invent. Math. **105** (1991), 547–569.

[H] G. Huisken, *Flow by mean curvature of convex surfaces into spheres*, J. Differential Geom. **20** (1984), 237–266.

[LSU] O. A. Ladyzhenskaya, V. A. Solonnikov, and N. N. Uraltseva, *Linear and quasilinear equations of parabolic type*, Amer. Math. Society, Providence, RI, 1988.

[OU1] V. Oliker and N. Uraltseva, *Evolution of nonparametric surfaces with speed depending on curvature* II: *The mean curvature case*, Comm. Pure Appl. Math. **46** (1993), 97–135.

[OU2] ———, *Evolution of nonparametric surfaces with speed depending on curvature.* III. *Some remarks on mean curvature and anisotropic flows*, Degenerate Diffusions (Wei-Ming Ni et al., editors), IMA Vols. Math. Appl., no. 47, Springer-Verlag, New York, 1993, pp. 141–156.

[S] J. Simons, *Minimal varieties in Riemannian manifolds*, Ann. of Math. (2) **88** (1968), 62–105.

[W] M. Wiegner, *On the asymptotic behaviour of solutions of nonlinear parabolic equations*, Math. Z. **188** (1984), 3–22.

Depatment of Mathematics, Emory University, Atlanta, GA 30322

Department of Mathematics and Mechanics, St. Petersburg State University

Amer. Math. Soc. Transl.
(2) Vol. **164**, 1995

Bifurcation Problem for Nonlinear Second Order Equations in Variable Regions

V. G. Osmolovskiĭ and A. V. Sidorov

We consider the Dirichlet problem for a scalar second order differential equation $F\left(D_x^2 u, D_x u, u, x\right) = 0$ in a region $\omega \subset \mathbb{R}^m$. Assume that $u \equiv 0$ solves the problem for an arbitrary choice of the region ω and the linearization of F corresponding to this solution is given by an elliptic operator L. We then prove the existence of nontrivial solutions branching from the trivial one under small deformations of ω in the case when the Dirichlet problem for the operator L has a one-dimensional kernel.

1. Introduction

Let $\Omega \in \mathbb{R}^m$, $m \geqslant 2$. Define a scalar function $F(r, p, z, x)$ on the space $\mathbb{R}^m \times \mathbb{R}^1 \times \mathbb{R}^m \times \mathbb{R}^n$, $n = m(m+1)/2$, and assume that

$$(1.1) \qquad \begin{gathered} D_r^\alpha D_p^\beta D_x^\gamma F(r, p, z, x) \in C^{k,\varepsilon}(\mathbb{R}^n \times \mathbb{R}^m \times \mathbb{R}^1 \times \Omega) \\ |\alpha| + |\beta| + |\gamma| \leqslant s + 1, \qquad \varepsilon \in (0, 1), \end{gathered}$$

for some positive integer s and k and multi-indices α, β, γ. In addition, suppose that

$$(1.2) \qquad \begin{gathered} F(0,0,0,x) = 0, \qquad F_{ij}(0,0,0,x)\xi_i\xi_j \geqslant \gamma\|\xi\|^2, \\ \gamma > 0, \qquad F_{ij} = \partial F/\partial r_{ij}. \end{gathered}$$

Let ω be a strictly inner subregion in Ω and $\partial\omega \in C^{k+2,\varepsilon}$. Consider the problem

$$(1.3) \qquad \begin{gathered} F(u) = F\left(D_x^2 u, D_x u, u, x\right) = 0, \qquad x \in \omega, \\ u(x) = 0, \qquad x \in \partial\omega, \qquad u \in C^{k+2,\varepsilon}(\overline{\omega}), \end{gathered}$$

where $D_x u$ denotes the gradient of the function u and $D_x^2 u$ denotes the matrix of second derivatives. Under assumptions (1.2), the function $u \equiv 0$ is a solution of (1.3) for any choice of ω.

1991 *Mathematics Subject Classification*. Primary 35G30.

The research was partially supported by the Russian Fund for Fundamental Research, grant no. 93-011-1696.

0065-9290/95/\$1.00 + \$.25 per page

Given $h(x) \in C^{k+2,\varepsilon}(\overline{\omega}, \mathbb{R}^m)$ with $C^{k+2,\varepsilon}(\overline{\omega}, \mathbb{R}^m)$-norm $|h|_{k+2,\varepsilon}$ small enough, the mapping

$$x \longmapsto y(x) = x + h(x) \tag{1.4}$$

is a diffeomorphism from the subregion ω to a strictly inner subregion $\omega_h \subset \Omega$.

Definition. A region ω is said to be a *bifurcation region* for (1.3) if there exist a function $h \in C^{k+2,\varepsilon}(\overline{\omega}, \mathbb{R}^m)$, $|h|_{k+2,\varepsilon} < \delta$, and a nontrivial solution $u(y)$, $y \in \omega_h$, of the problem

$$\begin{gathered} F\big(D_y^2 u, D_y u, u, y\big) = 0, \quad y \in \omega_h, \qquad u = 0, \quad y \in \partial\omega_h; \\ u \in C^{k+2,\varepsilon}(\overline{\omega}_h), \qquad |u|_{k+2,\varepsilon} < \delta \end{gathered} \tag{1.5}$$

for any positive sufficiently small δ.

In this article we state sufficient conditions for ω to be a bifurcation region.

2. Auxiliary assertions

Let X and Y be Banach spaces, and let $U \subset X \times \mathbb{R}^1$ be a neighborhood of the point $(0, 0)$. Assume $f : U \to Y$ to be s times continuously differentiable in the Fréchet sense and $f[0, \lambda] = 0$ for any $\lambda \in U \times \mathbb{R}^1$.

Definition. A point $\lambda = 0$ is said to be a *bifurcation point* for the equation

$$f[v, \lambda] = 0, \qquad (v, \lambda) \in U, \tag{2.1}$$

if there exists a solution of (2.1) with $v \neq 0$ in any sufficiently small neighborhood $V \subset U \subset X \times \mathbb{R}^1$ of the point $(0, 0)$.

We now formulate several statements related to the bifurcation problem [**N**, Chapter 3] which will be needed below.

a) *Necessary condition for bifurcation.* If $\lambda = 0$ is a bifurcation point, then a differential $f_v[0, 0]$ is not a diffeomorphism from X to Y.

b) *Sufficient condition for bifurcation.* Let $s > 2$, and let $Nf_v[0, 0]$ be the kernel of the operator $f_v[0, 0]$. We assume that it is one-dimensional and generated by the vector $\widehat{v}$ while the image of $f_v[0, 0]$, denoted by $Rf_v[0, 0] = Y_1$, is closed and of codimension one. Then, under the condition

$$f_{v\lambda}[0, 0]\widehat{v} \notin Y_1, \tag{2.2}$$

the point $\lambda = 0$ is a bifurcation point.

c) *Description of a branching solution.* Assume that b) holds. Let the space $Y = Y_1 + Y_1'$ be an orthogonal sum, and let P and P' be the projections on Y_1 and Y_1' respectively. If

$$P' f_{vv}[0, 0](\widehat{v}, \widehat{v}) \neq 0, \tag{2.3}$$

then the branching solution is a curve $v = v[\lambda]$ of class C^{s-2} such that

$$v[0] = 0, \qquad v'[0] = -2\big[P' f_{v\lambda}[0, 0]\widehat{v}\big]\big[P' f_{vv}[0, 0](\widehat{v}, \widehat{v})\big]^{-1}\widehat{v} \tag{2.4}$$

for any $|\lambda|$ small enough.

We also formulate the implicit function theorem, since it will be needed in the sequel.

IMPLICIT FUNCTION THEOREM [**N**, Chapter 2]. *Let X, H, Y be a triple of Banach spaces, $U \subset X \times H$ a neighborhood of the point $(0,0)$, and $f : U \to Y$ an s times continuously differentiable mapping such that $f[v,h]=0$ for $v=0$, $h=0$. Let $f_v[0,0]$ be an isomorphism from X to Y. Then the equation*

$$f[v,h]=0 \tag{2.5}$$

has a unique solution $v=v[h]$ possessing s continuous derivatives in a sufficiently small neighborhood U of $(0,0)$ and such that $v(0)=0$.

3. Bifurcations for equations in variable regions

In this section we derive both necessary and sufficient bifurcation conditions for the problem stated in §1 assuming that (1.1) and (1.2) hold.

THEOREM 3.1. (a) *If ω is a bifurcation region, then the problem*

$$\begin{gathered} Lu \equiv F_{ij} u_{x_i x_j} + F_{p_i} u_{x_i} + F_z u = 0, \\ u\big|_{\partial\omega} = 0, \qquad u \in C^{k+2,\varepsilon}(\overline{\omega}), \qquad F \equiv F(0,0,0,x) \end{gathered} \tag{3.1}$$

has a nontrivial solution.

(b) *Let (3.1) have a unique (up to a factor) nontrivial solution, and let $s>2$. Then ω is a bifurcation region.*

PROOF. (a) If ω is a bifurcation region, then for any positive sufficiently small δ there exists $h \in C^{k+2,\varepsilon}(\overline{\omega}, \mathbb{R}^m)$, $|h|_{k+2,\varepsilon} < \delta$, such that (1.5) has a nontrivial solution. In the x-variables, (1.5) has the form

$$\begin{gathered} F\left(\widehat{D}_x^2 v, \widehat{D}_x v, v, y(x)\right) = 0, \qquad v\big|_{\partial\omega} = 0, \\ v \in C^{k+2,\varepsilon}(\overline{\omega}), \qquad |v|_{k+2,\varepsilon} < C\delta. \end{gathered} \tag{3.2}$$

Here the function $v(x)$, the vector $\widehat{D}_x v$, and the matrix $\widehat{D}_x^2 v$ are given respectively by

$$\begin{gathered} v(x) = u(y(x)), \qquad \left(\widehat{D}_x v\right)_i = v_{x_k} x_{y_i}^k, \qquad x_{y_i}^k = \frac{\partial x_k}{\partial y_i}, \\ \left(\widehat{D}_x^2 v\right)_{ij} = v_{x_k x_l} x_{y_i}^k x_{y_j}^l + v_{x_k} x_{y_i y_j}^k, \end{gathered} \tag{3.3}$$

while $C \neq C(\delta)$ is a positive constant. The Jacobian matrix of the mapping (1.4) has the form $\dot{y}(x) = I + \dot{h}(x)$. Hence the Jacobian matrix of the inverse mapping is $\dot{x}(y) = \left(I + \dot{h}(x)\right)^{-1}$. It defines a mapping of a small neighborhood of the zero point in the space $C^{k+2,\varepsilon}(\overline{\omega}, \mathbb{R}^m)$ into the space of matrices with components belonging to $C^{k+1,\varepsilon}(\overline{\omega})$. This mapping is infinitely differentiable in the Fréchet sense. Therefore the left-hand side of equation (3.2) may be treated as a mapping $F[v,h]$ from a small neighborhood U of the zero point in $X \times H$ to Y, where

$$X = \{v \in C^{k+2,\varepsilon}(\overline{\omega}),\ v|_{\partial\omega} = 0\},$$
$$H = \{h \in C^{k+2,\varepsilon}(\overline{\omega}, \mathbb{R}^m)\}, \qquad Y = C^{k,\varepsilon}(\overline{\omega}). \tag{3.4}$$

Due to (1.1), this mapping has continuous Fréchet derivatives up to order s. Up to first order terms with respect to h, we have $\dot{x}(y) = I - \dot{h}(x)$, and the relations

$$\dot{x}_{ki} = x^k_{y_i} = \delta_{ki} - h^k_{x_i}, \qquad x^k_{y_i y_i} = -h^k_{x_i x_l} x^l_{y_j} = -h^k_{x_i x_l}(\delta_{jl} - h^l_{x-j}) = -h^k_{x_i x_j}$$

are valid up to the same order of approximation. So the quadratic approximations of $\widehat{D}_x v$ and $\widehat{D}^2_x v$ with respect to both v and h are

$$(\widehat{D}_x v)_i = v_{x_i} - v_{x_l} h^l_{x-i}, \qquad (\widehat{D}^2_x v)_{ij} = v_{x_i x_j} - v_{x_i x_l} h^l_{xj} - v_{x_j x_l} h^l_{x_i} - v_{x_l} h^l_{x_i x_j}.$$

Taking these formulas and (1.2) into account, one can obtain

$$F_v[0,0]v = Lv, \qquad F_h[0,0]h = F_{x_l} h_l = 0, \tag{3.5}$$

and, if $s \geqslant 2$,

$$\begin{aligned} &F_{vv}[0,0](v,v) = F''[0](v,v),\\ &F_{hh}[0,0](h,h) = F_{x_r x_s} h_r h_s = 0,\\ &F_{vh}[0,0](v,h) = 2\big[-F_{ij}\big(v_{x_i x_l} h^l_{x_j} + v_{x_j x_l} h^l_{x_i} + v_{x-l} h^l_{x_i x_j}\big)\\ &\qquad - F_{p^i} v_{x_l} h^l_{x-i} + F_{ijx_r} v_{x_i x_j} h^r + F_{p^i x_r} v_{x_i} h^r + F_{zx_r} v h_r\big]. \end{aligned} \tag{3.6}$$

Here $F''[0](v,v)$ stands for the quadratic part with respect to v of the expansion of $F[v]$ defined by (1.3) at the zero point, while the derivatives of the function $F(r,p,z,x)$ are calculated at the point $(0,0,0,x)$.

The equation (3.2) may be rewritten as

$$F[v,h] = 0, \qquad (v,h) \in U. \tag{3.7}$$

Since $F[0,h] \equiv 0$ and the region ω is a bifurcation region, the operator $F_v[0,0]$ cannot be an isomorphism from X to Y. The operator $F_v[0,0] = L$ is elliptic and its coefficients belong to $C^{k,\varepsilon}(\overline{\omega})$. Therefore the fact that L is not an isomorphism is equivalent to the existence of a nontrivial solution of (3.1).

(b) The problem dual to (3.1) has the form

$$\begin{aligned} &L^* u \equiv (F_{ij} u)_{x_i x_j} - (F_{p^i} u)_{x_i} + F_z u,\\ &u|_{\partial\omega} = 0, \qquad u \in C^{k+2,\varepsilon}(\overline{\omega}). \end{aligned} \tag{3.8}$$

The operator L^* is elliptic with coefficients from $C^{k,\varepsilon}(\overline{\omega})$. Let $\widehat{v}(x)$ be a solution of (3.1) with the unit $L_2(\omega)$-norm. Since by assumption a nontrivial solution of (3.1) is unique up to a factor, the same is true for (3.8). Denote by $\widehat{u}(x)$ a solution of (3.8) with the unit $L_2(\omega)$-norm. Given spaces X and Y defined in (3.4), denote $Y_1 = LX$.

Then Y_1 is a closed subspace of Y of codimension one. Denote by $Y_1' = \{\widehat{u}\}$ the one-dimensional subspace spanning $\widehat{u}$. The decomposition $Y = Y_1 + Y_1'$ is orthogonal in $L_2(\omega)$. A projection operator P' onto Y_1' can be represented in the form

$$P'w = \widehat{u} \int_{\omega} w\widehat{u}\, dx, \qquad w \in Y. \tag{3.9}$$

Let both the function $\widehat{h}(\lambda, x)$ and its partial derivatives with respect to λ up to order $s+1$ belong to $C^{k+2,\varepsilon}(\overline{\omega}, \mathbb{R}^m)$, and, in addition, let $\widehat{h}(0,x) \equiv 0$ and $\widehat{h}_{\lambda}(0,x) \equiv h(x)$. Then, for all sufficiently small $|\lambda|$, the mapping $x \longmapsto y(\lambda, x) = x + \widehat{h}(\lambda, x)$ is a diffeomorphism from ω to $\omega_\lambda = \omega_{\widehat{h}(\lambda,\cdot)}$, while $\widehat{h}(\lambda) = \widehat{h}(\lambda, x)$ can be treated as an s times differentiable mapping from a small neighborhood of the origin of the real axis to the space $C^{k+2,\varepsilon}(\overline{\omega}, \mathbb{R}^m)$. To prove part (b) of the theorem, it is enough to check that, given $h(x)$, the point $\lambda = 0$ is a bifurcation point for the problem

$$f[v, \lambda] \equiv F\left[v, \widehat{h}[\lambda]\right] = 0. \tag{3.10}$$

The mapping $f[v, \lambda]$ is defined in a sufficiently small zero neighborhood of the origin of $X \times \mathbb{R}^1$ and has $s > 2$ continuous Fréchet derivatives in it. A list of its first and second order differentials is given by

$$f_v[0,0]v = Lv, \qquad f_\lambda[0,0] = 0, \qquad f_{\lambda\lambda}[0,0] = 0,$$
$$f_{vv}[0,0](v,v) = F_{vv}[0,0](v,v), \qquad f_{v\lambda}[0,0](v,\lambda) = \lambda F_{vh}[0,0](v,h).$$

It follows that all sufficient bifurcation conditions for (3.10) are fulfilled except (2.2). Condition (2.2) can be rewritten in the form $P' f_{v\lambda}[0,0]\widehat{v} \neq 0$ using the projection operator P' of the form (3.9). This relation is equivalent to

$$\begin{aligned} \int_{\omega} \widehat{u}\big\{ & \left[F_{ijx_r}\widehat{v}_{x_i x_j} + F_{p^i x_r}\widehat{v}_{x_i} + F_{zx_r}\widehat{v}\right] h_r \\ & - F_{ij}\left[\widehat{v}_{x_i x_l} h^l_{x_j} + \widehat{v}_{x_j x_l} h^l_{x_i} + \widehat{v}_{x_l} h^l_{x_i x_j}\right] - F_{p^i}\widehat{v}_{x_l} h^l_{x_i}\big\} \neq 0. \end{aligned} \tag{3.11}$$

Let us transform the left-hand side of (3.11). Since $L\widehat{v} = 0$, we have $(L\widehat{v})_{x_r} = 0$ as well. Hence,

$$F_{ijx_r}\widehat{v}_{x_i x_j} + F_{p^i x_r}\widehat{v}_{x_i} + F_{zx_r}\widehat{v} = -L\widehat{v}_{x_r}. \tag{3.12}$$

Moreover,

$$\widehat{v}_{x_l x_i} h^l_{x_j} + \widehat{v}_{x_l x_j} h^l_{x_i} + \widehat{v}_{x_l} h^l_{x_i x_j} = \left(\widehat{v}_{x_l} h_l\right)_{x_i x_j} - \widehat{v}_{x_l x_i x_j} h_l,$$
$$\widehat{v}_{x_l} h_l = \left(\widehat{v}_{x_l} h_l\right)_{x_i} - \widehat{v}_{x_l x_i} h_l.$$

The last two formulas yield

$$\begin{aligned} - F_{ij} & \left[\widehat{v}_{x_i x_l} h^l_{x_j} + \widehat{v}_{x_j x_l} h^l_{x_i} + \widehat{v}_{x_l} h^l_{x_i x_j}\right] - F_{p^i}\widehat{v}_{x_l} h^l_{x_i} \\ &= -\left[F_{ij}\left(\widehat{v}_{x_l} h_l\right)_{x_i x_j} + F_{p^i}\left(\widehat{v}_{x_l} h_l\right)_{x_i} + F_z\left(\widehat{v}_{x_l} h_l\right)\right] \\ &\quad + \left[F_{ij}\left(\widehat{v}_{x_l}\right)_{x_i x_j} + F_{p^i}\left(\widehat{v}_{x_l}\right)_{x_i} + F_z\widehat{v}_{x_l}\right] h_l \\ &= -L\left(\widehat{v}_{x_l} h_l\right) + L\left(\widehat{v}_{x_l}\right) h_l. \end{aligned} \tag{3.13}$$

It follows from (3.12) and (3.13) that the expression under the integral sign in (3.11) coincides with $-\widehat{u}L(\widehat{v}_{x_l}h_l)$. Since $\widehat{u}$ solves problem (3.8),

$$-\int_\omega \widehat{u}L(\widehat{v}_{x_l}h_l)\,dx = \int_{\partial\omega} F_{ij}\widehat{u}_{x_j}\widehat{v}_{x_l}h_l\nu_i\,dS, \tag{3.14}$$

where ν_i is the ith component of the outer normal to $\partial\omega$. Let e_j be a jth orthonormal vector. Given a point $x \in \partial\omega$, the vector e_j can be represented in the form $e_j = (\nu, e_j)\nu + \widehat{e}_j$, where e_j is a tangent vector to $\partial\omega$ at the point x. In this case

$$\widehat{v}_{x_j} = (\nabla\widehat{v}, e_j) = \nu_j(\nabla\widehat{v}, \nu) + (\nabla\widehat{v}, \widehat{e}_j).$$

Since $\widehat{v}\big|_{\partial\omega} = 0$, we have $(\nabla\widehat{v}, \widehat{e}_j) = 0$, and the relation $\widehat{v}_{x_j} = \nu_j(\partial\widehat{v}/\partial\nu_j)$ is valid. Similar relations hold for $\widehat{u}_{x_j}$. Substituting them into (3.14), one can rewrite (3.11) in the equivalent form

$$\int_{\partial\omega} h_\nu \frac{\partial\widehat{u}}{\partial\nu}\frac{\partial\widehat{v}}{\partial\nu}F_{ij}\nu_i\nu_j\,ds \neq 0, \tag{3.15}$$

where $h_\nu = (h, \nu)$ is a normal component of the vector function $h(x)$.

To complete the proof, it suffices to check that there exists a function $h(x) \in C^{k+2,\varepsilon}(\overline{\omega}, \mathbb{R}^m)$ such that the left-hand side of (3.15) is not equal to zero. Since by ellipticity we have $F_{ij}\nu_i\nu_j > 0$ and $\partial\omega \in C^{k+2,\varepsilon}$, the required function $h(x)$ can exist only if $\partial\widehat{u}/\partial\nu \cdot \partial\widehat{v}/\partial\nu$ does not vanish identically. Since $\widehat{u}$ and $\widehat{v}$ are nontrivial solutions of (3.1) and (3.8) respectively, both $\partial\widehat{u}/\partial\nu$ and $\partial\widehat{v}/\partial\nu$ can vanish only over a closed subset of $\partial\omega$ having no inner points [**L**]. Then the zero points of the function $\partial\widehat{u}/\partial\nu \cdot \partial\widehat{v}/\partial\nu$ in $\partial\omega$ constitute a set of the first category that cannot coincide with $\partial\omega$ [**Y**, p. 24]. □

REMARK 3.1. Under the assumptions of part (b) of Theorem 3.1, condition (3.15) ensures the existence of a branching nonzero solution under a deformation of the region by means of a family of diffeomorphisms of the form $y(\lambda, x) = x + \lambda h(x)$. If $h(x)$ does not satisfy (3.15), then a bifurcation cannot occur. To prove this assertion, consider the following example:

$$F[u] = \Delta u + \mu u + u^2 = 0, \qquad x \in \omega;$$
$$u(x) = 0, \qquad x \in \partial\omega,$$

where μ is the first eigenvalue of the operator $-\Delta$ with zero boundary condition. It is well known that there exists a unique eigenfunction $u_1(x)$ (up to a factor) which preserves the sign in ω. Let $y(\lambda, x) = x + \lambda h$, with $h \in \mathbb{R}^m$ a fixed vector. The family $y(\lambda, x)$ of diffeomorphisms shifts the region ω along the vector λh. Since the above equation is invariant with respect to shifts, nonzero solutions can exist in the region ω_λ if and only if they exist in ω. Let us integrate the identity $u_1(x)F[u] \equiv 0$ over ω. We obtain

$$\int_\omega u_1(x)u^2(x)\,dx = 0,$$

but this is possible only in the case $u(x) \equiv 0$. On the other hand, using an orthogonal normalization procedure, one can choose $u_1(x)$ having unit $L^2(\omega)$ norm, and hence $\widehat{u} = \widehat{v} = u_1$. Moreover,

$$\int_\omega \Delta u_1 u_{1x_i}\, dx = \int_{\partial\omega} \frac{\partial u_1}{\partial \nu} u_{1x_i}\, dS + \int_\omega \Delta u_{1x_i}\, dx.$$

Since $-\Delta u_1 = \mu u_1$, $x \in \omega$, and $u_{1x_i} = \partial u_1/\partial \nu \cdot \nu_i$, $x \in \partial\omega$, the last relation is equivalent to

$$\int_{\partial\omega} \left(\frac{\partial u_1}{\partial \nu}\right)^2 \nu_i\, dS = 0,$$

which yields (3.15) after multiplication by the components h_i of the vector h.

REMARK 3.2. Condition (2.3) and formula (2.4) corresponding to problem (3.10) have the following form:

$$\int_\omega \widehat{u} F''[0](\widehat{v}, \widehat{v})\, dx \neq 0, \tag{3.16}$$

$$v'[0] = -2\left[\int_{\partial\omega} h_\nu \frac{\partial \widehat{u}}{\partial \nu}\frac{\partial \widehat{v}}{\partial \nu} F_{ij}\nu_i\nu_j\, dS\right]\left[\int_\omega \widehat{u} F''[0](\widehat{v}, \widehat{v})\, dx\right]^{-1} \widehat{v}. \tag{3.17}$$

REMARK 3.3. If the function F in (1.3) does not depend explicitly on u, then no subregion $\omega \subset \Omega$ can play the role of a bifurcation region. Indeed, due to the maximum principle, (3.1) will have only the zero solution.

REMARK 3.4. Suppose we are given $\Omega \subset \mathbb{R}^m$ and a hypersurface $\Sigma \subset \mathbb{R}^{m+1}$, while $z(x)\colon \Omega \to \Sigma$ stands for its sufficiently smooth parametrization and obeys a nonlinear second order differential equation. One can use Theorem 3.1 to determine whether there exists a surface spanning a contour $z(\partial\omega)$, $\omega \subset \Omega$, which is different from $\Sigma \cap z(\omega)$ and has a parametrization obeying the same equation in ω. To reduce this problem to a scalar one, it suffices to look for a parametrization of the new surface in the form $\zeta(x) = z(x) + \mu(z(x))u(x)$, where $u(x)$, $x \in \omega$, is an unknown scalar function and $\mu(z)$ is a normal vector to Σ at the point $z(x)$. This problem has been studied in **[Bo]** for minimal surfaces in $\mathbb{R}^3$ and regions ω of rectangular forms.

4. Stability of a branching solution of a variational problem

Given points $x \in \Omega \subset \mathbb{R}^m$, numbers $z \in \mathbb{R}^1$ and vectors $p \in \mathbb{R}^m$, define a scalar function $H(p, z, x)$ such that

$$\begin{gathered} D_p^\beta D_x^\mu D_z^\delta H(p, z, x) \in C^{k,\varepsilon}\left(\mathbb{R}^m \times \mathbb{R}^1 \times \Omega\right) \\ |\beta| + |\mu| + |\delta| \leqslant s + 3, \qquad \varepsilon \in (0, 1), \end{gathered} \tag{4.1}$$

for arbitrary multi-indices β, μ, δ and some integers s, k. Moreover, let

$$\begin{gathered} -H_{p^i p^j}(0, 0, x)\xi_i\xi_j \geqslant \gamma|\xi|^2, \qquad \gamma > 0, \\ -H_{p^i x_i}(0, 0, x) + H_z(0, 0, x) \equiv 0. \end{gathered} \tag{4.2}$$

Consider the functional

$$J[u, \omega] = \int_\omega H(D_x u, u, x)\, dx, \qquad u \in C^{k+2,\varepsilon}(\overline{\omega}), u\big|_{\partial\omega} = 0, \tag{4.3}$$

in a strictly inner subregion $\omega \subset \Omega$ with $\partial\Omega \in C^{k+2,\varepsilon}$. The Lagrange equation for this functional has the form

$$(4.4)\qquad \begin{aligned} F[u] \equiv &- H_{p^i p^j}(D_x u, u, x) u_{x_i x_j} - H_{p^i z}(D_x u, u, x) u_{x_i} \\ &- H_{p^i x_i}(D_x u, u, x) + H_z(D_x u, u, x) = 0, \qquad u\big|_{\partial\omega} = 0, \end{aligned}$$

while a linearization of its left-hand side at the zero point is given by

$$(4.5)\qquad Lv = -\frac{d}{dx_i}\left[H_{p^i p^j}(0,0,x) v_{x_j}\right] + \left[H_{zz}(0,0,x) - H_{p^i z x_i}(0,0,x)\right] v.$$

The operator L is obviously symmetric. Under the above assumptions, the function $u \equiv 0$ is an extremal of the functional (4.3), i.e., for a solution of (4.4) in any subregion ω.

Assume that the nonzero solution of the problem

$$(4.6)\qquad Lv = 0, \qquad v\big|_{\partial\omega} = 0, \qquad v \in C^{k+2,\varepsilon}(\overline{\omega}),$$

in a subregion ω is unique up to a factor. Denote by $\widehat{v}$ a solution of (4.6) having unit $L_2(\omega)$ norm. Let

$$\omega \ni x \longmapsto y(\lambda, x) = x + \widehat{h}(\lambda, x) \in \omega_\lambda, \qquad \widehat{h}(0,x) \equiv 0, \quad \widehat{h}_\lambda(0,x) \equiv h(x),$$

be a family of diffeomorphisms of the region ω defined in §3. If the condition

$$(4.7)\qquad \int_{\partial\omega} h_\nu \left(\frac{\partial \widehat{v}}{\partial \nu}\right)^2 H_{p^i p^j}(0,0,x) \nu_i \nu_j \, dS \neq 0$$

holds, then, according to Theorem 3.1(b), ω will be a bifurcation region.

LEMMA 4.1. *Assume that $F[u]$ is defined by (4.4) and for any $v \in C^{k+2,\varepsilon}(\overline{\omega})$ with $v\big|_{\partial\omega} = 0$ the equality*

$$(4.8)\qquad \int_\omega vF''[0](v,v)\, dx = \int_\omega H'''[0](v,v,v)\, dx$$

holds, where $6H'''[0](v,v,v)$ denotes cubic terms with respect to both $D_x v$ and v in the expansion of the increment $H(D_x v, v, x) - H(0,0,x)$.

PROOF. Define the left-hand side of (4.4) in the equivalent form

$$F[u] = -\frac{d}{dx_i} H_{p^i}(D_x u, u, x) + H_z(D_x u, u, x).$$

Then, for any v belonging to the above-mentioned class,

$$F''[0](v,v) = -\frac{d}{dx_i} H''_{p^i}[0](v,v) + H''_z[0](v,v),$$

where $2H''_{p^i}[0](v,v)$ and $2H''_z[0](v,v)$ are quadratic terms with respect to both $D_x v$ and v in the expansion of the increments

$$H_{p^i}(D_x v, v, x) - H_{p^i}(0,0,x) \quad \text{and} \quad H_z(D_x v, v, x) - H_z(0,0,x)$$

respectively. Multiplying the last equality by v and integrating over ω, we obtain

$$\int_\omega vF''[0](v,v)\,dx = \int_\omega \left[H''_{p^i}[0](v,v)v_{x_i} + H''_z[0](v,v)v\right] dx$$
$$= \int_\omega H'''[0](v,v,v)\,dx. \qquad \square$$

Along with (4.1), (4.2) and the requirement that the solution space be one-dimensional, assume that

$$\int_\omega H'''[0](\widehat{v},\widehat{v},\widehat{v})\,dx \neq 0. \tag{4.9}$$

Then, for any sufficiently small $|\lambda| \neq 0$, problem (4.4) has a nonzero solution $u(\lambda, y)$, $y \in \omega_\lambda$ in the region ω_λ, while the function $v[\lambda](x) \equiv u(\lambda, y(\lambda, x))$ determines an $s-2$ times differentiable curve in the space X, defined by (3.4) and, besides,

$$v[0](x) \equiv 0, \qquad v'[0](x) \equiv c\widehat{v}(x),$$
$$c = 2\left[\int_{\partial\omega} h_\nu \left(\frac{\partial \widehat{v}}{\partial \nu}\right)^2 H_{p^i p^j}(0,0,x)\nu_i\nu_j\,dS\right]\left[\int_\omega H'''[0](\widehat{v},\widehat{v},\widehat{v})\,dx\right]^{-1}. \tag{4.10}$$

The sign of the function $\Phi(\lambda) = J\left[u(\lambda,y),\omega_\lambda\right] - J\left[0,\omega_\lambda\right]$, interpreted as the difference between values of the energy of the two extremals $u \equiv u(\lambda, y)$ and $u \equiv 0$, can indicate the stability of the branching extremal.

THEOREM 4.1. *Under conditions* (4.1) *with* $s \geqslant 5$ *and* (4.2), *the assumption that the set of solutions of* (4.6) *is one-dimensional, and the assumption that* (4.7), (4.9) *hold for all sufficiently small* $|\lambda|$,

$$\operatorname{sgn}\Phi(\lambda) = -\operatorname{sgn}\lambda \int_{\partial\omega} h_\nu \left(\frac{\partial \widehat{v}}{\partial \nu}\right)^2 H_{p^i p^j}(0,0,x)\nu_i\nu_j\,dS. \tag{4.11}$$

PROOF. Given λ, let us change coordinates using the diffeomorphism inverse to $x \mapsto y(\lambda, x)$ and represent the function $\Phi(\lambda)$ in the form

$$\Phi(\lambda) = \int_\omega \left[H\left(\widehat{D}_x v[\lambda](x), v[\lambda](x), y(\lambda,x)\right) - H\left(0,0,y(\lambda,x)\right)\right] \det \dot{y}(\lambda,x)\,dx$$

with the operator $\widehat{D}_x$ given by (3.3). If $s \geqslant 5$, then the X-valued function $v[\lambda]$ has at least three Fréchet derivatives. Then by (4.1) there exist at least three derivatives of $\Phi(\lambda)$. Now we want to prove that

$$\Phi(0) = \Phi'(0) = \Phi''(0) = 0,$$
$$\Phi'''(0) = -c^2 \int_{\partial\omega} h_\nu \left(\frac{\partial \widehat{v}}{\partial \nu}\right)^2 H_{p^i p^j}(0,0,x)\nu_i\nu_j\,dS, \tag{4.12}$$

with the constant c defined in (4.10).

Given $v(x) \in X$ and sufficiently small $|\lambda|$, define a function $G[v,\lambda]$ by

$$G[v,\lambda] = \int_\omega \left[H\left(\widehat{D}_x v(x), v(x), y(\lambda,x)\right) - H\left(0,0,y(\lambda,x)\right)\right] \det \dot{y}(\lambda,x)\,dx$$

Due to (4.1), the function $G[v,\lambda]$ possesses $s \geqslant 3$ continuous Fréchet derivatives in a neighborhood of zero in $X \times \mathbb{R}^1$. Besides, $\Phi(\lambda) = G\left[v(\lambda),\lambda\right]$. Since the Lagrange equation is invariant with respect to coordinate transformations [**Bu**, p. 205], the functions $v \equiv v[\lambda]$ and $v \equiv 0$ are extremals of the functional $G[v, \lambda]$ for sufficiently small $|\lambda|$. This assertion along with the definition of $G[v, \lambda]$ yields

$$G[0, \lambda] \equiv 0, \qquad G_v[0, \lambda] \equiv 0, \qquad G_v[v[\lambda], \lambda] \equiv 0. \tag{4.13}$$

The following formulas for $\Phi(0)$, $\Phi'(0)$, $\Phi''(0)$, $\Phi'''(0)$ are evident

$$\begin{aligned}
\Phi(0) &= 0, \\
\Phi'(0) &= G_v[0,0]v'[0] + G_\lambda[0,0] = 0, \\
\Phi''(0) &= \frac{d}{d\lambda}\big(G_v[v(\lambda), \lambda]v'(\lambda)\big)\big|_{\lambda=0} + G_{v\lambda}[0,0]v'[0] + G_{\lambda\lambda}[0,0], \\
\Phi'''(0) &= \frac{d^2}{d\lambda^2}\big(G_v[v(\lambda), \lambda]v'(\lambda)\big)\big|_{\lambda=0} + G_{vv\lambda}[0,0](v'[0], v'[0]) \\
&\quad + G_{v\lambda}[0,0]v''[0] + 2G_{v\lambda\lambda}[0,0]v'[0] + G_{\lambda\lambda\lambda}[0,0].
\end{aligned}$$

Since $G_v[0, \lambda]w = 0$ for all $w \in X$, it follows that

$$G_{v\lambda}[0,0]w = 0, \qquad G_{v\lambda\lambda}[0,0]w = 0. \tag{4.14}$$

Hence,

$$G_{v\lambda}[0,0]v'[0] = G_{v\lambda}[0,0]v''[0] = G_{v\lambda\lambda}[0,0]v'[0] = 0,$$

and so $\Phi''(0) = 0$ and

$$\Phi'''(0) = G_{vv\lambda}[0,0]\big(v'[0], v'[0]\big).$$

Let $w[\lambda](x)$ be a continuously differentiable curve in the space X. Let us differentiate the identity $G_v[v[\lambda], \lambda]w[\lambda] \equiv 0$ with respect to λ. We obtain

$$G_{vv}[v[\lambda], \lambda]\big(v'[\lambda], w[\lambda]\big) + G_{v\lambda}[v[\lambda], \lambda]w[\lambda] + G_v[v[\lambda], \lambda]w'[\lambda] \equiv 0. \tag{4.15}$$

One more differentiation with respect to λ gives

$$\begin{aligned}
G_{vvv}[0,0]\big(v'[0], v'[0], v'[0]\big) + 2G_{vv\lambda}[0,0]\big(v'[0], v'[0]\big)& \\
+ 2G_{vv}[0,0]\big(v''[0], v'[0]\big) = 0&,
\end{aligned} \tag{4.16}$$

if we take (4.13) into account and put $w[\lambda] \equiv v'[\lambda]$. If we take $w[\lambda] \equiv v''[\lambda]$ and $\lambda = 0$, then (4.15) implies $G_{vv}[0,0]\big(v'[0], v''[0]\big) = 0$. An expression for $\Phi'''(0)$ can be derived from (4.16):

$$\Phi'''(0) = -\tfrac{1}{2}G_{vvv}[0,0]\big(v'[0], v'[0], v'[0]\big). \tag{4.17}$$

Since

$$G[v, 0] = \int_\omega \big[H(D_x v, v, x) - H(0,0,x)\big]\, dx,$$

it follows that $G_{vvv}[0,0]\big(v'[0], v'[0], v'[0]\big)$ can be represented in the form

$$G_{vvv}[0,0]\big(v'[0], v'[0], v'[0]\big) = \int_\omega H'''[0]\big(v'[0], v'[0], v'[0]\big)\, dx. \tag{4.18}$$

Combining (4.17), (4.18), and (4.10), we complete the proof of (4.12). The relation (4.11) follows immediately from (4.12).

Thanks to (4.11), one may see that if the parameter λ crosses the point O, a minimal (out of two possible) value of the functional J changes one extremal to another. The next theorem permits us to judge how the extremum type changes for two extremals while the parameter λ passes through the point O. The type of extremum is determined by the sign of the second variation of $J[u,\omega]$ at $u\equiv 0$, $\omega=\omega_\lambda$, and $u\equiv u(\lambda,y)$, $\omega=\omega_\lambda$. □

THEOREM 4.2. *Let the assumptions of Theorem* 4.1 *hold. Then there exists a family of functions* $g(\lambda,y)\in C^{k+2,\varepsilon}(\overline{\omega}_\lambda)$, $g(\lambda,y)\big|_{\partial\omega_\lambda}=0$, *not identically zero,*

$$J_{uu}[0,\omega]\big(g(0.x),g(0,x)\big)=0$$

and for all sufficiently small $|\lambda|\neq 0$ *we have*

$$\begin{aligned}\operatorname{sgn} J_{uu}[u(\lambda,y),\omega_\lambda]\big(g(\lambda,y),g(\lambda,y)\big)&=-\operatorname{sgn} J_{uu}[0,\omega_\lambda]\big(g(\lambda,y),g(\lambda,y)\big)\\&=\operatorname{sgn}\lambda\int_{\partial\omega}h_\nu\left(\frac{\partial\widehat{v}}{\partial\nu}\right)^2 H_{p^ip^j}(0,0,x)\nu_i\nu_j\,dS.\end{aligned}$$

PROOF. The relation

$$\begin{aligned}&J_{uu}\left[w(\lambda,y,\omega_\lambda\right]\big(g(\lambda,y),g(\lambda,y)\big)\\&\qquad=G_{vv}\left[w(\lambda,y(\lambda,x)),\lambda\right]\big(g(\lambda,y(\lambda,x)),g(\lambda,y(\lambda,x))\big)\end{aligned}$$

holds for all $w(\lambda,y),g(\lambda,y)\in C^{k+2,\varepsilon}(\overline{\omega}_\lambda)$, $w(\lambda,y),g(\lambda,y)\big|_{\partial\omega_\lambda}=0$.

Recall that $G[v,\lambda]$ was defined in the proof of Theorem 4.1, while the family of diffeomorphisms $y(\lambda,x)$ was defined in Theorem 3.1. Choose $w(\lambda,y)$ in this equality to be one of the two extremals of J in ω_λ, either $w(\lambda,y)\equiv 0$ or $w(\lambda,y)\equiv u(\lambda,y)$. Then $w(\lambda,y(\lambda,x))\equiv 0$ or $w(\lambda,y(\lambda,x))\equiv v[\lambda](x)$. For $g(\lambda,y)$ take a function such that $g(\lambda,y(\lambda,x))\equiv v'[0](x)$. The family of functions $g(\lambda,y)$ described above possesses the properties required by the theorem. Thus the following two relations are valid:

$$\begin{aligned}J_{uu}[0,\omega_\lambda]\big(g(\lambda,y),g(\lambda,y)\big)&=G_{vv}[0,\lambda]\big(v'[0],v'[0]\big),\\J_{uu}\left[u(\lambda,y),\omega_\lambda\right]\big(g(\lambda,y),g(\lambda,y)\big)&=G_{vv}\left[v[\lambda],\lambda\right]\big(v'[0],v'[0]\big).\end{aligned}\tag{4.19}$$

Let us prove the following system of equalities:

$$\begin{aligned}G_{vv}[0,0]\big(v'[0],v'[0]\big)&=0,\\\frac{d}{d\lambda}G_{vv}[0,\lambda]\big(v'[0],v'[0]\big)\big|_{\lambda=0}&=-\frac{1}{2}G_{vvv}[0,0]\big(v'[0],v'[0],v'[0]\big),\\\frac{d}{d\lambda}G_{vv}\left[v[\lambda],\lambda\right]\big(v'[0],v'[0]\big)\big|_{\lambda=0}&=\frac{1}{2}G_{vvv}[0,0]\big(v'[0],v'[0],v'[0]\big).\end{aligned}\tag{4.20}$$

It results from (4.15) for $\lambda=0$, (4.13) and (4.14) that

$$G_{vv}[0,0]\big(v'[0],w[0]\big)=0.\tag{4.21}$$

Putting $w[\lambda]\equiv v'[\lambda]$ in (4.21), we derive the first equality in (4.20). The second one follows from (4.16) and (4.21) with $w[\lambda]\equiv v''[\lambda]$, since the total derivative with respect to λ on the left-hand side can be replaced by a partial one. The left-hand side of the last equality in (4.20) coincides with

$$G_{vv\lambda}[0,0]\big(v'[0],v'[0]\big)+G_{vvv}[0,0]\big(v'[0],v'[0],v'[0]\big).$$

Using (4.21) with $w[\lambda]\equiv v''[\lambda]$ and (4.16), we obtain

$$2G_{vv\lambda}[0,0]\big(v'[0],v'[0]\big)+G_{vvv}[0,0]\big(v'[0],v'[0],v'[0]\big)=0,$$

which completes the proof of the last equality in (4.20).

Using (4.18) and (4.10), we can derive

$$G_{vvv}[0,0]\big(v'[0],v'[0],v'[0]\big)=2c^2\int_{\partial\omega}h_\nu\left(\frac{\partial\widehat{v}}{\partial\nu}\right)^2 H_{p^ip^j}(0,0,x)\nu_i\nu_j\,dS.$$

Combining (4.19), (4.20) and the last formula, we can obtain the relations

$$\begin{aligned}&J_{uu}[0,\omega_\lambda]\big(g(\lambda,y),g(\lambda,y)\big)\\&\qquad=-c^2\lambda\int_{\partial\omega}h_\nu\left(\frac{\partial\widehat{v}}{\partial\nu}\right)^2 H_{p^ip^j}(0,0,x)\nu_i\nu_j\,dS+o(\lambda),\\&J_{uu}\big[u(\lambda,y),\omega_\lambda\big]\big(g(\lambda,y),g(\lambda,y)\big)\\&\qquad=c^2\lambda\int_{\partial\omega}h_\nu\left(\frac{\partial\widehat{v}}{\partial\nu}\right)^2 H_{p^ip^j}(0,0,x)\nu_i\nu_j\,dS+o(\lambda)\end{aligned}$$

which are valid for $\lambda\to0$, and thus complete the proof of the theorem. □

5. Bifurcations in variable regions with nonsmooth boundaries

If $\partial\omega$ or the function F does not possess the required smoothness, it becomes impossible to use Hölder spaces. In the setting of Sobolev spaces, a nonlinear object does not have, as a rule, the required number of Fréchet derivatives. Thus we cannot use the sufficient conditions for bifurcation stated in §2. Nevertheless, if we are dealing with weak nonlinearity, we can treat it as a compact perturbation and try to apply a technique based on degree theory to the bifurcation problem in variable regions.

Assume that the coefficients $a_{ij}(x)$, $i,j=1,\dots,m$, of a symmetric positive definite matrix and the function $a(x)$ belong to the space $C^2(\overline{\Omega})$, while the function $\varphi(x,v)\in C^1(\overline{\Omega}\times\mathbb{R}^1)$ obeys the inequality

$$|\varphi(x,v)|+|\varphi_x(x,v)|+|v\varphi_v(x,v)|\leqslant C|v|^q,\qquad q\in\left(1,\frac{m+2}{m-2}\right),\qquad c>0.\tag{5.1}$$

Define the operator $F[u]$ by

$$F[u]=F\left(D_x^2u,D_xu,u,x\right)\equiv Lu+\varphi(x,u),\qquad Lu\equiv-\frac{\partial}{\partial x_i}\left(a_{ij}(x)\frac{\partial u}{\partial x_j}\right)+a(x)u.\tag{5.2}$$

Suppose ω is a strictly inner subregion of Ω and $\partial\omega$ consists of a finite number of pieces belonging to the class $C^{2,\varepsilon}$. Consider the problem

$$F[u]=0,\qquad u\big|_{\partial\omega}=0\tag{5.3}$$

in the region ω. The relation (5.3) means that the integral identity

$$\int_\omega\big[a_{ij}(x)u_{x_i}(x)\psi_{x_j}(x)+a(x)u(x)\psi(x)+\varphi(x,u(x))\psi(x)\big]\,dx=0\tag{5.4}$$

holds for a function $u\in\overset{\circ}{W}{}_2^1(\omega)$ and an arbitrary function $\psi\in\overset{\circ}{W}{}_2^1(\omega)$. Given $h(x)\in C^1(\overline{\omega},\mathbb{R}^m)$ with sufficiently small $C^1(\overline{\omega},\mathbb{R}^m)$-norm $|h|_1$, the mapping (1.4) is a diffeomorphism from the region ω to a strictly inner subregion $\omega_h\subset\Omega$. The region ω is called a *bifurcation region* for (5.3) if for sufficiently small $\delta>0$ there exist a function $h \in C^1(\overline{\omega}, \mathbb{R}^m)$, $|h|_1 < \delta$, and a solution $u(y)$, $y \in \omega_h$, of (5.3) in the sense of the integral identity

$$\int_{\omega_h} \left[a_{ij}(y)u_{y_i}(y)\psi_{y_j}(y) + a(y)u(y)\psi(y) + \varphi(y, u(y))\psi(y)\right] dy = 0. \tag{5.5}$$

Here $u(y) \in \overset{\circ}{W}{}_2^1(\omega_h)$ is the function we are looking for, which must not be identically equal to zero, while $\psi(y)$ is an arbitrary element of $\overset{\circ}{W}{}_2^1(\omega_h)$. Let us change coordinates in (5.5) using the diffeomorphism inverse to (1.4). Then (5.5) can be rewritten in the form

$$\begin{aligned}\int_\omega \{a_{ij}(x)v_{x_i}\psi_{x_j} &+ \left[a_{ij}(x+h(x))x^k_{y_i}x^l_{y_j}\det[I+\dot h(x)] - a_{ij}(x)\delta_{ki}\delta lj\right]v_{x_k}\psi_{x_l} \\ &+ (a(x+h(x))\det[I+\dot h(x)]v\psi \\ &+ \varphi\big(x+h(x), v(x)\big)\det[I+\dot h(x)]\psi(x)\}\,dx = 0\end{aligned} \tag{5.6}$$

where $v(x) \in \overset{\circ}{W}{}_2^1(\omega)$ is the function we are looking for and $\psi(x) \in \overset{\circ}{W}{}_2^1(\omega)$ is an arbitrary function. Introduce an inner product in the space $\overset{\circ}{W}{}_2^1(\omega)$:

$$[u, v] = \int_\omega a_{ij}(x)u_{x_i}(x)v_{x_j}(x)\,dx,$$

and apply the imbedding theorem of $\overset{\circ}{W}{}_2^1(\omega)$ in $L_p(\omega)$, along with (5.1), for the function φ. This permits us to rewrite (5.6) in equivalent form,

$$F[v, h] \equiv v + B(h)v - K[h]v + g[v, h] = 0. \tag{5.7}$$

The four terms on the left-hand side of (5.7) correspond to the four terms under the integral sign in (5.6) in the same order. For each $h \in C^1(\overline{\omega}, \mathbb{R}^m)$ with C^1-norm small enough, $B[h]$ is a bounded linear operator, and $K[h]$ is a compact linear operator. Moreover, both operators are selfadjoint and $B[0] = 0$. Since both coefficients a_{ij} and a are twice differentiable, the operators $B[h]$ and $K[h]$ are continuously differentiable in the Fréchet sense if they are treated as operator-valued functions of the argument $h \in C^1$ valued in the space of operators bounded in $\overset{\circ}{W}{}_2^1(\omega)$. For any $u, v \in \overset{\circ}{W}{}_2^1(\omega)$ and $A[h] = I + B[h] - K[h]$, the equality

$$\begin{aligned}\left[(A'[0]h)v, u\right] = \int \{ \left[a_{ij}v_{x_i}u_{x_j} + avu\right]\operatorname{div} h &+ \left[a_{ijx_k}v_{x_i}u_{x_j} + a_{x_k}vu\right]h^k \\ &- a_{ij}h^k_{x_j}v_{x_i}u_{x_k} - a_{ij}h^k_{x_i}v_{x_k}u_{x_j}\}\,dx\end{aligned} \tag{5.8}$$

holds. Let us study various properties of the nonlinear operator $g[v, h]$. Due to assumptions (5.1) and the Newton-Leibniz formula

$$\varphi(x+h_1, v_1) - \varphi(x+h_2, v_2) = \int_0^1 \frac{d}{dt}\varphi\big(x+h_2+t(h_1-h_2), v_2+t(v_1-v_2)\big)\,dt,$$

we have the estimate

$$\begin{aligned}&\left|\varphi(x+h_1, v_1) - \varphi(x+h_2, v_2)\right| \\ &\quad \leqslant C\left[|v_1|^q + |v_2|^q\right]|h_1-h_2| + C\left[|v_1|^{q-1} + |v_2|^{q-1}\right]|v_1-v_2|,\end{aligned}$$

which yields the inequality

$$\begin{aligned}\big|\det[I+\dot h_1]\varphi(x+h_1,v_1)-\det[I+\dot h_2]\varphi(x+h_2,v_2)\big|\\ \leqslant C|h_1-h_2|_1\big[|v_1|^q+|v_2|^q\big]+C|v_1-v_2|\big[|v_1|^{q-1}+|v_2|^{q-1}\big].\end{aligned}\tag{5.9}$$

It follows from (5.9) and the definition of $g[v,h]$ that

$$\begin{aligned}\big|[g[v_1,h_1]-g[v_2,h_2],\psi]\big| \leqslant C\|\psi\|_p\big[\|v_1^q\|_{p'}+\|v_2^q\|_{p'}\big]|h_1-h_2|_1\\ +C\|\psi\|_p\big[\|v_1^{q-1}\|_r+\|v_2^{q-1}\|_r\big]\|v_1-v_2\|_p\end{aligned}\tag{5.10}$$

where $\|\cdot\|_s$ stands for the norm in the space $L_s(\omega)$, $1/p+1/p'=1$, $2/p+1/r=1$, while p is chosen so as to ensure that the imbedding of $\overset{\circ}{W}{}_2^1(\omega)$ into $L_p(\omega)$ is compact. Since

$$\|v^q\|_{p'}=\|v\|_{qp'}^q,\qquad qp'=\frac{qp}{p-1},$$

$$\|v^{q-1}\|_r=\|v\|_{(q-1)r}^{q-1},\qquad (q-1)r=\frac{(q-1)p}{p-2},$$

it suffices to require the restrictions $qp'\leqslant p$, $(q-1)r\leqslant p$ in order to ensure that the right-hand side of (5.10) be bounded. The last two inequalities are equivalent to the requirement $q\leqslant p-1$, which is fulfilled due to assumptions on q from (5.1) and the restriction on p: $p<2m/(m-2)$. Hence (5.10) implies

$$\begin{aligned}\|g[v_1,h_1]-g[v_2,h_2]\|_{1,2}\leqslant C\big[\|v_1\|_{1,2}^q+\|v_2\|_{1,2}^q\big]|h_1-h_2|_1\\ +C\big[\|v_1\|_{1,2}^{q-1}+\|v_2\|_{1,2}^{q-1}\big]\|v_1-v_2\|_p\end{aligned}\tag{5.11}$$

where $\|\cdot\|_{1,2}$ is the norm in the space $\overset{\circ}{W}{}_2^1(\omega)$. The inequality (5.11) implies that g is continuous with respect to its arguments, compact with respect to v, and satisfies

$$\|g[v,h]\|_{1,2}=o\big(\|v\|_{1,2}\big)\qquad\text{for } \|v\|_{1,2}\to 0,\tag{5.12}$$

uniformly with respect to $|h|_1\leqslant\delta$.

The above constructions permit us to reduce the initial problem on bifurcations in a variable region to a problem on bifurcations for the operator equation (5.7).

To study this question let us start with some abstract statements. Assume that $B(\lambda)$ and $K(\lambda)$, $\lambda\in\mathbb{R}^1$, are families of selfadjoint bounded linear operators acting in a real Hilbert space H. Both families are assumed to be continuously differentiable with respect to λ, the operators $K(\lambda)$ are assumed compact, and $B(0)=0$. Let the set of solutions of the equation

$$v-K(0)v=0\tag{5.13}$$

be one-dimensional, and let $\widehat{v}$ be a solution of (5.13) such that $\|\widehat{v}\|=1$. Consider the bifurcation problem for the operator

$$f[v,\lambda]=A(\lambda)v+g(v,\lambda),\qquad A(\lambda)\equiv I+B(\lambda)-K(\lambda)$$

where $g(v,\lambda)$ is a mapping from $H\times\mathbb{R}^1$ to H continuous with respect to its arguments, compact for each λ, and such that

$$\|g(v,\lambda)\|=o(\|v\|)\qquad\text{uniformly with respect to } |\lambda|\leqslant\delta.\tag{5.14}$$

Obviously, $f[0,\lambda] = 0$. Below we need a version of a well-known theorem of M. A. Krasnoselskiĭ [**N**, p. 72].

THEOREM 5.1. *If, along with the above assumptions, the inequality*

$$(A'[0]\widehat{v},\widehat{v}) \neq 0 \tag{5.15}$$

holds, then $\lambda = 0$ is a bifurcation point for the equation $f[v,\lambda] = 0$.

PROOF. Let us start with some preliminary constructions. The operator $I + B(\lambda)$ is invertible if $|\lambda|$ is small enough. Hence, one may investigate bifurcations for the equivalent equation

$$\begin{gathered}\widehat{f}[w,\lambda] \equiv \widehat{A}(\lambda)w + \widehat{g}(w,\lambda) = 0,\\ \widehat{A}(\lambda) = I - \widehat{K}(\lambda), \qquad \widehat{K}(\lambda) = \big(I + B(\lambda)\big)^{-1/2}K(\lambda)\big(I + B(\lambda)\big)^{-1/2},\\ \widehat{g}(w,\lambda) = \big(I + B(\lambda)\big)^{-1/2}g\big(\big(I + B(\lambda)\big)^{-1/2}w,\lambda\big), \qquad w = \big(I + B(\lambda)\big)^{1/2}v.\end{gathered} \tag{5.16}$$

Since $\widehat{K}(0) = K(0)$, $\mu = 1$ is a simple eigenvalue of the operator $\widehat{K}(0)$ corresponding to the eigenfunction $\widehat{v}$. Since $\widehat{K}(\lambda)$ is a continuously differentiable family of selfadjoint operators for $|\lambda|$ small enough, the eigenvalue $\mu(\lambda)$, $\mu(0) = 1$, will be differentiable with respect to λ for sufficiently small $|\lambda|$ and, in addition [**F**, p. 17],

$$\mu'(0) = \big(\widehat{K}'(0)\widehat{v},\widehat{K}'(0)\widehat{v}\big).$$

Taking into account the equalities

$$\widehat{K}'(0) = \tfrac{1}{2}B'(0)K(0) + K'(0) - \tfrac{1}{2}K(0)B'(0), \qquad K(0)\widehat{v} = \widehat{v},$$

the inequality (5.15), and the fact that $K(0)$ is selfadjoint, we obtain

$$\mu'(0) = -\big(A'(0)\widehat{v},\widehat{v}\big) \neq 0. \tag{5.17}$$

Since $\mu = 1$ is not a condensation point of the spectrum of a compact operator, it results from (5.17) that for all sufficiently small $|\lambda| \neq 0$

$$1 \notin \sigma[\widehat{K}(\lambda)], \tag{5.18}$$

where $\sigma(L)$ is the spectrum of L.

The operator $\widehat{K}(0)$ has no more than a finite number of eigenvalues $\mu_1(0) > \cdots > \mu_l(0) > 1$. For $|\lambda|$ small enough, simple eigenvalues belonging to this set will shift a little bit; multiple ones will shift as well. Nevertheless, under this perturbation, the above-mentioned part of the spectrum will still be located to the right of 1 and the total multiplicity of the spectrum remains the same. Denote by $\beta(\lambda)$ the total multiplicity of the eigenvalues of the operator $K(\lambda)$ that are greater than 1. Then the above considerations and inequalities (5.17) lead to

$$|\beta(-\lambda) - \beta(\lambda)| = 1 \tag{5.19}$$

for all sufficiently small $|\lambda| \neq 0$.

Now let us turn to the proof of the theorem. Assume that there is no bifurcation, that is:

(5.20) For a sufficiently small $\delta > 0$, the set of all solutions w, λ, $\|w\| \leqslant \delta$, $|\lambda| \leqslant \delta$, of problem (5.16) consists of the points $w = 0$, λ only.

Then in a ball $B_\delta(0) \subset H$ one may define the degree of the mapping $\widehat{f}[w, \lambda]$, and this degree, $\deg\big(\widehat{f}[w, \lambda], B_\delta(0), 0\big)$, does not depend on λ; see [**N**]. Define the family of mappings $\widehat{f}_t[w, \lambda]$ by

$$\widehat{f}_t[w, \lambda] = \widehat{A}(\lambda)w + \frac{1}{t}\widehat{g}(tw, \lambda).$$

The family $\widehat{f}_t[w, \lambda]$ depends on $t \in [0, 1]$ in a continuous way, and the equation $\widehat{f}_t[w, \lambda] = 0$ has no solutions w, $\|w\| = \delta$, for all $0 \neq |\lambda| \leqslant \delta$ and $t \in [0, 1]$. The absence of those solutions for $t \neq 0$ is a consequence of (5.20), while for $t = 0$ it follows from (5.18). Hence for $\lambda \neq 0$ the degree $\deg\big(\widehat{f}_t[w, \lambda], B_\delta(0), 0\big)$ is defined and does not depend on t. Therefore

$$\deg\big(\widehat{f}[w, \lambda], B_\delta(0), 0\big) = \deg\big(\widehat{A}(\lambda)w, B_\delta(0), 0\big), \qquad \lambda \neq 0.$$

The right-hand side of the last equality may be easily calculated [**N**, p. 61] if (5.18) holds. As the result, we get

$$\deg\big(\widehat{f}[w, \lambda], B_\delta(0), 0\big) = (-1)^{p(\lambda)}, \qquad \lambda \neq 0.$$

By (5.19), the last formula contradicts the fact that the degree of $\widehat{f}[w, \lambda]$ does not depend on λ. □

REMARK 5.1. In the standard version of the Krasnoselskiĭ theorem, $B(\lambda) \equiv 0$ and $K(\lambda) = (1 + \lambda)K$. In this case the operator $K(\lambda)$ depends on the parameter λ in a much simpler way. This permits to drop the assumption that K is a selfadjoint operator, as well as (5.15) and the condition that the eigenvalue $\mu = 1$ is simple. The latter condition may be replaced by the assumption that $\mu = 1$ has odd multiplicity.

Now we can state the main result of this section. We assume that all the above restrictions on the coefficients a_{ij}, a, the function φ, and $\partial\omega$ are valid.

THEOREM 5.2. (a) *Let ω be a bifurcation region for equation* (5.3). *Then the problem*

$$Lv = 0, \qquad v \in \overset{\circ}{W}{}^1_2(\omega), \tag{5.21}$$

has a nonzero solution.

(b) *Assume that the set of solutions of* (5.21) *is a one-dimensional space. Then ω is a bifurcation region for the equation* (5.3).

PROOF. (a) Suppose that (5.21) has only the zero solution. Then the operator $I + B[0] - K[0]$ corresponding to this problem is invertible. Hence, for all $h(x)$ with sufficiently small C^1-norm, the operator $I + B[h] - K[h]$ is invertible as well and the norm of the inverse operator is bounded uniformly with respect to h in a sufficiently small $C^1(\overline{\omega}, \mathbb{R}^m)$ neighborhood of the zero point. Now the estimate $\|v\|_{1,2} = o\big(\|v\|_{1,2}\big)$ follows from (5.7) and (5.12) for each solution of (5.7), uniformly with respect to h. The last relation means that, locally, $v \equiv 0$ is a unique solution of the problem (5.7) for all h having sufficiently small C^1-norms. This contradicts the fact that ω is a bifurcation region.

(b) Assume that the function $\widehat{h}(\lambda, x)$ and its partial derivatives up to the second order with respect to λ belong to the space $C^1(\overline{\omega}, \mathbb{R}^m)$ and, in addition, $\widehat{h}(0, x) \equiv 0$, $\widehat{h}_\lambda(0, x) \equiv h(x)$. Then for all sufficiently small $|\lambda|$, the mapping $x \mapsto y(\lambda, x) = x + \widehat{h}(\lambda, x)$ is a diffeomorphism of ω onto the region ω_λ, while the function $\widehat{h}[\lambda] \equiv \widehat{h}(\lambda, x)$ may be treated as a continuously differentiable curve in the space $C^1(\overline{\omega}, \mathbb{R}^m)$. To prove part (b) of the theorem, it suffices to check that $\lambda = 0$ is a bifurcation point for the problem $f[v, \lambda] \equiv F\left[v, \widehat{h}[\lambda]\right] = 0$, with $F[v, h]$ defined in (5.7). Since Theorem 5.1 may be applied to the function $f[v, \lambda]$, it remains to verify that (5.15) is valid. Due to (5.8), we must prove that for some $h(x)$

$$
\begin{aligned}
(5.22)\qquad \int_\omega \{ & \left[a_{ij}\widehat{v}_{x_i}\widehat{v}_{x_j} + a\widehat{v}^2\right] \operatorname{div} h \\
& + \left[a_{ijx_k}\widehat{v}_{x_i}\widehat{v}_{x_j} + a_{x_k}\widehat{v}^2\right]h^k - a_{ij}h^k_{x_j}\widehat{v}_{x_i}\widehat{v}_{x_k} - a_{ij}h^k_{x_i}\widehat{v}_{x_k}\widehat{v}_{x_j}\} \, dx \neq 0,
\end{aligned}
$$

where $\widehat{v}$ is a solution of (5.21) with a unit energy norm. Given twice continuously differentiable functions v and u, denote

$$
l[v, u] = a_{ij}v_{x_i}u_{x_j} + avu, \qquad l_{x_k}[v, u] = a_{ijx_k}v_{x_i}u_{x_j} + a_{x_k}vu.
$$

Since

$$
h^k_{x_i}v_{x_k} = \left(h^k v_{x_k}\right)_{x_i} - h^k v_{x_k x_i}, \qquad h^k_{x_j}u_{x_k} = \left(h^k u_{x_k}\right)_{x_j} - h^k u_{x_k x_j},
$$

we get

$$
\begin{aligned}
-a_{ij}h^k_{x_j}v_{x_i}u_{x_k} &= -a_{ij}\left(h^k u_{x_k}\right)_{x_j}v_{x_i} + a_{ij}\left(u_{x_k}\right)_{x_j}v_{x_i}h^k, \\
-a_{ij}h^k_{x_i}v_{x_k}u_{x_j} &= -a_{ij}\left(h^k v_{x_k}\right)_{x_i}u_{x_j} + a_{ij}\left(v_{x_k}\right)_{x_i}u_{x_j}h^k.
\end{aligned}
$$

Therefore

$$
\begin{aligned}
-a_{ij}&h^k_{x_j}v_{x_i}u_{x_k} - a_{ij}h^k_{x_i}v_{x_k}u_{x_j} \\
&= -l[v, h^k u_{x_k}] - l[h^k v_{x_k}, u] + l[v, u_{x_k}]h^k + l[v_{x_k}, u]h^k.
\end{aligned}
$$

Thus the expression under the integral sign in (5.8) has the form

$$
(5.23)\qquad \operatorname{div}\left(hl[v, u]\right) - l\left[h^k v_{x_k}, u\right] - l\left[v, h^k u_{x_k}\right]
$$

if u and v are twice continuously differentiable functions. Since $\widehat{v}$ is a solution of the elliptic problem (5.21) with coefficients possessing the required smoothness, we have $\widehat{v}(x) \in C^{2,\varepsilon}$ both at inner points of the region ω and at points of those pieces of $\partial\omega$ that belong to $C^{2,\varepsilon}$. Therefore, the expression under the integral sign in (5.22) coincides with (5.23) for $u = v = \widehat{v}$.

Let $\omega(\rho) \subset \omega$, let $\partial\omega(\rho)$ consist of a finite number of pieces belonging to the class $C^{2,\varepsilon}$, let $\partial\omega(\rho) \cap \partial\omega$ be located on smooth (class $C^{2,\varepsilon}$) pieces of $\partial\omega$, and let $\operatorname{meas}(\omega \setminus \omega(\rho)) \to 0$ if $\rho \to 0$. Then (5.22) is equivalent to the relation

$$\lim_{\rho\to 0} \int_{\omega(\rho)} \left[\operatorname{div}\left(hl[\widehat{v}, \widehat{v}]\right) - 2l[h^k \widehat{v}_{x_k}, \widehat{v}]\right] dx \neq 0. \tag{5.24}$$

Applying integration by parts to (5.24) and taking (5.21) into account at the inner points of ω, we can rewrite (5.24) as follows:

$$\lim_{\rho\to 0} \int_{\partial\omega(\rho)} \left[l[\widehat{v}, \widehat{v}]h_\nu - 2a_{ij}\widehat{v}_{x_i}(h^k \widehat{v}_{x_k})\nu_j\right] dS \neq 0. \tag{5.25}$$

Choose $h(x)$ to be identically equal to zero in a small neighborhood of the nonsmooth part of $\partial\omega$. Since the boundary condition $\widehat{v} = 0$ is satisfied on smooth parts of $\partial\omega$, the limit as $\rho \to 0$ may be easily calculated and, given such an h, (5.25) will aquire the form

$$\int_{\partial\omega} a_{ij}\nu_i\nu_j \left(\frac{\partial\widehat{v}}{\partial\nu}\right)^2 h_\nu \, dS \neq 0. \tag{5.26}$$

Since $\partial\widehat{v}/\partial\nu$ cannot vanish on open sets in $\partial\omega$ **[L]**, there exists a function $h(x)$ such that (5.26) holds. □

REMARK 5.2. Relation (5.26) coincides with (3.15) in the selfadjoint case.

References

[Bo] A. Yu. Borisovich, *Lyapunov-Schmidt method and types of singularities of critical points of key function in the problem of bifurcations of minimal surfaces*, Global Analysis—Studies and Applications, Vol. IV, Lecture Notes in Math., vol. 1453, Springer-Verlag, Berlin, 1990, pp. 201–210.

[Bu] V. S. Buslaev, *Calculus of variations*, Leningrad. Gos. Univ., Leningrad, 1980. (Russian)

[F] K. Friedrichs, *Perturbation of spectra in Hilbert space*, Amer. Math. Soc., Providence, RI, 1965.

[L] E. M. Landis, *On some properties of solutions of elliptic equations*, Dokl. Akad. Nauk SSSR **107** (1956), no. 5, 460–463. (Russian)

[N] L. Nirenberg, *Topics in nonlinear functional analysis*, Courant Inst. Math. Sci., New York Univ., New York, 1974.

[Y] K. Yosida, *Functional analysis*, Springer-Verlag, Berlin, 1965.

DEPARTMENT OF MATHEMATICS AND MECHANICS, ST. PETERSBURG STATE UNIVERSITY

Translated by YA. I. BELOPOL'SKAYA

Amer. Math. Soc. Transl.
(2) Vol. **164**, 1995

Existence of a Weak Solution of the Minimax Problem Arising in Coulomb-Mohr Plasticity

S. Repin and G. Seregin

Introduction

In the last decade much progress has been made in the mathematical theory of plasticity. From the mathematical point view, problems arising in perfect plasticity turn out to be highly interesting and difficult. The existence of a weak solution for variational problems in Hencky plasticity, was proved by a number of mathematicians (see, for example, [**AG1, AG2, KT, S1**]). The main feature of such problems is the linear growth of the variational functional with respect to some or all components of the strain tensor. For this reason, the variational problem that follows from the classical formulation in a "natural" way is ill-posed in the sense of the existence of a solution, and should be relaxed. The most acceptable class in which one can prove existence theorems in Hencky plasticity is the space BD of vector-valued functions of bounded deformation, introduced in [**MSCh, Su**] and systematically studied in [**TStr**].

In the present paper we consider similar problems for plasticity theory with the Coulomb-Mohr yield condition. As in the case of Hencky plasticity, functionals of the corresponding variational problems are not coercive with respect to a norm of some reflexive Sobolev space, and that is why it is impossible to prove existence theorems in such spaces. Besides, as compared to the case of Hencky plasticity, the integrand of the Coulomb-Mohr variational functional has a more complicated structure, which is explained, from the physical point of view, by unlike behavior of the elastic-plastic media under expansion and compression. Limiting cases of such a plasticity model are Hencky plasticity and, for instance, free-flowing media that resist compression and do not resist expansion.

The classical setting of the problem, describing the equilibrium of an elastic-plastic media, allows us to give a "natural" functional formulation, which may be reduced to a minimax problem for the pair of functions (σ, u), where σ is the stress tensor and u is the displacement field (Problem A). Unfortunately, such a formulation, or more precisely, such a choice of admissible class of displacement fields does not ensure the existence of a solution, as illustrated by the example considered in §1.

In this paper we define weak solutions of the classical problem and prove some statements to justify our definition. In particular, replacing the Lagrangian of Problem A by a relaxed one and extending the space of admissible displacement fields

1991 *Mathematics Subject Classification*. Primary 49J35, 49J52; Secondary 73E.

so that the new Lagrangian makes sense, we obtain a relaxed minimax problem (Problem A$^+$) which has at least one solution under very general conditions. Using known procedures, we can assign to each Lagrangian of Problems A and A$^+$ a pair of variational problems in duality. These are Problems B and C, B$^+$ and C$^+$. Solutions of Problems B or B$^+$, C or C$^+$ may be interpreted as the stress tensor and displacement vector respectively. It turns out that Problems B and B$^+$ are identical, while Problem C$^+$ is a variational extension of Problem C.

A very interesting and important problem is to investigate the smoothness of weak solutions of the corresponding variational problems with convex functionals (see, for instance, [**AG3, G, LU**]). For plasticity theory with the von Mises yield condition, this question was studied, for example, in the papers [**EK, HKi, S2–S7**]. We intend to devote a separate paper to the analysis of regularity in the case of Coulomb-Mohr plasticity.

The outline of the paper is as follows. The classical problem and its "natural" functional formulation are posed in §1. In §2 we study the coercivity of the Coulomb-Mohr variational functional. In §3 we define weak solutions and prove the existence theorem. In §4 we construct another variational extension of Problem C (Problem C*) which preserves its greatest lower bound, but, in general, does not ensure the existence of a solution. Such an extension may be used to generate new variational-difference schemes based on discontinuous conforming finite elements. Many questions connected with numerical analysis of these schemes in application to problems of Hencky plasticity and other variational problems for functionals with linear growth were studied in the papers [**JS, R1–R6, S1, S9, StT**]. Finally, in Appendix 1, we prove that the kinematic and static coefficients of the limit loads are equal and, moreover, the maximizing stress tensor exists if the value of the parameter α characterizing the shape of the yield surface in the Coulomb-Mohr condition is small enough.

§1. The classical problem and its "natural" functional formulation

We denote by $\mathbb{M}^{m\times n}$ the space of real $m\times n$ matrices and by $\mathbb{M}^{n\times n}_s$ the subspace of symmetric matrices in $\mathbb{M}^{n\times n}$. Next, let $\varepsilon^D=\varepsilon-\frac{1}{n}\operatorname{tr}\varepsilon\,\mathbb{1}$ be the deviator of a matrix $\varepsilon\in\mathbb{M}^{n\times n}_s$, where $\operatorname{tr}\varepsilon$ is its trace and $\mathbb{1}$ is the unit matrix in $\mathbb{M}^{n\times n}_s$. In addition, we use the notation

$$u\cdot v=u_iv_i,\qquad u\otimes v=(u_iv_i),\qquad u=(u_i),\ v=v_i\in\mathbb{R}^n,$$

$$\varepsilon:\sigma=\operatorname{tr}((\varepsilon^\top\sigma)=\varepsilon_{ij}\sigma_{ij},\qquad |\varepsilon|^2=\varepsilon:\varepsilon,\qquad \varepsilon=(\varepsilon_{ij}),\ \sigma=(\sigma_{ij})\in\mathbb{M}^{n\times n}.$$

Here we have adopted the convention of summation over repeated indices, and $\sigma^\top$ is the transpose of σ.

Let Ω be a bounded domain in $\mathbb{R}^n$ ($n=2$ or 3) whose boundary $\partial\Omega$ is Lipschitz continuous, and let $\partial_1\Omega$ and $\partial_2\Omega$ be measurable parts of $\partial\Omega$ such that $\partial_1\Omega\cap\partial_2\Omega=\varnothing$ and $\partial\Omega=\overline{\partial_1\Omega}\cup\overline{\partial_2\Omega}$.

We consider a homogeneous and isotropic elastoplastic body in equilibrium under the action of the given loads. The corresponding boundary value problem for the stress tensor σ and the displacement field u has the following form (see, for example, [**DuL**]):

$$\text{(1.1 a)}\qquad \operatorname{div}\sigma(x)+f(x)=0,\qquad \mathscr{F}(\sigma(x))\leqslant 0\quad\text{for } x\in\Omega,$$

$$\text{(1.1 b)}\qquad \Big(\varepsilon(u)(x)-\frac{1}{n^2K_0}\operatorname{tr}\sigma(x)\mathbb{1}-\frac{1}{2\mu}\sigma^D(x)\Big):(\tau-\sigma(x))\leqslant 0$$

for $x \in \Omega$ and all $\tau \in \mathbb{M}_s^{n\times n}$ such that $\mathscr{F}(\tau) \leqslant 0$,

$$\text{(1.1 c)} \qquad \sigma(x)\nu(x) = F(x) \quad \text{for } x \in \partial_2\Omega, \qquad u(x) = u_0(x) \quad \text{for } x \in \partial_1\Omega.$$

Here $\varepsilon_{ij}(u) = \frac{1}{2}(u_{i,j} + u_{j,i})$ is the strain tensor corresponding to the displacement vector u, $u_{i,j} = \partial u_i/\partial x_j$, $\operatorname{div}\sigma = (\sigma_{ij,j})$, $f = (f_i)$ and $F = (F_i)$ are the given volume and surface loads, ν is the unit outward normal to $\partial\Omega$, the function $\mathscr{F} : \mathbb{M}_s^{n\times n} \to \mathbb{R}$ is given by

$$\mathscr{F}(\tau) := |\tau^D| + \alpha \operatorname{tr}\tau - \sqrt{2}k_*, \tag{1.2}$$

and, finally, the positive constants K_0, μ, α, k_* describe elastic and plastic properties of media.

In this case the condition $\mathscr{F}(\tau) \leqslant 0$ conforms to the Coulomb-Mohr yield law (see **[DP, N]**), which is also called the Drucker-Prager yield condition (see, for example, **[MO, SiDe]**). It may be remarked that if $\alpha = 0$, the Coulomb-Mohr condition becomes the von Mises condition.

Let us give the "natural" functional formulation of the classical problem (1.1)

Problem A. Find the stress tensor $\sigma \in K$ and the displacement field $u = u_0 + V_0$ such that

$$l(u,\tau) \leqslant l(u,\sigma) \leqslant l(v,\sigma) \tag{1.3}$$

for all $v \in u_0 + V_0$ and $\tau \in K$.

Here the Lagrangian l is introduced by means of the expression

$$l(v,\tau) := \int_\Omega \Big(\varepsilon(v) : \tau - \tfrac{1}{2}a(\tau,\tau) - f \cdot v\Big)\, dx - \int_{\partial_2\Omega} F \cdot v \, dl$$

in which we use the bilinear form

$$a(\tau,\sigma) = \frac{1}{n^2 K_0} \operatorname{tr}\tau \operatorname{tr}\sigma + \frac{1}{2\mu}\tau^D : \sigma^D$$

and the set K of admissible stress tensors τ together with the set $u_0 + V_0$ of admissible displacement fields v are defined as follows:

$$K := \big\{\, \tau \in \Sigma : \mathscr{F}(\tau(x)) \leqslant 0 \text{ for a.e. } x \text{ in } \Omega \,\big\},$$
$$V_0 := \big\{\, v \in V : v = 0 \text{ on } \partial_1\Omega \,\big\}.$$

In these formulas we have used the Banach spaces $\Sigma := L_2(\Omega; \mathbb{M}_s^{n\times n})$ and $V := W_2^1(\Omega;\mathbb{R}^n)$, where $L_p(\Omega;\mathbb{R}^m)$ and $W_p^l(\Omega;\mathbb{R}^m)$ denote, as usual, the Lebesgue and Sobolev spaces of vector-valued functions from Ω into $\mathbb{R}^m$. We also assume that

$$f \in L_n(\Omega;\mathbb{R}^n), \qquad F \in L_\infty(\partial_2\Omega;\mathbb{R}^n), \qquad u_0 \in V. \tag{1.4}$$

It is not difficult to show that if there are smooth functions σ and u that satisfy (1.1), then they are a solution of Problem A. Moreover, if a pair of functions σ, u is a saddle point of the Lagrangian l on the set $(u_0 + V_0) \times K$, then it is also a solution of (1.1).

The Lagrangian l generates a couple of variational problems, which are in duality.

Problem B. Find the stress tensor $\sigma \in K \cap Q_f$ such that

$$R(\sigma) = \sup\{ R(\tau) : \tau \in K \cap Q_f \}.$$

Here

$$R(\tau) = -\frac{1}{2}\int_\Omega a(\tau,\tau)\,dx + \int_\Omega \varepsilon(u_0) : \tau\,dx - M(u_0),$$

$$M(u) = \int_{\partial_2\Omega} F \cdot u\,dl + \int_\Omega f \cdot u\,dx$$

and $Q_f = \{\sigma \in \Sigma : \operatorname{div}\sigma + f = 0,\ \sigma\nu = F \text{ on } \partial_2\Omega\}$ is the set of the tensor fields satisfying the equilibrium equations. In this definition we understand the relation $\operatorname{div}\sigma + f = 0$ in the sense of distributions, while the condition $\sigma\nu = F$ on $\partial_2\Omega$ for $\sigma \in \Sigma$ with $\operatorname{div}\sigma \in L_2(\Omega;\mathbb{R}^n)$ means that

$$\int_\Omega \big(\sigma : \varepsilon(v) + v \cdot \operatorname{div}\sigma\big)\,dx = \int_{\partial_2\Omega} F \cdot v\,dl \quad \forall v \in V_0.$$

Problem C. Find the displacement vector $u \in u_0 + V_0$ such that

$$J(u) = \inf\{ J(v) : v \in u_0 + V_0 \}, \tag{1.5}$$

where

$$J(v) = \sup\{ l(\tau,v) : \tau \in K \}. \tag{1.6}$$

According to minimax theory, the following theorem holds.

THEOREM 1.1. *A pair (σ,u) is a solution of Problem* A *if and only if the vector u is a solution of Problem* C *and the tensor σ is a solution of Problem* B.

The functional $-R(\sigma)$ is continuous, and coercive on the set Σ, and $Q_f \cap K$ is a convex closed subset in Σ; this implies the following assertion.

THEOREM 1.2. *If $Q_f \cap K \neq \varnothing$, then Problem* B *is uniquely solvable.*

Here and later on we assume that $Q_f \cap K$ is not empty.

Hence the solvability of A is equivalent to the solvability of C. Calculating the supremum in (1.6) when the set K is determined by means of convex function $\mathscr{F}$ (see (1.2)), we obtain the following form of the functional J:

$$J(u) = J_0(u) - M(u), \qquad J_0(u) = \int_\Omega g(\varepsilon(u))\,dx,$$

where

$$g(\varepsilon) = \sup\Big\{ \varepsilon : \tau - \tfrac{1}{2}a(\tau,\tau) : \tau \in \mathbb{M}^{n\times n}_s,\ \mathscr{F}(\tau) \leqslant 0 \Big\}.$$

If we introduce the convex function $\varphi\colon \mathbb{M}^{n\times n}_s \to \mathbb{R}$ as follows:

$$\varphi(\varepsilon) = \begin{cases} \dfrac{K_0}{2}\left(\operatorname{tr}\varepsilon + \dfrac{\sqrt{2}k_*}{\alpha K_0 n}\right)^2 + \mu|\varepsilon^D|^2 & \text{if } 2\mu|\varepsilon^D| + \alpha K_0 n \operatorname{tr}\varepsilon \leqslant 0, \\ \dfrac{k_*^2}{\alpha^2 K_0 n^2} + \dfrac{\sqrt{2}k_*}{\alpha n}\operatorname{tr}\varepsilon + D\Big[\Big(|\varepsilon^D| - \dfrac{1}{\alpha n}\operatorname{tr}\varepsilon\Big)_+\Big]^2 & \\ \qquad\qquad \text{if } 2\mu|\varepsilon^D| + \alpha K_0 n \operatorname{tr}\varepsilon > 0, & \end{cases} \tag{1.7 a}$$

where

$$D = \frac{1}{2}\left(\frac{1}{2\mu} + \frac{1}{\alpha^2 K_0 n^2}\right)^{-1},$$

then

$$g(\varepsilon) = \varphi\left(\varepsilon - \frac{\sqrt{2}k_*}{\alpha K_0 n^2}\mathbb{1}\right). \tag{1.7 b}$$

In (1.7 a) and later on $(z)_\pm$ is the positive (negative) part of z, i.e., $(z)_\pm = \frac{1}{2}(|z| \pm z)$.

It is easy to see that the function φ satisfies the inequality

$$\beta\frac{\sqrt{2}k_*}{\alpha\sqrt{n}} \leqslant \frac{\varphi(\varepsilon)}{|\varepsilon|} - \frac{k_*^2}{\alpha^2 n^2 K_0 |\varepsilon|} \leqslant \frac{\sqrt{2}k_*}{\alpha\sqrt{n}}, \qquad \beta = \alpha\sqrt{n}(1+\alpha^2 n)^{-1/2},$$

when $|\varepsilon^D| \leqslant \frac{1}{\alpha n}\operatorname{tr}\varepsilon$.

That is why it is impossible to establish the coercivity of the functional J on the affine manifold $u_0 + V_0$ of the reflexive space V and use known theorems of convex analysis to prove the existence of the solution. Moreover, an example set forth below shows that Problem C (and thus Problem A too) may have no solution.

Let $n = 2$, $\alpha = 1/\sqrt{2}$ and $\Omega = \{(r,\varphi) : r_1 < r < r_2,\ -\pi < \varphi \leqslant \pi\}$, where (r,φ) is a polar coordinate system. We shall consider the elastoplastic problem for the annulus Ω when

$$\begin{gathered} f = 0, \qquad \partial_1\Omega = \partial\Omega, \qquad u = (u_r, u_\varphi), \\ u_\varphi = 0 \quad \text{on } \partial\Omega, \qquad u_r = 0 \quad \text{if } r = r_1, \qquad u_r = U \quad \text{if } r = r_2 \end{gathered} \tag{1.8}$$

and

$$U_* := \frac{k_*(r_2^2 - r_1^2)}{2(K_0+\mu)r_2} < U < U_{**} := \frac{k_*}{2K_0}r_2. \tag{1.9}$$

Now our aim is to show that in this case Problems A and C are unsolvable.

To do this we introduce the tensor

$$\sigma^* = \begin{pmatrix} \sigma^*_{rr}, \sigma^*_{r\varphi} \\ \sigma^*_{r\varphi}, \sigma^*_{\varphi\varphi} \end{pmatrix}$$

with components

$$\sigma^*_{rr} = 2\left(K_0 C_1 - \frac{\mu C_2}{r^2}\right), \qquad \sigma^*_{\varphi\varphi} = 2\left(K_0 C_1 + \frac{\mu C_2}{r^2}\right), \qquad \sigma^*_{r\varphi} = 0, \tag{1.10}$$

where

$$C_1 = \frac{Ur_2 - \Delta r_1}{r_2^2 - r_1^2}, \qquad C_2 = r_1 r_2\frac{\Delta r_2 - Ur_1}{r_2^2 - r_1^2}$$

and Δ is some positive constant.

First we prove that $\sigma^* \in Q_f \cap K$. Indeed, it is easy to check that the equilibrium equations

$$\frac{\partial \sigma_{rr}}{\partial r} + \frac{1}{r}\frac{\partial \sigma_{r\varphi}}{\partial \varphi} + \frac{\sigma_{rr} - \sigma_{\varphi\varphi}}{r} = 0, \qquad \frac{\partial \sigma_{r\varphi}}{\partial r} + \frac{1}{r}\frac{\partial \sigma_{\varphi\varphi}}{\partial \varphi} + 2\frac{\sigma_{r\varphi}}{r} = 0$$

are fulfilled, and thus $\sigma^* \in Q_f$. In addition, we have

$$\mathcal{F}(\sigma^*) = \frac{1}{\sqrt{2}}\left(\frac{4\mu|C_2|}{r^2} + 4K_0C_1 - 2k_*\right).$$

Consequently if the conditions

$$(1.11) \qquad C_2 < 0, \qquad -\mu C_2/r_1^2 + K_0C_1 = k_*/2$$

hold, then $\mathcal{F}(\sigma^*) \leqslant 0$ for all $r \in [r_1, r_2]$ and, moreover, we have $\mathcal{F}(\sigma^*) < 0$ if $r \in]r_1, r_2]$.

If we suppose that

$$(1.12) \qquad \Delta = \left(K_0r_1^2 + \mu r_2^2\right)^{-1}\left(Ur_1r_2(\mu + K_0) - k_*r_1(r_2^2 - r_1^2)/2\right),$$

while the value of boundary condition is chosen in accordance with (1.9), then conditions (1.11) are satisfied and

$$0 < \Delta < U_{**}r_1/r_2.$$

Now let us show that σ^* is the solution of Problem B. Using well-known arguments, it is easy to establish that σ^* minimizes $-R$ on the set $Q_f \cap K$ if and only if it satisfies the variational inequality

$$(1.13) \qquad \int_\Omega \left(\varepsilon(u_0) : (\tau - \sigma^*) - a(\sigma^*, \tau - \sigma^*)\right) dx \leqslant 0 \quad \forall \tau \in Q_f \cap K,$$

where u_0 is some sufficiently smooth function subject to the boundary conditions (1.8).

Introduce the function u as follows:

$$(1.14) \qquad u = (u_r, u_\varphi), \qquad u_r = C_1r + C_2r^{-1}, \qquad u_\varphi = 0.$$

Due to the choice of constants C_1, C_2, and Δ, this function obeys the boundary conditions $u_r(r_1) = \Delta$ and $u_r(r_2) = U$, and the corresponding strain tensor

$$\varepsilon(u) = \begin{pmatrix} \varepsilon_{rr}(u), \varepsilon_{r\varphi}(u) \\ \varepsilon_{r\varphi}(u), \varepsilon_{\varphi\varphi}(u) \end{pmatrix}$$

has the components

$$\varepsilon_{rr}(u) = \frac{\partial u_r}{\partial r} = C_1 - C_2r^{-2}, \qquad \varepsilon_{\varphi\varphi}(u) = \frac{1}{r}\frac{\partial u_\varphi}{\partial \varphi} + \frac{u_r}{r} = C_1 + C_2r^{-2},$$

$$\varepsilon_{r\varphi}(u) = \frac{1}{2}\left(\frac{\partial u_\varphi}{\partial r} - \frac{u_\varphi}{r} + \frac{1}{r}\frac{\partial u_r}{\partial \varphi}\right) = 0.$$

For any $\tau \in Q_f \cap K$ (see Appendix 2) there is a sequence $\tau^m \in C^\infty(\overline{\Omega}; \mathbb{M}_s^{n\times n}) \cap K$ such that $\tau^m \to \tau$ in Σ and $\operatorname{div}\tau^m \to \operatorname{div}\tau$ in $L_2(\Omega; \mathbb{R}^n)$. Setting

$$I_m(\tau) = \int_\Omega \{\varepsilon(u_0) : (\tau^m - \sigma^*) + (u_0 - u) \cdot \operatorname{div}(\tau^m - \sigma^*) - a(\sigma^*, \tau^m - \sigma^*)\}\, dx,$$

we find that

$$(1.15)\qquad I_m(\tau) \longrightarrow \int_\Omega \big(\varepsilon(u_0):(\tau-\sigma^*) - a(\sigma^*,\tau-\sigma^*)\big)\,dx \quad \forall\tau\in Q_f\cap K.$$

All functions forming the integrand of $I_m(\tau)$ are smooth. Hence we can integrate by parts and obtain

$$I_m(\tau) = \int_{\partial\Omega}((\tau^m-\sigma^*)\nu)\cdot(u_0-u)\,dl + \int_\Omega \big(\varepsilon(u):(\tau^m-\sigma^*) - a(\sigma^*,\tau^m-\sigma^*)\big)\,dx.$$

Taking the definition of the tensors $\varepsilon(u)$ and σ^* into account, we observe that $\varepsilon(u):(\tau^m-\sigma^*) - a(\sigma^*,\tau^m-\sigma^*) = 0$. Hence

$$I_m(\tau) = \int_{\partial\Omega}((\tau^m-\sigma^*)\nu)\cdot(u_0-u)\,dl = \big(\tau^m_{rr}(r_1) - \sigma^*_{rr}(r_1)\big)\Delta.$$

Next, it is easy to check that owing to the chosen value of parameter Δ we have $\sigma^*_{rr}(r_1) = k_*$. As soon as $\tau^m \in K$, we establish that

$$|\tau^m_{rr} - \tau^m_{\varphi\varphi}| + \tau^m_{rr} + \tau^m_{\varphi\varphi} \leqslant 2k_* \implies \tau^m_{rr} \leqslant k_*.$$

Therefore (see (1.12)), we have $I_m(\tau) \leqslant 0$. This, together with (1.15) and the fact that $\tau \in Q_f \cap K$ is arbitrary, leads us to the variational inequality (1.13). So, the tensor σ^* is the solution of Problem B.

Now let us assume that Problem A has a solution $u^* \in u_0 + V_0$. Because the tensor σ^* is smooth and $\mathscr{F}(\sigma^*) < 0$ for $r \in \,]r_1, r_2]$, using the first inequality in (1.3), we can uniquely determine the strain tensor $\varepsilon(u^*)$ for $r \in \,]r_1, r_2]$ and find that $\varepsilon(u^*) = \varepsilon(u)$ on this interval. Since $u^*_r(r_2) = u_r(r_2) = U$, we conclude that $u^* = u$ for $r \in \,]r_1, r_2]$. Thus, $u^*(r_1) = \Delta > 0$ and, consequently, $u^* \notin u_0 + V_0$. This means that Problems C and A are unsolvable. Nevertheless, as will be shown later, the pair u^*, σ^* is a weak solution of the classical elastoplastic problem in the sense of the definition to be given in §3.

§2. On the coercivity of the functional J

On the space $\mathbb{M}^{n\times n}_s$ let us define the functional $\varepsilon \mapsto [\varepsilon]$ by putting

$$[\varepsilon] := \left(((\operatorname{tr}\varepsilon)_-)^2 + \left(\left(|\varepsilon^D| - \frac{1}{\alpha n}\operatorname{tr}\varepsilon\right)_+\right)^2\right)^{1/2}.$$

LEMMA 2.1. *The functional* $[\varepsilon]$ *has the following properties*:

$$(2.1)\qquad [\lambda\varepsilon] = \lambda[\varepsilon] \quad \forall\lambda\in\mathbb{R}_+,\ \forall\varepsilon\in\mathbb{M}^{n\times n}_s,$$

$$(2.2)\qquad [\varepsilon_1+\varepsilon_2] \leqslant [\varepsilon_1] + [\varepsilon_2] \quad \forall\varepsilon_1,\varepsilon_2\in\mathbb{M}^{n\times n}_s.$$

PROOF. The property (2.1) is evident. To prove (2.2), we can use the well-known inequalities $(z_1+z_2)_+ \leqslant (z_1)_+ + (z_2)_+$ and $(z_1+z_2)_- \leqslant (z_1)_- + (z_2)_-$. Setting $\theta(\varepsilon) := |\varepsilon^D| - \frac{1}{\alpha n}\operatorname{tr}\varepsilon$, we obtain

$$\big(\theta(\varepsilon_1+\varepsilon_2)\big)_+ \leqslant \big(\theta(\varepsilon_1)+\theta(\varepsilon_2)\big)_+ \leqslant \big(\theta(\varepsilon_1)\big)_+ + \big(\theta(\varepsilon_2)\big)_+$$

and, therefore,

$$\begin{aligned}[\varepsilon_1+\varepsilon_2] &\leqslant \Big(\big((\operatorname{tr}\varepsilon_1+\operatorname{tr}\varepsilon_2)_-\big)^2+\big((\theta(\varepsilon_1))_+ + (\theta(\varepsilon_2))_+\big)^2\Big)^{1/2}\\ &\leqslant \Big(\big((\operatorname{tr}\varepsilon_1)_-\big)^2+\big((\theta(\varepsilon_1))_+\big)^2\Big)^{1/2}+\Big(\big((\operatorname{tr}\varepsilon_2)_-\big)^2+\big((\theta(\varepsilon_2))_+\big)^2\Big)^{1/2}\\ &= [\varepsilon_1]+[\varepsilon_2].\end{aligned}$$

□

COROLLARY. *It follows from* (2.1) *and* (2.2) *that the functional* $\varepsilon \mapsto [\varepsilon]$ *is convex.*

LEMMA 2.2. *If the function* φ *is defined according to* (1.7 a), *then*

$$\varphi(\varepsilon) \geqslant C_1|\varepsilon| + C_2[\varepsilon]^2 + C_3, \quad \forall \varepsilon \in \mathbb{M}_s^{n\times n}, \tag{2.3}$$

where the constants C_1, C_2, C_3 *do not depend on* ε *and* $C_1, C_2 > 0$.

PROOF. First we prove that

$$\varphi(\varepsilon) \geqslant C_4\big((\operatorname{tr}\varepsilon)_-\big)^2 + C_5, \quad \forall \varepsilon \in \mathbb{M}_s^{n\times n}, \tag{2.4}$$

where the constants C_4, C_5 do not depend on ε.

Indeed, if $\operatorname{tr}\varepsilon \geqslant 0$, then $(\operatorname{tr}\varepsilon)_- = 0$ and (see (1.7))

$$\begin{aligned}\varphi(\varepsilon) &= \frac{k_*^2}{\alpha^2 n^2 K_0} + \frac{\sqrt{2}k_*}{\alpha n}\operatorname{tr}\varepsilon + D\Big(\Big(|\varepsilon^D| - \frac{1}{\alpha n}\operatorname{tr}\varepsilon\Big)_+\Big)^2\\ &\geqslant \Big(\frac{D}{2\alpha^2 n^2} + \frac{K_0}{4}\Big)\big((\operatorname{tr}\varepsilon)_-\big)^2.\end{aligned} \tag{2.5}$$

If $\operatorname{tr}\varepsilon < 0$, then $(\operatorname{tr}\varepsilon)_- = -\operatorname{tr}\varepsilon$ and we have to consider two cases. In the first one $2\mu|\varepsilon^D| + \alpha n K_0 \operatorname{tr}\varepsilon \leqslant 0$ and, consequently,

$$\varphi(\varepsilon) = \frac{K_0}{2}\Big(\operatorname{tr}\varepsilon + \frac{\sqrt{2}k_*}{\alpha n K_0}\Big) + \mu|\varepsilon^D|^2 \geqslant \frac{K_0}{2}\Big(\frac{(\operatorname{tr}\varepsilon)^2}{2} - \frac{2k_*^2}{\alpha^2 n^2 K_0^2}\Big). \tag{2.6}$$

In the second case $2\mu|\varepsilon^D| + \alpha n K_0 \operatorname{tr}\varepsilon > 0$ and

$$\begin{aligned}\varphi(\varepsilon) &= \frac{k_*^2}{\alpha^2 n^2 K_0} + \frac{\sqrt{2}k_*}{\alpha n}\operatorname{tr}\varepsilon + D\Big(|\varepsilon^D| - \frac{1}{\alpha n}\operatorname{tr}\varepsilon\Big)^2\\ &\geqslant \frac{k_*^2}{\alpha^2 n^2 K_0} + \frac{\sqrt{2}k_*}{\alpha n}\operatorname{tr}\varepsilon + \frac{D}{\alpha^2 n^2}(\operatorname{tr}\varepsilon)^2\\ &\geqslant \frac{D}{2\alpha^2 n^2}\big((\operatorname{tr}\varepsilon)_-\big)^2 + \frac{k_*^2}{\alpha^2 n^2 K_0} - \frac{k_*^2}{D}.\end{aligned} \tag{2.7}$$

Now, (2.4) follows from (2.5)–(2.7) if we choose suitable values of the constants C_4, C_5.

The estimate

$$\varphi(\varepsilon) \geqslant C_6\Big(\Big(|\varepsilon^D| - \frac{1}{\alpha n}\operatorname{tr}\varepsilon\Big)_+\Big)^2 + C_7, \quad \forall \varepsilon \in \mathbb{M}_s^{n\times n}, \tag{2.8}$$

where C_6, C_7 do not depend on ε and $C_6 > 0$, can be established with the help of similar considerations.

If $|\varepsilon^D| - \frac{1}{\alpha n}\operatorname{tr}\varepsilon > 0$, then there are two possibilities. In the first one $2\mu|\varepsilon^D| + \alpha n K_0 \operatorname{tr}\varepsilon \leqslant 0$, and we have

$$\begin{aligned}\varphi(\varepsilon) &= \frac{K_0}{2}\Big(\operatorname{tr}\varepsilon + \frac{\sqrt{2}k_*}{\alpha n K_0}\Big)^2 + \mu|\varepsilon^D|^2 \geqslant \frac{K_0}{2}\Big(\frac{(\operatorname{tr}\varepsilon)^2}{2} - \frac{2k_*^2}{\alpha^2 n^2 K_0^2}\Big) + \mu|\varepsilon^D|^2 \\ &\geqslant \min\Big\{\frac{K_0\alpha^2 n^2}{4}, \mu\Big\}\Big(\frac{(\operatorname{tr}\varepsilon)^2}{\alpha^2 n^2} + |\varepsilon^D|^2\Big) - \frac{k_*^2}{\alpha^2 n^2 K_0} \\ &\geqslant \frac{1}{2}\min\Big\{\frac{K_0\alpha^2 n^2}{4}, \mu\Big\}\Big(\Big(|\varepsilon^D| - \frac{1}{\alpha n}\operatorname{tr}\varepsilon\Big)_+\Big)^2 - \frac{k_*^2}{\alpha^2 n^2 K_0}.\end{aligned} \tag{2.9}$$

In the second case $2\mu|\varepsilon^D| + \alpha n K_0 \operatorname{tr}\varepsilon > 0$ and

$$\begin{aligned}\varphi(\varepsilon) &= \frac{k_*^2}{\alpha^2 n^2 K_0} + \frac{\sqrt{2}k_*}{\alpha n}\operatorname{tr}\varepsilon + D\Big(|\varepsilon^D| - \frac{1}{\alpha n}\operatorname{tr}\varepsilon\Big)^2 \\ &= \frac{k_*^2}{\alpha^2 n^2 K_0} + \sqrt{2}k_*|\varepsilon^D| - \sqrt{2}k_*\Big(|\varepsilon^D| - \frac{1}{\alpha n}\operatorname{tr}\varepsilon\Big) + D\big(|\varepsilon^D| - \frac{1}{\alpha n}\operatorname{tr}\varepsilon\big)^2 \\ &\geqslant \frac{D}{2}\Big(\Big(|\varepsilon^D| - \frac{1}{\alpha n}\operatorname{tr}\varepsilon\Big)_+\Big)^2 + \frac{k_*^2}{\alpha^2 n^2 K_0} - \frac{k_*^2}{D}.\end{aligned} \tag{2.10}$$

Finally, for $|\varepsilon^D| - \frac{1}{\alpha n}\operatorname{tr}\varepsilon \leqslant 0$, we have $\operatorname{tr}\varepsilon \geqslant 0$, $\big(|\varepsilon^D| - \frac{1}{\alpha n}\operatorname{tr}\varepsilon\big)_+ = 0$, and thus

$$\begin{aligned}\varphi(\varepsilon) &= \frac{k_*^2}{\alpha^2 n^2 K_0} + \frac{\sqrt{2}k_*}{\alpha n}\operatorname{tr}\varepsilon \\ &\geqslant \frac{1}{2}\Big(D + \min\Big\{\frac{K_0\alpha^2 n^2}{4}, \mu\Big\}\Big)\Big(\Big(|\varepsilon^D| + \frac{1}{\alpha n}\operatorname{tr}\varepsilon\Big)_+\Big)^2 + \frac{k_*^2}{\alpha^2 n^2 K_0}.\end{aligned} \tag{2.11}$$

Now, choosing suitable constants, we obtain (2.8) from (2.9)–(2.11).

The proof will be finished if we establish the inequality

$$\varphi(\varepsilon) \geqslant C_8|\varepsilon| + C_9, \quad \forall \varepsilon \in \mathbb{M}_s^{n\times n}. \tag{2.12}$$

Let us set $\beta = \sqrt{2}k_*/(1 + \alpha\sqrt{n})$. If $\tau \in \mathbb{M}_s^{n\times n}$, $|\tau| \leqslant \beta$, then we have $|\tau^D| \leqslant \beta$, $|\operatorname{tr}\tau| \leqslant \sqrt{n}\beta$, and, consequently, $|\tau^D| + \alpha\operatorname{tr}\tau \leqslant \beta(1 + \alpha\sqrt{n}) = \sqrt{2}k_*$. Hence

$$\begin{aligned}g(\varepsilon) &= \sup\Big\{\varepsilon : \tau - \tfrac{1}{2}a(\tau,\tau) : \mathscr{F}(\tau) \leqslant 0\Big\} \\ &\geqslant \sup\Big\{\varepsilon : \tau - \tfrac{1}{2}a(\tau,\tau) : |\tau| \leqslant \beta\Big\} \geqslant \sup\{\varepsilon : \tau : |\tau| \leqslant \beta\} = \beta|\varepsilon|.\end{aligned}$$

Since

$$\varphi(\varepsilon) = g\Big(\varepsilon + \frac{\sqrt{2}k_*}{\alpha K_0 n^2}\mathbb{1}\Big) \geqslant \beta\Big|\varepsilon + \frac{\sqrt{2}k_*}{\alpha K_0 n^2}\mathbb{1}\Big| \geqslant \beta|\varepsilon| - \beta\frac{\sqrt{2}k_*}{\alpha K_0 n\sqrt{n}} \quad \forall \varepsilon \in \mathbb{M}_s^{n\times n}$$

we have proved (2.12). Lemma 2.2 follows from (2.4), (2.8), and (2.12). □

As soon as $[\mathbb{1}] = 0$ we can easily obtain the estimate

$$J_0(u) \geqslant C_1 \int_\Omega |\varepsilon(u)|\, dx + C_2 \int_\Omega [\varepsilon(u)]^2\, dx + C_3 - \frac{C_1\sqrt{2}k_*}{\alpha n\sqrt{n}K_0} \operatorname{meas}\Omega. \tag{2.13}$$

Indeed,

$$\begin{aligned} J_0(u) &\geqslant \int_\Omega \varphi\left(\varepsilon(u) - \frac{\sqrt{2}k_*}{\alpha n^2 K_0}\mathbb{1}\right) dx \\ &\geqslant C_1 \int_\Omega \left|\varepsilon(u) - \frac{\sqrt{2}k_*}{\alpha n^2 K_0}\mathbb{1}\right| dx + C_2 \int_\Omega \left[\varepsilon(u) - \frac{\sqrt{2}k_*}{\alpha n^2 K_0}\mathbb{1}\right]^2 dx + C_3. \end{aligned}$$

But

$$[\varepsilon(u)] \leqslant \left[\varepsilon(u) - \frac{\sqrt{2}k_*}{\alpha n^2 K_0}\mathbb{1}\right] + \frac{\sqrt{2}k_*}{\alpha n^2 K_0}[\mathbb{1}] = \left[\varepsilon(u) - \frac{\sqrt{2}k_*}{\alpha n^2 K_0}\mathbb{1}\right]$$

and thus (2.13) follows from the last two inequalities.

Unfortunately an estimate analogous to (2.13) cannot be established for the functional J. Moreover, for some loads F and f the greatest lower bound of Problem C may be equal to $-\infty$. This situation is typical of perfect plasticity problems and is related to the limit load phenomenon. That is why Problem C is meaningful only if certain supplementary conditions are added. We shall call these conditions the "safe load hypothesis" and suppose that loads F, f are "safe" if for some $\delta > 0$ there is a stress tensor $\sigma^1 \in \Sigma$ such that

$$\operatorname{div}\sigma^1 + f = 0 \quad \text{in } \Omega, \qquad \sigma^1\nu = F \quad \text{on } \partial_2\Omega, \tag{2.14 a}$$

$$|\sigma^{1D}| + \alpha \operatorname{tr}\sigma^1 \leqslant \sqrt{2}k_* - \delta \quad \text{a.e. in } \Omega. \tag{2.14 b}$$

THEOREM 2.1. *If conditions* (1.4) *and* (2.14) *are satisfied, then*

$$J(u) \geqslant C_{10} \int_\Omega |\varepsilon(v)|\, dx + C_{11} \int_\Omega [\varepsilon(v)]^2\, dx + C_{12}, \quad \forall v \in u_0 + V_0 \tag{2.15}$$

where the constants C_{10}, C_{11}, C_{12} do not depend on v and $C_{10}, C_{11} > 0$.

PROOF. For any $\varkappa \in \Sigma$ satisfying the condition $|\varkappa^D| + \alpha \operatorname{tr}\varkappa \leqslant \delta/2$ a.e. in Ω we find that $\sigma^1 + \varkappa \in K$ and

$$|\sigma^{1D} + \varkappa^D| + \alpha \operatorname{tr}(\sigma^1 + \varkappa) \leqslant \sqrt{2}k_* - \delta/2 \quad \text{a.e. in } \Omega.$$

According to (2.14 a) we have the identity

$$\int_{\partial_2\Omega} F \cdot w\, dl = \int_\Omega (\sigma^1 : \varepsilon(w) - f \cdot w)\, dx \quad \forall w \in V_0.$$

Hence

$$M(v) = M(u_0) + \int_\Omega \sigma^1 : (\varepsilon(v) - \varepsilon(u_0))\, dx \quad \forall v \in u_0 + V_0$$

and

$$J(v) = \sup\Big\{ \int_\Omega \Big(\tau : \varepsilon(v) - \tfrac{1}{2}a(\tau,\tau)\Big)\,dx - M(v) : \tau \in K \Big\}$$
$$= \sup\Big\{ \int_\Omega \Big((\tau - \sigma^1) : \varepsilon(v) - \tfrac{1}{2}a(\tau,\tau)\Big)\,dx : \tau \in K \Big\}$$
$$- M(u_0) + \int_\Omega \sigma^1 : \varepsilon(u_0)\,dx$$
$$\geqslant \int_\Omega \sigma^1 : \varepsilon(u_0)\,dx$$
$$+ \sup\Big\{ \int_\Omega \Big(\varkappa : \varepsilon(v) - \tfrac{1}{2}a(\sigma^1 + \varkappa, \sigma^1 + \varkappa)\Big)\,dx :$$
$$\varkappa \in \Sigma, |\varkappa^D| + \alpha \operatorname{tr} \varkappa \leqslant \delta/2 \text{ a.e. in } \Omega \Big\} - M(u_0).$$

Since

$$a(\sigma^1 + \varkappa, \sigma^1 + \varkappa) \leqslant (1 + \beta^2)a(\sigma^1, \sigma^1) + \Big(1 + 1/\beta^2\Big)a(\varkappa, \varkappa)$$

for any $\beta \neq 0$, we arrive at the inequality

$$J(v) \geqslant \sup\Big\{ \int_\Omega \Big(\varkappa : \varepsilon(u) - \frac{1}{2}\Big(1 + \frac{1}{\beta^2}\Big)a(\varkappa,\varkappa)\Big)\,dx : \varkappa \in \Sigma,$$
$$|\varkappa^D| + \alpha \operatorname{tr} \varkappa \leqslant \frac{\delta}{2} \text{ a.e. in } \Omega \Big\} + C_{13}(\beta, u_0),$$

where

$$C_{13}(\beta, u_0) = \int_\Omega \Big(\sigma^1 : \varepsilon(u_0) - \frac{1 + \beta^2}{2}a(\sigma^1, \sigma^1)\Big)\,dx - M(u_0).$$

Next, let us set $\varkappa = \beta^2\eta/(1 + \beta^2)$. Then $\eta \in \Sigma$, $|\eta^D| + \alpha \operatorname{tr} \eta \leqslant \delta(\beta^2 + 1)/2\beta^2$ a.e. in Ω, and, consequently,

$$J(v) \geqslant \frac{\beta^2}{1 + \beta^2} \sup\Big\{ \int_\Omega \Big(\eta : \varepsilon(v) - \tfrac{1}{2}a(\eta,\eta)\Big)\,dx : \eta \in \Sigma,$$
$$|\eta^D| + \alpha \operatorname{tr} \eta \leqslant \frac{\delta(\beta^2 + 1)}{2\beta^2} \text{ a.e. in } \Omega \Big\} + C_{13}(\beta, u_0).$$

Finally, choosing β^2 so that $\sqrt{2}k_* = \delta(\beta^2 + 1)/(2\beta^2)$, we obtain

$$J(v) \geqslant \frac{\delta}{2\sqrt{2}k_*} J_0(v) + C_{14}(\delta, u_0) \quad \text{for all } v \in u_0 + V_0,$$

which, together with (2.13), gives us (2.15).

Now we shall describe the particular case when the tensor satisfying (2.14) exists if only $Q_f \cap K \neq \varnothing$. Let

$$\partial_1 \Omega = \partial \Omega \tag{2.16}$$

and $\sigma^0 \in Q_f \cap K$. Introduce the tensor $\sigma^1 = \sigma^0 + p\mathbb{1}$, where p is some negative constant. In this case

$$\operatorname{div} \sigma^1 + f = 0 \quad \text{in } \Omega, \qquad |\sigma^{1D}| + \alpha \operatorname{tr} \sigma^1 \leqslant \sqrt{2}k_* + \alpha p \quad \text{a.e. in } \Omega$$

and, consequently, (2.14) is fulfilled if we take $p < -\delta/\alpha$.

But in the general case it may be very difficult to check conditions (2.14). Thus, it is more efficient to use formulations related to the limit load. By means of the Lagrangian

$$l_0(v,\tau) := \int_\Omega \tau : \varepsilon(v)\, dx$$

we can define two numbers: the static and kinematic coefficients of the limit load:

$$\begin{aligned} \lambda_s &:= \sup\{\, r(\tau) : \tau \in K \,\}, \qquad & r(\tau) &:= \inf\{\, l_0(v,\tau) : v \in \overline{V}_0 \,\},\\ \lambda_k &:= \inf\{\, j(v) : v \in \overline{V}_0 \,\}, \qquad & j(v) &:= \sup\{\, l_0(v,\tau) : \tau \in K \,\}, \end{aligned} \tag{2.17}$$

where $\overline{V}_0 = \{\, v \in V_0 : M(v) = 1 \,\}$.

Directly from these definitions it follows that

$$\lambda_s \leqslant \lambda_k, \qquad j(v) = \int_\Omega g_0\big(\varepsilon(v)\big)\, dx, \tag{2.18}$$

where $g_0 \colon \mathbb{M}^{n\times n}_s \longrightarrow \mathbb{R}$ can be represented in the form

$$\begin{aligned} g_0(\varepsilon) &:= \sup\{\, \tau : \varepsilon : \tau \in \mathbb{M}^{n\times n}_s,\ \mathscr{F}(\tau) \leqslant 0 \,\}\\ &= \begin{cases} \dfrac{\sqrt{2}k_*}{\alpha n}\operatorname{tr}\varepsilon & \text{if } |\varepsilon^D| \leqslant \frac{1}{n\alpha}\operatorname{tr}\varepsilon,\\ +\infty & \text{otherwise.} \end{cases} \end{aligned} \tag{2.19}$$

Moreover,

$$r(\tau) > -\infty \implies \int_\Omega \tau : \varepsilon(v)\, dx = r(\tau) M(v) \quad \forall v \in V_0, \tag{2.20}$$

which shows that $\lambda_s < 1$ implies $Q_f \cap K = \varnothing$. Thus, due to the assumptions made above, $\lambda_s \geqslant 1$.

Next, it is easy to see that if $\lambda_s > 1$, then (2.14) holds.

Indeed, in accordance with the definition of λ_s, there exists a tensor τ^* such that $r(\tau^*) \in [(\lambda_s+1)/2, \lambda_s]$. Setting $\sigma^1 = \tau^*/r(\tau^*)$, we obtain (see (2.20)) that $\sigma^1 \in Q_f$ and

$$|\sigma^{1D}| + \alpha \operatorname{tr}\sigma^1 \leqslant \sqrt{2}k_*(1-\delta_*) \quad \text{a.e. in } \Omega$$

if $\delta_* = \big(r(\tau^*) - 1\big)/r(\tau^*)$.

Note that in the case (2.16) we have $\lambda_s = +\infty$. To show this, we introduce the sequence

$$\sigma^m = m\sigma^0 + \frac{\sqrt{2}k_*}{n\alpha}(1-m)\mathbb{1}, \quad m \in \mathbb{R}_+.$$

Evidently $\operatorname{div}\sigma^m = mf$, and

$$|\sigma^{mD}| + \alpha\operatorname{tr}\sigma^m = m\big(|\sigma^{0D}| + \alpha\operatorname{tr}\sigma^0\big) + \sqrt{2}k_*(1-m) \leqslant \sqrt{2}k_*$$

so that $\sigma^m \in K$. Because $\lambda_s \geqslant r(\sigma^m) = m$, where m is arbitrary large, we conclude that $\lambda_s = +\infty$. Inequality (2.18) shows that in this case λ_k is also infinite, because in this case $\operatorname{dom} j \cap \overline{V}_0 = \varnothing$. Indeed, if $v \in \operatorname{dom} j$, then $v = 0$ on $\partial\Omega$ and so

$$0 = \int_\Omega \operatorname{div} v\, dx \geqslant \frac{1}{\alpha n}\int_\Omega |\varepsilon^D(v)|\, dx \implies v \equiv 0 \text{ in } \Omega.$$

This means that λ_k and λ_s are finite only if $\partial_2\Omega \neq \varnothing$. It is clear that from the computational point of view the functional j is more suitable than r. That is why it would be important to know when the equality

$$\lambda_s = \lambda_k \tag{2.21}$$

holds. In Appendix 1 we show that (2.21) is correct if $\operatorname{dom} j \cap \overline{V}_0 \neq \varnothing$ and the parameter α is sufficiently small.

Finally, let us demonstrate an example showing that the value of λ_k (and, consequently, the value of λ_s too) may be finite. Assume that $\partial_2\Omega = \partial\Omega$, while the loads are self-balanced, i.e.,

$$M(v) = 0, \quad \text{for all } v \in E(\Omega) := \big\{ v \in V : \varepsilon(v) = 0 \text{ a.e. in } \Omega \big\} \tag{2.22}$$

and $M(e) = \delta > 0$, where $e = (x_1, 0, 0, \dots, 0)$. Evidently $M(u) = 1$ if $u = \delta^{-1} e \in V$. Note that the function u can be represented in the form $u = \widetilde{u} + \widehat{u}$, where $\widehat{u} \in E(\Omega)$ and $\widetilde{u} \in V_0 := \{ v \in V : \int_\Omega v \cdot w \, dx = 0 \ \forall w \in E(Om)\}$ and this representation is unique. In accordance with the assumptions made above, $M(\widetilde{u}) = M(u) - M(\widehat{u}) = 1$, and hence $\widetilde{u} \in \overline{V}_0$. Furthermore,

$$\varepsilon(\widetilde{u}) = \varepsilon(u), \qquad \operatorname{div} u = \delta^{-1}, \qquad |\varepsilon^D(u)| = \Big(1 - 1/n\Big)^{1/2} \delta^{-1}$$

so that $\widetilde{u} \in \operatorname{dom} j$ if $\alpha \leqslant (n^2 - n)^{-1/2}$. In this case

$$\lambda_k \leqslant \frac{\sqrt{2} k_*}{\alpha n} \delta^{-1} \operatorname{meas} \Omega < +\infty. \qquad \square$$

§3. Weak solution of the classical problem

On the set $U \times (Q \cap K)$, where

$$Q := \big\{ \tau \in \Sigma : \operatorname{div} \tau \in L_n(\Omega; \mathbb{R}^n),\ \tau \nu = F \text{ on } \partial_2\Omega \big\}, \qquad U := L_{\frac{n}{n-1}}\big(\Omega; \mathbb{R}^n\big),$$

we can define the extended Lagrangian

$$L(v, \tau) := \int_\Omega \Big\{ \tau : \varepsilon(u_0) + (u_0 - v) \cdot \operatorname{div} \tau - \tfrac{1}{2} a(\tau, \tau) - f \cdot v \Big\} \, dx - \int_{\partial_2\Omega} F \cdot u_0 \, dl$$

and formulate the following minimax problem.

Problem A$^+$. Find $u \in U$ and $\sigma \in Q \cap K$ such that

$$L(u, \tau) \leqslant L(u, \sigma) \leqslant L(v, \sigma) \tag{3.1}$$

for all $\tau \in Q \cap K$ and $v \in U$.

Definition. We shall say that the pair $(u, \sigma) \in U \times (Q \cap K)$ is a *weak solution of the classical problem* if it satisfies (3.1), or, in other words, if it is the saddle-point of the extended Lagrangian.

The assertions proved below show the validity of this definition.

Using the Lagrangian L, we introduce the extended functional $\Phi\colon U \longrightarrow \overline{\mathbb{R}}$ as follows:

$$\Phi(v) := \sup\{\, L(v,\tau) : \tau \in Q \cap K \,\}, \qquad v \in U.$$

Let us establish some simple properties of the functional Φ. It follows directly from the definition of the set Q and of the Lagrangians L and l that

$$l(w,\tau) = L(w,\tau) \quad \forall w \in u_0 + V_0,\ \forall \tau \in Q \cap K. \tag{3.2}$$

The identity (3.2) implies the estimate

$$\Phi(v) \leqslant J(v) \quad \forall v \in u_0 + V_0. \tag{3.3}$$

Now our aim is to obtain some auxiliary results and then to prove the main assertion of this section, stated below in Theorem 3.1.

LEMMA 3.1. *Suppose that $u_m \in u_0 + V_0$ and $u_m \rightharpoonup u$ in $L_{\frac{n}{n-1}}(\Omega;\mathbb{R}^n)$; then*

$$\liminf_{m\to\infty} J(u_m) \geqslant \Phi(u).$$

(Recall that the symbols $\rightarrow$ and $\rightharpoonup$ denote strong and weak convergences respectively.)

PROOF OF LEMMA 3.1. By the definition of the functional J and by (3.2) we have

$$J(u_m) \geqslant l(u_m,\tau) = L(u_m,\tau) \quad \forall \tau \in Q \cap K.$$

Due to the continuity of the Lagrangian L with respect to the weak convergence in $L_{\frac{n}{n-1}}(\Omega;\mathbb{R}^n)$, by passing to the limit in the last inequality we obtain

$$\liminf_{m\to\infty} J(u_m) \geqslant L(u,\tau) \quad \forall \tau \in Q \cap K.$$

Taking into account the definition of the functional Φ, we arrive at the required statement. $\square$

LEMMA 3.2. *Suppose that there is a tensor function σ^2 such that*

$$\sigma^2 \in Q \cap K \cap L_n(\Omega;\mathbb{M}_s^{n\times n}). \tag{3.4}$$

Then

$$\Phi(v) = J(v) \quad \forall v \in u_0 + V_0.$$

REMARK. The condition (3.4) is evidently satisfied for $n = 2$ if only $Q_f \cap K \neq \varnothing$.

PROOF OF LEMMA 3.2. By the definition of the functional J, for any $\varepsilon > 0$ one can find a tensor $\sigma^\varepsilon \in K$ such that

$$J(v) \leqslant l(v, \sigma^\varepsilon) + \varepsilon \tag{3.5}$$

for some function $v \in u_0 + V_0$. Let us choose a sequence $\widetilde{\sigma}^m \in C^\infty(\overline{\Omega}; \mathbb{M}_s^{n\times n}) \cap K$ (see Appendix 2) so that

$$\widetilde{\sigma}^m \longrightarrow \sigma^\varepsilon \quad \text{in } \Sigma. \tag{3.6}$$

Now define $\sigma^m = \varphi_m \widetilde{\sigma}^m + (1 - \varphi_m)\sigma^2$, where the sequence φ_m possesses the following properties:

$$\varphi_m \in C_0^\infty(\Omega), \qquad 0 \leqslant \varphi_m \leqslant 1 \quad \text{in } \Omega, \qquad \varphi_m \longrightarrow 1 \quad \text{a.e. in } \Omega.$$

Since

$$\operatorname{div} \sigma^m = \operatorname{div} \sigma^2 + \varphi_m \operatorname{div}(\widetilde{\sigma}^m - \sigma^2) + (\widetilde{\sigma}^m - \sigma^2)\nabla\varphi_m,$$

the condition (3.4) implies that $\operatorname{div} \sigma^m \in L_n(\Omega; \mathbb{R}^n)$ for any number m. Moreover,

$$\begin{aligned}
\int_\Omega \left(\varepsilon(w) : \sigma^m + w \cdot \operatorname{div} \sigma^m\right) dx &= \int_\Omega \left(\varepsilon(w) : \sigma^2 + w \cdot \operatorname{div} \sigma^2\right) dx \\
&\quad + \int_\Omega \left[\varepsilon(w) : ((\widetilde{\sigma}^m - \sigma^2)\varphi_m) + w \cdot \left(\varphi_m \operatorname{div}(\widetilde{\sigma}^m - \sigma^2) + (\widetilde{\sigma}^m - \sigma^2)\nabla\varphi_m\right)\right] dx \\
&= \int_{\partial_2\Omega} F \cdot w \, dl + \int_\Omega \left[(\widetilde{\sigma}^m - \sigma^2) : \varepsilon(w\varphi_m) + (\varphi_m w) \cdot \operatorname{div}(\widetilde{\sigma}^m - \sigma^2)\right] dx \\
&= \int_{\partial_2\Omega} F \cdot w \, dl.
\end{aligned}$$

Therefore $\sigma^m \in K \cap Q$, and consequently (see (3.2))

$$l(v, \sigma^m) = L(v, \sigma^m) \leqslant \Phi(v). \tag{3.7}$$

On the other hand,

$$\sigma^m \longrightarrow \sigma^\varepsilon \quad \text{in } \Sigma. \tag{3.8}$$

Indeed,

$$\begin{aligned}
\left(\int_\Omega |\sigma^m - \sigma^\varepsilon|^2 \, dx\right)^{1/2} &\leqslant \left(\int_\Omega \varphi_m |\widetilde{\sigma}^m - \sigma^\varepsilon|^2 \, dx\right)^{1/2} \\
&\quad + \left(\int_\Omega (1 - \varphi_m)|\sigma^\varepsilon|^2 \, dx\right)^{1/2} + \left(\int_\Omega (1 - \varphi_m)|\sigma^2|^2 \, dx\right)^{1/2}.
\end{aligned}$$

Applying the Lebesgue theorem to the second and third integrals in the right-hand side of this inequality and condition (3.6) to the first one, we get (3.8). Because $\sigma^m \in K$, it follows from (3.8) that $l(v, \sigma^m) \longrightarrow l(v, \sigma^\varepsilon)$. Taking (3.5) and (3.7) into account, we obtain $J(v) \leqslant \Phi(v) + \varepsilon$. To finish the proof it suffices to use (3.3) and the fact that the parameter $\varepsilon > 0$ was arbitrary. □

Now we shall consider the following variational problem.

Problem C$^+$. Find a function $u \in U$ such that $\Phi(u) = \inf\{\, \Phi(v) : v \in U \,\}$.

Since for any $\tau \in Q \cap K$ we have

$$\inf\{\, L(v,\tau) : v \in U \,\} = \begin{cases} R(\tau) & \text{if } \tau \in Q_f, \\ -\infty & \text{if } \tau \notin Q_f, \end{cases}$$

by using the definition of the functional Φ and known results of minimax theory, we get the following analog of Theorem 1.1.

THEOREM 3.1. *The pair* $(u,\sigma) \in U \times (Q \cap K)$ *is the solution of Problem* A$^+$ *if and only if the vector* $u \in U$ *is the solution of Problem* C$^+$ *and the tensor* σ *is the solution of Problem* B.

This theorem shows that the solvability of Problem A$^+$ is equivalent to the solvability of Problem C$^+$.

THEOREM 3.2. *If conditions* (1.4) *and* (2.14) *are satisfied, then Problem* A$^+$ *has at least one solution* $u \in U$, $\sigma \in Q \cap K$, *and, moreover,*

$$\Phi(u) = R(\sigma) = \inf\{\, J(v) : v \in u_0 + V_0 \,\}. \tag{3.9}$$

REMARK 3.1. Relation (3.9) shows that variational Problem C$^+$ preserves the greatest lower bound of the basic Problem C and thus can be treated as a variational extension of this problem.

Before proving Theorem 3.2, it is necessary to establish some additional results. Let us define the regularized functional

$$J^\delta(v) = \frac{\delta}{2}\int_\Omega |\varepsilon(v)|^2\,dx + J(v), \qquad v \in V,\ 0 < \delta \leqslant 1,$$

and consider the corresponding variational problem: Find $u^\delta \in u_0 + V_0$ such that

$$J^\delta(u^\delta) = \inf\{\, J^\delta(v) : v \in u_0 + V_0 \,\}. \tag{3.10}$$

Evidently problem (3.10) always has a unique solution satisfying the Euler equation

$$\int_\Omega \sigma^\delta : \varepsilon(w)\,dx = M(w) \quad \forall w \in V_0, \tag{3.11}$$

where

$$\sigma^\delta = \delta\varepsilon(u^\delta) + \frac{\partial g}{\partial \varepsilon}(\varepsilon(u^\delta))$$

and the derivative $\frac{\partial g}{\partial \varepsilon}(\varkappa)$ in accordance with (1.7 a) has the representation

$$\frac{\partial g}{\partial \varepsilon}(\varkappa) = \frac{\partial \varphi}{\partial \varepsilon}\left(\varkappa - \frac{\sqrt{2}k_*}{\alpha n^2 K_0}\mathbb{1}\right) \quad \forall \varkappa \in \mathbb{M}^{n\times n}_s,$$

$$\frac{\partial \varphi}{\partial \varepsilon}(\varkappa) = \begin{cases} K_0\left(\operatorname{tr}\varkappa + \dfrac{\sqrt{2}k_*}{\alpha n K_0}\right)\mathbb{1} + 2\mu\varkappa^D & \text{if } 2\mu|\varkappa^D| + \alpha n K_0 \operatorname{tr}\varkappa \leqslant 0, \\ \dfrac{\sqrt{2}k_*}{\alpha n}\mathbb{1} + 2D\left(\left(|\varkappa^D| - \dfrac{1}{\alpha n}\operatorname{tr}\varkappa\right)_+\right)\left(\dfrac{\varkappa^D}{|\varkappa^D|} - \dfrac{1}{\alpha n}\mathbb{1}\right) & \\ & \text{if } 2\mu|\varkappa^D| + \alpha n K_0 \operatorname{tr}\varkappa > 0. \end{cases}$$

LEMMA 3.3. *If the assumptions of Theorem 3.2 are satisfied, then the expressions*

$$\int_\Omega [\varepsilon(u^\delta)]^2\,dx, \quad \int_\Omega |\varepsilon(u^\delta)|\,dx, \quad \delta\int_\Omega |\varepsilon(u^\delta)|^2\,dx, \quad \int_\Omega |\sigma_0^\delta|^2\,dx, \tag{3.12}$$

where

$$\sigma_0^\delta := \sigma^\delta - \delta\varepsilon(u^\delta) = \frac{\partial g}{\partial \varepsilon}(\varepsilon(u^\delta)),$$

are uniformly bounded for $\delta \in]0,1]$.

PROOF. It is obvious that

$$J(u^\delta) \leqslant J^\delta(u^\delta) \leqslant J^\delta(u_0) \leqslant J^1(u_0)$$

for $\delta \in]0,1]$. These inequalities together with (2.15) show that the first three integrals in (3.12) are uniformly bounded. Next, if $2\mu|\varkappa^D| + \alpha n K_0 \operatorname{tr}\varkappa \leqslant 0$, we have $\operatorname{tr}\varkappa = -(\operatorname{tr}\varkappa)_-$ and hence

$$\left|\frac{\partial\varphi}{\partial\varepsilon}(\varkappa)\right| \leqslant \frac{\sqrt{2}k_*}{\alpha\sqrt{n}} + K_0(1+\alpha n)(\operatorname{tr}\varkappa)_- \leqslant \frac{\sqrt{2}k_*}{\alpha\sqrt{n}} + K_0(1+\alpha n)[\varkappa].$$

In the opposite case

$$\left|\frac{\partial\varphi}{\partial\varepsilon}(\varkappa)\right| \leqslant \frac{\sqrt{2}k_*}{\alpha\sqrt{n}} + 2D\left(1+\frac{1}{\alpha\sqrt{n}}\right)[\varkappa].$$

Therefore

$$\begin{aligned}\left|\frac{\partial g}{\partial\varepsilon}(\varepsilon(u^\delta))\right| &\leqslant \frac{\sqrt{2}k_*}{\alpha\sqrt{n}} + \max\Big\{2D\big(1+\frac{1}{\alpha\sqrt{n}}\big), K_0(1+\alpha n)\Big\}\left[\varepsilon(u^\delta) - \frac{\sqrt{2}k_*}{\alpha n^2 K_0}\mathbb{1}\right] \\ &\leqslant \frac{\sqrt{2}k_*}{\alpha\sqrt{n}} + \max\Big\{2D\big(1+\frac{1}{\alpha\sqrt{n}}\big), K_0(1+\alpha n)\Big\}\left([\varepsilon(u^\delta)] + \left[-\frac{\sqrt{2}k_*}{\alpha n^2 K_0}\mathbb{1}\right]\right).\end{aligned}$$

Thus, using the previous estimates, we obtain the required result. □

PROOF OF THEOREM 3.2. It follows from Lemma 3.3 that passing, if necessary, to subsequences (still denoted u^δ, σ^δ), we have

$$\begin{aligned} &u^\delta \rightharpoonup u \quad \text{in } U, && u^\delta \to u \quad \text{in } L_q(\Omega;\mathbb{R}^n) \text{ if } q \in [1, n/n-1[, \\ &\sigma_0^\delta \rightharpoonup \sigma \quad \text{in } \Sigma, && \sigma^\delta \rightharpoonup \sigma \quad \text{in } \Sigma. \end{aligned} \tag{3.13}$$

We shall prove that the pair u,σ is the solution of Problem A^+ and that (3.9) holds.

Relation (3.11) shows that $\sigma^\delta \in Q_f$ and, moreover, taking into account (3.13), we conclude that

$$\sigma^\delta \in Q_f \implies \sigma \in Q_f \implies \sigma \in Q. \tag{3.14}$$

Besides, the inequality

$$\left|\left(\frac{\partial g}{\partial\varepsilon}(\varkappa)\right)^D\right| + \alpha\operatorname{tr}\left(\frac{\partial g}{\partial\varepsilon}(\varkappa)\right) \leqslant \sqrt{2}k_* \quad \forall\varkappa \in \mathbb{M}_s^{n\times n}$$

implies $\sigma_0^\delta \in K$ for $\delta \in]0,1]$ and hence

$$\sigma \in K. \tag{3.15}$$

Next, according to the definition of σ_0^δ and well-known theorems of convex analysis, we have

$$\sigma_0^\delta : \varepsilon(u^\delta) - g(\varepsilon(u^\delta)) - \tfrac{1}{2} a(\sigma_0^\delta, \sigma_0^\delta) = 0 \quad \text{a.e. in } \Omega.$$

Thus

$$\begin{aligned} J^\delta(u^\delta) &= \int_\Omega \Big(\frac{\delta}{2} |\varepsilon(u^\delta)|^2 + \sigma_0^\delta : \varepsilon(u^\delta) - \frac{1}{2} a(\sigma_0^\delta, \sigma_0^\delta) \Big)\, dx - M(u^\delta) \\ &= -\frac{\delta}{2} \int_\Omega |\varepsilon(u^\delta)|^2\, dx + \int_\Omega \sigma^\delta : \varepsilon(u^\delta)\, dx - \frac{1}{2} \int_\Omega a(\sigma_0^\delta, \sigma_0^\delta)\, dx - M(u^\delta) \\ &\quad \text{(see (3.11))} \\ &= -\frac{\delta}{2} \int_\Omega |\varepsilon(u^\delta)|^2\, dx + \int_\Omega \sigma^\delta : \varepsilon(u_0)\, dx - M(u_0) - \frac{1}{2} \int_\Omega a(\sigma_0^\delta, \sigma_0^\delta)\, dx, \end{aligned}$$

and we finally obtain

$$J(u^\delta) \leqslant J^\delta(u^\delta) = R(\sigma_0^\delta) - \frac{\delta}{2} \int_\Omega |\varepsilon(u^\delta)|^2\, dx + \delta \int_\Omega \varepsilon(u^\delta) : \varepsilon(u_0)\, dx. \tag{3.16}$$

Since the convex functional $-R$ is continuous on Σ, we find with the help of (3.13) and Lemma 3.1 that

$$\Phi(u) \leqslant R(\sigma) - \liminf_{\delta \to 0} \frac{\delta}{2} \int_\Omega |\varepsilon(u^\delta)|^2\, dx \leqslant R(\sigma). \tag{3.17}$$

Using the implication (3.14), we have

$$L(v, \sigma) = R(\sigma) \quad \forall v \in U, \tag{3.18}$$

$$l(v, \sigma) = R(\sigma) \quad \forall v \in u_0 + V_0, \tag{3.19}$$

and, therefore

$$\begin{aligned} \Phi(v) &\geqslant R(\sigma) \quad \forall v \in U, \\ J(v) &\geqslant R(\sigma) \quad \forall v \in u_0 + V_0. \end{aligned} \tag{3.20}$$

Now the inequalities

$$L(u, \tau) \leqslant \Phi(u) \leqslant R(\sigma) = L(v, \sigma) \quad \forall v \in U, \forall \tau \in Q \cap K$$

follow from (3.17), (3.18). This means that the pair $u \in U$, $\sigma \in Q \cap K$ is the solution of Problem A^+ satisfying the equality $\Phi(u) = R(\sigma)$. Moreover, using (3.17), we find that

$$\liminf_{\delta \to 0} \frac{\delta}{2} \int_\Omega |\varepsilon(u^\delta)|^2\, dx = 0.$$

Finally, passing to the limit in (3.16) with respect to (3.20), it is easy to establish the inequality

$$\liminf_{\delta \to 0} J(u^\delta) \leqslant R(\sigma) \leqslant \inf\{\, J(v) : v \in u_0 + V_0 \,\},$$

which proves (3.9). □

As a matter of fact, it is not difficult to show that the weak solution of Problem A possesses some additional differential properties.

Indeed, introducing the set

$$\Sigma_{00} = \left\{ \tau \in C_0^\infty(\Omega; \mathbb{M}_s^{n\times n}) : |\tau| \leqslant 1 \text{ in } \Omega \right\}$$

and setting $\beta = \varepsilon/2(1+\alpha\sqrt{n})$, we obtain $\sigma^1 \beta\tau \in K \cap Q\ \forall \tau \in \Sigma_{00}$. Evidently

$$L(u, \sigma^1 + \beta\tau) \leqslant L(u, \sigma) = \Phi(u) \quad \forall \tau \in \Sigma_{00},$$

and thus

$$-\int_\Omega u \cdot \operatorname{div} \tau \, dx \leqslant \frac{1}{2}\beta \int_\Omega a(\tau, \tau)\, dx + \int_\Omega a(\sigma^1, \tau)\, dx + \frac{1}{\beta}\Phi(u) + \text{const.}$$

Passing to the least upper bound on $\tau \in \Sigma_{00}$, we find that

$$\int_\Omega |\varepsilon(u)| < +\infty \implies u \in BD(\Omega),$$

where $BD(\Omega)$ is the space of $\mathbb{R}^n$-valued functions $u \in L_1(\Omega, \mathbb{R}^n)$ whose linearized strains $\varepsilon_{ij}(u) = \frac{1}{2}(u_{i,j} + u_{j,i})$ are measures (see **[MSCh, Su, TStr]**).

§4. Other relaxed formulations for Problem C

In many cases it is useful to examine, together with Problem C$^+$ described above, some other versions of the relaxed formulations when the relaxed functional is defined on functions having jumps of discontinuity along some surfaces. These problems may be also regarded as relaxations of Problem C.

Let Ω be divided into subdomains Ω_+ and Ω_- by means of a Lipschitz continuous surface Γ so that $\Omega_+ \cap \Omega_- = \varnothing$, $\overline{\Omega} = \overline{\Omega}_+ \cup \overline{\Omega}_-$. Vectors of the outward unit normals to the boundaries $\partial\Omega_+$, $\partial\Omega_-$ are denoted ν_+ and ν_-. On the set

$$V_\Gamma(\Omega) := \left\{ \{v, w\} : w \in u_0 + V_0,\ v = v^\pm \text{ in } \Omega_\pm,\ v^\pm \in W_2^1(\Omega_\pm; \mathbb{R}^n) \right\}$$

we can define the functional

$$\begin{aligned} \Phi_\Gamma(v, w) = &\int_{\Omega_+} g(\varepsilon(v^+))\, dx + \int_{\Omega_-} g(\varepsilon(v^-))\, dx \\ &+ \int_\Gamma g_0(-S(\nu^+, v^+) - S(\nu^-, v^-))\, dl \\ &+ \int_{\partial\Omega} g_0(S(\nu, w - v))\, dl - \int_\Omega f \cdot v\, dx - \int_{\partial_2\Omega} F \cdot w\, dl, \end{aligned}$$

where $S(a, b) = \frac{1}{2}(a \otimes b + b \otimes a)$, $a, b \in \mathbb{R}^n$, and the functional g_0 was introduced in §2 (see (2.19)).

Now it is possible to formulate a new variational problem.

Problem $\mathbf{C}^*$. Find a pair of functions $\{u, w\} \subset V_\Gamma(\Omega)$ such that

$$\Phi_\Gamma(u, w) = \inf\{\, \Phi_\Gamma(v; \widetilde{w}) : \{v, \widetilde{w}\} \in V_\Gamma(\Omega)\,\}.$$

It is easy to see that

$$J(v) = \Phi_\Gamma(v, u_0) \quad \forall v \in u_0 + V_0 \tag{4.1}$$

and, consequently, in general

$$\inf\{\, \Phi_\Gamma(v, \widetilde{w}) : \{v, \widetilde{w}\} \in V_\Gamma(\Omega)\,\} \leqslant \inf\{\, J(v) : v \in u_0 + V_0\,\}. \tag{4.2}$$

The question of interest is when (4.2) is an equality. The answer is given by the following theorem:

THEOREM 4.1. *If conditions* (1.4) *and* (2.14) *are satisfied, then*:
(i) *We have*

$$\inf\{\, \Phi_\Gamma(v, \widetilde{w}) : \{v, \widetilde{w}\} \in V_\Gamma(\Omega)\,\} = \inf\{\, J(v) : v \in u_0 + V_0\,\}. \tag{4.3}$$

(ii) *If Problem* C *has a solution* $u \in u_0 + V_0$, *then the pair* $\{u, u_0\}$ *is the solution of Problem* C^*

(iii) *If the pair* $(u, u_0) \in V_\Gamma(\Omega)$ *is the solution of Problem* C^*, *then the function* u *is the solution of Problem* C^+.

PROOF. We denote

$$T\big(F, \tau; w, v\big) = -\int_{\partial_2\Omega} F \cdot w\, dl + \int_\Omega \big(\tau : \varepsilon(w) + (w - v) \cdot \operatorname{div} \tau\big)\, dx$$

for functions $v \in L_2\big(\Omega; \mathbb{R}^n\big)$, $w \in u_0 + V_0$, and

$$\tau \in \Sigma_2 := \big\{\, \tau \in \Sigma : \operatorname{div} \tau \in L_2(\Omega; \mathbb{R}^n)\,\big\}.$$

It follows from the definition of the set Q that

$$\begin{aligned} &T\big(F, \tau; w, v\big) = T\big(F, \tau; u, v\big) \\ &\qquad \forall \tau \in \Sigma_2,\ \forall v \in L_2\big(\Omega; \mathbb{R}^n\big),\ \forall u, w \in u_0 + V_0. \end{aligned} \tag{4.4}$$

Using the functional T, we can introduce the extended functional in the form

$$L(v, \tau) = T\big(F, \tau; u_0, v\big) - \int_\Omega \Big(\frac{1}{2} a(\tau, \tau) - f \cdot v\Big)\, dx, \tag{4.5}$$

where $v \in L_2\big(\Omega; \mathbb{R}^n\big)$ and $\tau \in \Sigma_2$.

For any tensor $\tau \in K \cap \Sigma_2$, there is a sequence (see Appendix 2) such that $\tau^m \in C^\infty\big(\overline{\Omega}; \mathbb{M}_s^{n\times n}\big) \cap K$ and

$$\tau^m \longrightarrow \tau \quad \text{in } \Sigma, \qquad \operatorname{div}\tau^m \longrightarrow \operatorname{div}\tau \quad \text{in } L_2\big(\Omega; \mathbb{R}^n\big).$$

This means that

$$T\big(F, \tau^m; w, v\big) \xrightarrow[m\to\infty]{} T\big(F, \tau; w, v\big) \tag{4.6}$$

if $(v, w) \in V_\Gamma(\Omega)$ and $\tau \in \Sigma_2$. On the other hand, integrating by parts, we obtain

$$\begin{aligned} T\big(F, \tau^m; w, v\big) &= -\int_{\partial_2\Omega} F \cdot w\, dl + \int_{\Omega_+} \tau^m : \varepsilon(v^+)\, dx + \int_{\Omega_-} \tau^m : \varepsilon(v^-)\, dx \\ &\quad - \int_\Gamma \tau^m : \big(S(v^+, v^+) + S(v^-, v^-)\big)\, dl + \int_{\partial\Omega} \tau^m : S(v, w - v)\, dl. \end{aligned} \tag{4.7}$$

Let us consider the sequence of real numbers

$$I_m(v, w) := T\big(F, \tau^m; w, v\big) - \int_\Omega \Big(\frac{1}{2} a(\tau^m, \tau^m) - f \cdot v\Big)\, dx$$

which in accordance with (4.7) is bounded from above:

$$I_m(v, w) \leqslant \Phi_\Gamma(v, w), \quad \{v, w\} \in V_\Gamma(\Omega). \tag{4.8}$$

It is also follows from (4.4)–(4.6) that

$$L(v, \tau) \leqslant \Phi_\Gamma(v, w) \quad \forall \{v, w\} \in V_\Gamma(\Omega),\ \forall \tau \in \Sigma_2 \cap K$$

and, consequently,

$$\Phi(v) = \sup\big\{\, L(v, \tau) : \tau \in Q \cap K \,\big\} \leqslant \Phi_\Gamma(v, w) \quad \forall \{v, w\} \in V_\Gamma(\Omega). \tag{4.9}$$

Evidently, if the pair $\{v, w\}$ belongs to the set $V_\Gamma(\Omega)$, then $v \in U$, so that (4.9) implies

$$\inf\big\{\, \Phi(\widetilde{v}) : \widetilde{v} \in U \,\big\} \leqslant \Phi_\Gamma(v, w) \quad \forall \{v, w\} \in V_\Gamma(\Omega).$$

This estimate together with (4.2) and Theorem 3.2 proves equality (4.3), i.e., part (i) of the theorem. Part (ii) follows from (4.1) and part (i). Finally, (4.9) combined with (4.3) gives part (iii). Theorem 4.1 is proved. □

REMARK 4.1. The functional $\Phi_\Gamma(v, w)$ is defined on the set $V_\Gamma(\Omega)$ containing functions that may have jumps of discontinuity along the surface Γ or the boundary $\partial\Omega$. An important particular case of the variational formulation C* is the following problem:

Find a pair of functions $u \in V$, $w \in u_0 + V_0$ such that

$$\Phi(u, w) = \inf\big\{\, \Phi(v, \widetilde{w}) : v \in V,\ \widetilde{w} \in u_0 + V_0 \,\big\}, \tag{4.10}$$

where

$$\Phi(v, \widetilde{w}) = \int_\Omega g\big(\varepsilon(v)\big)\, dx - M(v) + \int_{\partial\Omega} g_0\big(S(v, \widetilde{w} - v)\big)\, dl - \int_{\partial_2\Omega} F \cdot (\widetilde{w} - v)\, dl.$$

REMARK 4.2. Variational formulations similar to Problem C* have been used for constructing conforming discontinuous finite element schemes (see **[JS, R1–S1, S9, StT]**).

Now let us return to the problem considered in §1 concerning the elasto-plastic deformation of the annulus with boundary conditions (1.8). Our aim is to show that the pair $u^* = u$, σ^* defined by (1.9), (1.10), (1.11), (1.14) is a weak solution of Problem A, i.e., it is a solution of A^+.

It should be noted that the fact that the function u^* does not satisfy the boundary condition on the inner side of the annulus may be interpreted as a jump of discontinuity on the boundary $\partial\Omega$.

To prove that u^* is a weak solution of Problem A, we shall demonstrate that the pair of functions $\{u^*, u_0\}$ is the solution of C^*. Then, according to Theorem 4.1, u^* would be a solution of Problem C^+ and, as soon as σ^* is the solution of Problem B, we could obtain the necessary result with the help of Theorem 3.1.

Indeed, in this case

$$\Phi_\Gamma(u^*, u_0) = 2\pi\left[\int_{r_1}^{r_2} (2K_0C_1^2 + \frac{2\mu C_2^2}{r^4})r\,dr + r_1 k_* \Delta\right],$$

$$R(\sigma^*) = 2\pi\left[-\int_{r_1}^{r_2} (2K_0C_1^2 + \frac{2\mu C_2^2}{r^4})r\,dr + \sigma_{rr}^*(r_2)r_2 U\right].$$

Taking into account (1.11) together with the relations

$$\Delta = C_1r_1 + C_2r_1^{-1}, \qquad U = C_1r_2 + C_2r_2^{-1}$$

we find that

$$\Phi_\Gamma(u^*, u_0) - R(\sigma^*) = 8\pi\int_{r-1}^{r_2} (rK_0C_1^2 + \frac{\mu C_2^2}{r^3})\,dr$$

$$+ 4\pi\left[(C_1r_1^2 + C_2)\left(K_0C_1 - \frac{\mu C_2}{r_1^2}\right) - (C_1r_2^2 + C_2)\left(K_0C_1 - \frac{\mu C_2}{r_2^2}\right)\right] = 0.$$

Using the arguments given at the end of §1, we can affirm that the $\{u^*, \sigma^*\}$ is the unique solution of Problem A^+.

§5. Appendix 1

In this section we shall assume that $\partial_2\Omega \neq \varnothing$. We shall also assume that in the particular case when $\partial_2\Omega = \partial\Omega$ the condition (2.22) is fulfilled and

$$V_0 = \left\{u \in V : \int_\Omega u \cdot v\,dx = 0,\ \forall v \in E(\Omega)\right\}, \quad u_0 = 0.$$

Besides this, we postulate the existence of a function such that

$$\int_\Omega \operatorname{div} w_*\,dx = 1. \tag{5.1}$$

Obviously, in the case when $\partial_2\Omega = \partial\Omega$, condition (5.1) holds. Define

$$L_0^2 := \left\{f \in L_2(\Omega) : \int_\Omega f\,dx = 0\right\}.$$

Then, as shown in [LS], for any $f \in L_0^2$ one can find a function $u \in W_0 := \overset{\circ}{W}{}_2^1(\Omega; \mathbb{R}^n)$ satisfying the relations

$$\operatorname{div} u = f \quad \text{in } \Omega, \qquad \||\nabla u|\|_{L_2(\Omega)} \leqslant C_* \|f\|_{L_2(\Omega)}$$

where the constant C_* does not depend on f. Consider the following variational problem:

Find $u_f \in W_0$ such that

$$\big\| |\nabla u_f| \big\|_{L_2(\Omega)} = \inf\{ \big\| |\nabla v| \big\|_{L_2(\Omega)} : v \in W_0,\ \operatorname{div} v = f \text{ in } \Omega \}. \tag{5.2}$$

Since (5.2) has a unique solution for any $f \in L_0^2$, we can introduce the operator $\pi \colon L_0^2 \to W_0$ by putting $\pi f = u_f$. Evidently the operator π is linear and bounded.

LEMMA 5.1. *Suppose that condition* (5.1) *is fulfilled and*

$$\begin{aligned}\Delta(\alpha, n, \|\pi\|, w_*) := 1 - \alpha n \Bigg[\|\pi\| + \operatorname{meas}^{1/2}\Omega \Bigg(\|\pi\| \Big\| \frac{1}{\operatorname{meas}\Omega} - \operatorname{div} w_* \Big\|_{L_2(\Omega)} \\ + \Big(\int_\Omega |\varepsilon^D(w_*)|^2\,dx \Big)^{1/2} \Bigg) \Bigg] > 0,\end{aligned} \tag{5.3}$$

and the tensor $\tau \in \Sigma$ satisfies the variational identity

$$\int_\Omega \tau : \varepsilon(v)\,dx = \int_\Omega \tau^0 : \varepsilon(v)\,dx + \int_\Omega f^0 \cdot v\,dx + \int_{\partial_2\Omega} F^0 \cdot v\,dl \quad \forall v \in V_0, \tag{5.4}$$

where

$$\tau^0 \in \Sigma, \qquad f^0 \in L_2(\Omega, \mathbb{R}^n), \qquad F^0 \in L_2(\partial_2\Omega; \mathbb{R}^n) \tag{5.5}$$

and in the case $\partial_2\Omega = \partial\Omega$ the loads are self-balanced:

$$\int_\Omega f^0 \cdot v\,dx + \int_{\partial_2\Omega} F^0 \cdot v\,dl = 0 \quad \forall v \in E(\Omega). \tag{5.6}$$

Then, if

$$\tau \in K, \tag{5.7}$$

the following estimate holds for the tensor τ:

$$\begin{aligned}\int_\Omega |\tau|^2\,dx \leqslant 4k_*^2 \operatorname{meas}\Omega + \frac{(2\alpha^2 + \frac{1}{n}) C_1(n, k_*, \Omega, \|\pi\|, w_*)}{\Delta^2} \\ \times \Big\{ 1 + \int_\Omega (|\tau^0|^2 + |f^0|^2)\,dx + \int_{\partial_2\Omega} |F^0|^2\,dl \Big\}.\end{aligned} \tag{5.8}$$

PROOF. Define the function

$$u_* := v_* + w_* \int_\Omega \operatorname{tr}\tau \implies u_* = 0 \quad \text{on } \partial_1\Omega, \tag{5.9}$$

where

$$v_* := \pi\Big(\operatorname{tr}\tau - \operatorname{div} w_* \int_\Omega \operatorname{tr}\tau\,dx \Big).$$

It follows from the definition of the operator π that

$$\operatorname{div} u_* = \operatorname{tr}\tau \quad \text{a.e. in } \Omega. \tag{5.10}$$

If $\partial_1\Omega \neq \varnothing$, then $u_* \in V_0$. If $\partial_1\Omega = \varnothing$, then, in general, we could use the representation $u_* = \widetilde{u}_* + \widehat{u}_*$, where $\widetilde{u}_* \in V_0$, $\widehat{u}_* \in E(\Omega)$, but by (5.6) and the fact that $\varepsilon(u_*) = 0$ this representation may be omitted.

Setting $v = u_*$, in (5.4) we have

$$\frac{1}{n}\int_\Omega \operatorname{tr}^2\tau\,dx = -\int_\Omega \tau^D : \varepsilon^D(u_*)\,dx + M^0(u_*). \tag{5.11}$$

Here

$$M^0(v) := \int_\Omega \left(\tau^0 : \varepsilon(v) + f\cdot v\right)dx + \int_{\partial_2\Omega} F\cdot v\,dl, \quad \forall v \in V.$$

Next, using (5.7), we find that

$$\begin{aligned}
\left|\int_\Omega \tau^D : \varepsilon^D(u_*)\,dx\right| &\leqslant \int_\Omega |\tau^D||\varepsilon^D(u_*)|\,dx \\
&\leqslant \int_\Omega \sqrt{2}k_*|\varepsilon^D(u_*)|\,dx - \alpha\int_\Omega \operatorname{tr}\tau|\varepsilon^D(u_*)|\,dx \\
&\leqslant \sqrt{2}k_*\int_\Omega |\varepsilon^D(u_*)|\,dx \\
&\quad + \alpha\int |\operatorname{tr}\tau|\left(|\varepsilon^D(u_*)| + |\varepsilon^D(w_*)|\left|\int_\Omega \operatorname{tr}\tau\,dx\right|\right)dx \\
&\leqslant \sqrt{2}k_*\int_\Omega |\varepsilon^D(u_*)|\,dx \\
&\quad + \alpha\left(\int_\Omega \operatorname{tr}^2\tau\,dx\right)^{1/2}\left(\int_\Omega |\varepsilon^D(u_*)|^2\,dx\right)^{1/2} \\
&\quad + \alpha\left|\int_\Omega \operatorname{tr}\tau\,dx\right|\left(\int_\Omega \operatorname{tr}^2\tau\,dx\right)^{1/2}\left(\int_\Omega |\varepsilon^D(w_*)|^2\,dx\right)^{1/2} \\
&\leqslant \sqrt{2}k_*\int_\Omega |\varepsilon^D(u_*)|\,dx \\
&\quad + \alpha\,\mathrm{meas}^{1/2}\,\Omega\left(\int_\Omega |\varepsilon^D(w_*)|^2\,dx\right)^{1/2}\int_\Omega \operatorname{tr}^2\tau\,dx \\
&\quad + \alpha\left(\int_\Omega \operatorname{tr}^2\tau\,dx\right)^{1/2}\|\pi\|\left\|\operatorname{tr}\tau - \operatorname{div} w_*\int_\Omega \operatorname{tr}\tau\,dx\right\|_{L_2(\Omega)}.
\end{aligned} \tag{5.12}$$

Substituting the estimate

$$\begin{aligned}
&\left\|\operatorname{tr}\tau - \operatorname{div} w_*\int_\Omega \operatorname{tr}\tau\,dx\right\|_{L_2(\Omega)} \\
&\quad\leqslant \left\|\operatorname{tr}\tau - \frac{1}{\mathrm{meas}\,\Omega}\int_\Omega \operatorname{tr}\tau\,dx\right\|_{L_2(\Omega)} + \left|\int_\Omega \operatorname{tr}\tau\,dx\right|\left\|\frac{1}{\mathrm{meas}\,\Omega} - \operatorname{div} w_*\right\|_{L_2(\Omega)} \\
&\quad\leqslant \|\operatorname{tr}\tau\|_{L_2(\Omega)}\left(1 + \mathrm{meas}^{1/2}\,\Omega\left\|\frac{1}{\mathrm{meas}\,\Omega} - \operatorname{div} w_*\right\|_{L_2(\Omega)}\right)
\end{aligned}$$

into (5.12), we obtain

$$\left|\int_\Omega \tau^D : \varepsilon^D(u_*)\,dx\right| \leqslant \sqrt{2}k_* \int_\Omega |\varepsilon^D(u_*)|\,dx + \alpha \int_\Omega \operatorname{tr}^2 \tau\,dx \left[\|\pi\| + (\operatorname{meas}\Omega)^{1/2}\left(\|\pi\|\left\|\frac{1}{\operatorname{meas}\Omega} - \operatorname{div} w_*\right\|_{L_2(\Omega)} + \left(\int_\Omega |\varepsilon^D(w_*)|^2\,dx\right)^{1/2}\right)\right].$$

Now (5.11) and (5.12) imply

$$\begin{aligned}\frac{\Delta(\alpha, \|\pi\|, w_*)}{n}\int_\Omega \operatorname{tr}^2 \tau\,dx &\leqslant \sqrt{2}k_* \int_\Omega |\varepsilon^D(u_*)|\,dx + M^0(u_*)\\ &\leqslant \sqrt{2}k_* \int_\Omega |\varepsilon^D(v_*)|\,dx + M^0(v_*)\\ &\quad + \sqrt{2}k_*\left|\int_\Omega \operatorname{tr}\tau\,dx\right| \int_\Omega |\varepsilon^D(w_*)|\,dx + \left(\int_\Omega \operatorname{tr}\tau\,dx\right) M^0(w_*).\end{aligned} \tag{5.13}$$

Taking into account the estimate of $\left\|\operatorname{tr}\tau - \operatorname{div} w_* \int_\Omega \operatorname{tr}\tau\,dx\right\|_{L_2(\Omega)}$, we have

$$\begin{aligned}\||\nabla v_*|\|_{L_2(\Omega)} &\leqslant \|\pi\|\left\|\operatorname{tr}\tau - \operatorname{div} w_* \int_\Omega \operatorname{tr}\tau\,dx\right\|_{L_2(\Omega)}\\ &\leqslant \|\pi\|\|\operatorname{tr}\tau\|_{L_2(\Omega)}\left(1 + \operatorname{meas}^{1/2}\Omega\left\|\frac{1}{\operatorname{meas}\Omega} - \operatorname{div} w_*\right\|_{l_2(\Omega)}\right).\end{aligned}$$

Indeed, transforming (5.13), we deduce that

$$\int_\Omega \operatorname{tr}^2 \tau\,dx \leqslant \frac{C_1(n, k_*, \Omega, \|\pi\|, w_*)}{\Delta^2} \times \left\{\int_\Omega (|\tau^0|^2 + |f^0|^2)\,dx + \int_{\partial_2\Omega} |F^0|^2\,dl + 1\right\}.$$

This together with the inequality

$$|\tau^D|^2 \leqslant 2\big((\sqrt{2}k_*)^2 + \alpha^2 \operatorname{tr}^2 \tau\big)$$

implies (5.8). □

REMARK 5.1. If $\partial_2\Omega = \partial\Omega$, then condition (5.3) can be simplified. Let us set $w_*(x) = x/(n \operatorname{meas}\Omega)$. Then $\operatorname{div} w_* = 1/\operatorname{meas}\Omega$, $\varepsilon^D(w_*) = 0$ and, consequently, (5.3) is equivalent to

$$\Delta(\alpha, n, \|\pi\|, w_*) = 1 - \alpha n\|\pi\| > 0. \tag{5.14}$$

The main result of this section is

THEOREM 5.1. *If conditions* (5.1), (5.3) *are satisfied and*

$$\exists u_0 \in \overline{V}_0 \cap \operatorname{dom} j, \tag{5.15}$$

where the linear manifold $\overline{V}_0$ was defined in §2, *then* (2.21) *is correct and, moreover, one can find a tensor $\sigma \in K$ such that $\lambda_s = r(\sigma)$.*

REMARK 5.2. Obviously (5.15) implies (5.1) with $w_* = u_0 / \int_\Omega \operatorname{div} u_0 \, dx$. (Note that the corresponding value of Δ in (5.3) may not be optimal.)

PROOF OF THEOREM 5.1. Here we use the same ideas as in the proof of Theorem 3.2. Consider the variational problem

Find $u^{\delta,N} \in \overline{V}_0$ such that

$$J_N^\delta(\delta, N) = \inf\{ J_N^\delta(v) : v \in \overline{V}_0 \}, \tag{5.16}$$

where

$$J_N^\delta(v) = \delta \int_\Omega |\varepsilon(v)|^2 \, dx + \int_\Omega g^N(\varepsilon(v)) \, dx, \quad N \in \mathbb{N}, \ \delta \in]0,1],$$

$$g^N(\varepsilon) = \varphi^N\left(\varepsilon - \frac{\sqrt{2}k_*}{\alpha N n} \mathbb{1}\right) \quad \forall \varepsilon \in \mathbb{M}_s^{n\times n},$$

$$\varphi^N(\varepsilon) := \begin{cases} \dfrac{N}{2}\left(\dfrac{1}{n}\left(\operatorname{tr}\varepsilon + \dfrac{\sqrt{2}k_*}{\alpha N}\right)^2 + |\varepsilon^D|\right), & \text{if } |\varepsilon^D| + \alpha \operatorname{tr}\varepsilon < 0, \\ \dfrac{k_*^2}{\alpha^2 N n} + \dfrac{\sqrt{2}k_*}{\alpha n}\operatorname{tr}\varepsilon + \dfrac{N}{2}\left(1 + \dfrac{1}{\alpha^2 n}\right)^{-1}\left[\left(|\varepsilon^D| - \dfrac{\operatorname{tr}\varepsilon}{\alpha n}\right)_+\right]^2, & \\ & \text{if } |\varepsilon^D| + \alpha\operatorname{tr}\varepsilon \geqslant 0. \end{cases}$$

It is easy to see that (5.16) possesses a unique solution. The corresponding Euler equations have the form

$$\int_\Omega \sigma^{\delta,N} : \varepsilon(v) \, dx = \lambda_N^\delta M(v), \quad \forall v \in V_0, \tag{5.17}$$

where

$$\sigma^{\delta,N} := \delta\varepsilon(u^{\delta,N}) + \frac{\partial g}{\partial \varepsilon}(\varepsilon(u^{\delta,N})), \qquad \lambda_N^\delta = \int_\Omega \sigma^{\delta,N} : \varepsilon(u_0) \, dx.$$

Now we can apply the results of §2 by setting $K_0 = N/n$ and $\mu = N/2$. Then Lemma 2.2 implies the estimate

$$\varphi^N(\varepsilon) \geqslant \frac{\sqrt{2}k_*}{1 + \alpha\sqrt{n}} |\varepsilon|^2 + C_2 N [\varepsilon]^2 + C_3/N, \quad \forall \varepsilon \in \mathbb{M}_s^{n\times n} \tag{5.18}$$

with constants $C_2 > 0$, C_3 that do not depend on N and ε.

Next, taking (5.17) into account, we establish the relation

$$J_N^\delta(u^{\delta,N}) = -\frac{\delta}{2}\int_\Omega |\varepsilon(u^{\delta,N})|^2 \, dx + \lambda_N^\delta - \frac{1}{2N}\int_\Omega |\sigma_0^{\delta,N}|^2 \, dx, \tag{5.19}$$

where

$$\sigma_0^{\delta,N} = \sigma^{\delta,N} - \delta\varepsilon(u^{\delta,N}). \tag{5.20}$$

To obtain the uniform estimates (with respect to N and δ) we use the inequality

$$J_N^\delta(u^{\delta,N}) \leqslant J_N^\delta(u_0) = \delta \int_\Omega |\varepsilon(u_0)|^2 \, dx + \int_\Omega \varphi^N\left(\varepsilon(u_0) - \frac{\sqrt{2}k_*}{\alpha N n}\mathbb{1}\right) dx.$$

If $|\varepsilon^D(u_0)| + \alpha \operatorname{div} u_0 - \sqrt{2}k_*/N < 0$, then

$$\varphi^N\left(\varepsilon(u_0) - \frac{\sqrt{2}k_*}{\alpha n N}\mathbb{1}\right) = \frac{N}{2}|\varepsilon(u_0)|^2 \leqslant \frac{N}{2}\left(\frac{\sqrt{2}k_*}{\alpha N}\right)^2 \frac{1}{\min\{1, \alpha^2 n\}}.$$

In the converse case

$$\begin{aligned}
&\varphi^N\left(\varepsilon(u_0)-\frac{\sqrt{2}k_*}{\alpha n N}\mathbb{1}\right)\\
&=\frac{\sqrt{2}k_*}{\alpha n}\operatorname{div}u_0+\frac{N}{2}\left(1+\frac{1}{\alpha^2 n}\right)^{-1}\left[\left(|\varepsilon(u_0)|-\frac{\operatorname{div}u_0}{\alpha n}+\frac{\sqrt{2}k_*}{\alpha^2 n N}\right)_+\right]-\frac{k_*^2}{\alpha^2 n N}\\
&\leqslant\frac{\sqrt{2}k_*}{\alpha n}\operatorname{div}u_0+\frac{N}{2}\left(1+\frac{1}{\alpha^2 n}\right)^{-1}\left(\frac{\sqrt{2}k_*}{\alpha^2 n N}\right)^2.
\end{aligned}$$

It follows from the last two inequalities that one can find a positive constant C_4 that does not depend on N and δ so that $J_N^\delta\left(u^{\delta,N}\right)\leqslant C_4$. This, together with (5.18), implies the estimates

$$\delta\int_\Omega\left|\varepsilon\left(u^{\delta,N}\right)\right|^2dx,\qquad N\int_\Omega\left[\varepsilon\left(u^{\delta,N}\right)\right]^2dx\leqslant C_5 \tag{5.21}$$

where the positive constant C_5 does not depend on N and δ.

It is not difficult to check that

$$\sigma_0^{\delta,N}\in K. \tag{5.22}$$

So, taking into consideration the assumptions of the theorem together with the relations (5.17), (5.20), we can apply Lemma 5.1 for $\tau^0=-\delta\varepsilon\left(u^{\delta,N}\right)$, $f^0=\lambda_N^\delta f$, $F^0=\lambda_N^\delta F$, and find the estimate

$$\begin{aligned}
\int_\Omega\left|\sigma_0^{\delta,N}\right|^2dx\leqslant C_6\Bigg(\delta^2\int_\Omega\left|\varepsilon\left(u^{\delta,N}\right)\right|^2dx\\
+\left(\lambda_N^\delta\right)^2\int_\Omega|f|^2dx+\left(\lambda_N^\delta\right)^2\int_{\partial_2\Omega}|F^2|\,dl+1\Bigg),
\end{aligned} \tag{5.23}$$

where the positive constant C_6 does not depend on δ and N.

Using (5.16), we get

$$\frac{\partial g^N(\varepsilon)}{\partial\varepsilon}=\frac{\partial\varphi^N}{\partial\varepsilon}\left(\varepsilon-\frac{\sqrt{2}k_*}{\alpha N n}\mathbb{1}\right),$$

$$\frac{\partial\varphi^N}{\partial\varepsilon}(\varkappa)=\begin{cases}\dfrac{N}{n}\left(\operatorname{tr}\varkappa+\dfrac{\sqrt{2}k_*}{\alpha N}\right)\mathbb{1}+N\varkappa^D, & \text{if } |\varkappa^D|+\alpha\operatorname{tr}\varkappa<0,\\[2ex] \dfrac{\sqrt{2}k_*}{\alpha\sqrt{n}}\mathbb{1}+N\left(1+\dfrac{1}{\alpha^2 n}\right)^{-1}\left(\dfrac{\varkappa^D}{|\varkappa^D|}-\dfrac{1}{\alpha n}\mathbb{1}\right)\\ \qquad\times\left(\left(|\varkappa^D|-\dfrac{1}{\alpha n}\operatorname{tr}\varkappa\right)_+\right), & \text{if } |\varkappa^D|+\alpha\operatorname{tr}\varkappa\geqslant 0.\end{cases}$$

It is easy to verify that

$$\left|\frac{\partial g}{\partial\varepsilon}\left(\varepsilon\left(u^{\delta,N}\right)\right)\right|\leqslant C_7+NC_8\left(\left[\varepsilon\left(u^{\delta,N}\right)\right]+\left[-\frac{\sqrt{2}k_*}{\alpha n^2}\mathbb{1}\right]\right),$$

where the constants $C_8>0$, C_7 do not depend on δ, N. Thus

$$\frac{1}{2N}\int_\Omega\left|\sigma_0^{\delta,N}\right|^2dx\leqslant\frac{C_9}{N}+C_{10}N\int_\Omega\left[\varepsilon\left(u^{\delta,N}\right)\right]^2dx. \tag{5.24}$$

Here the constants $C_{10} > 0$, C_9 also do not depend on δ, N. Estimates (5.21), (5.24), and (5.19) imply

$$\begin{aligned} 0 \leqslant \lambda_N^\delta &= \frac{1}{2N}\int_\Omega |\sigma_0^{\delta,N}|^2\,dx + \frac{\delta}{2}\int_\Omega |\varepsilon(u^{\delta,N})|^2\,dx + J_N^\delta(u^{\delta,N}) \\ &\leqslant \frac{C_9}{N} + C_{10}C_5 + C_4. \end{aligned} \tag{5.25}$$

Then, with the help of (5.21), (5.24), we immediately obtain

$$\int_\Omega |\sigma_0^{\delta,N}|^2\,dx \leqslant C_{11} \quad (C_{11} \text{ does not depend on } \delta,\ N). \tag{5.26}$$

Now, choosing subsequences if necessary (still denoted by $u^{\delta,N}$, $\sigma^{\delta,N}$), we get

$$u^{\delta,N} \xrightarrow[N\to\infty]{} u^\delta \in \overline{V}_0 \quad \text{in } V, \tag{5.27 a}$$

$$\int_\Omega [\varepsilon(u^{\delta,N})]^2\,dx \to 0 = \int_\Omega [\varepsilon(u^\delta)]^2\,dx, \tag{5.27 b}$$

$$[\varepsilon(u^{\delta,N})] \to [\varepsilon(u^\delta)] \quad \text{a.e. in } \Omega, \tag{5.27 c}$$

$$\sigma^{\delta,N} \rightharpoonup \sigma^\delta \quad \text{in } \Sigma, \qquad \sigma_0^{\delta,N} \rightharpoonup \sigma_0^\delta \quad \text{in } \Sigma, \tag{5.27 d}$$

where $\sigma_0^\delta \in K$, $\sigma^\delta = \delta\varepsilon(u^\delta) + \sigma_0^\delta$ and

$$\int_\Omega \sigma^\delta : \varepsilon(v)\,dx = \lambda^\delta M(v) \quad \forall v \in V_0, \qquad \lambda^\delta = \int_\Omega \sigma^\delta : \varepsilon(u_0)\,dx. \tag{5.28}$$

It follows from (5.27 b) that

$$u^\delta \in \operatorname{dom} j \cap \overline{V}_0. \tag{5.29}$$

Next, from the definition of the integrand g and (5.19) we deduce that

$$\int_\Omega \sigma^{\delta,N} : \varepsilon(u_0)\,dx = \lambda_N^\delta \geqslant J_N^\delta(u^{\delta,N}) \geqslant \int_\Omega \left(\frac{\sqrt{2}k_*}{\alpha n} \operatorname{div} u^{\delta,N} - \frac{2k_*^2}{\alpha^2 nN} \right) dx. \tag{5.30}$$

Passing to the limit in (5.30), we establish (5.29) together with the estimate

$$j(u^\delta) \leqslant j^\delta(u^\delta) \leqslant \lambda^\delta - \frac{\delta}{2}\int_\Omega |\varepsilon(u^\delta)|^2\,dx, \tag{5.31}$$

where

$$j^\delta(v) = \frac{\delta}{2}\int_\Omega |\varepsilon(v)|^2\,dx + j(v).$$

Since

$$\begin{aligned} \lambda^\delta - \frac{\delta}{2}\int_\Omega |\varepsilon(u^\delta)|^2\,dx &= \frac{\delta}{2}\int_\Omega |\varepsilon(u^\delta)|^2\,dx + \int_\Omega \sigma_0^\delta : \varepsilon(u^\delta)\,dx \\ &\geqslant \frac{\delta}{2}\int_\Omega |\varepsilon(u^\delta)|^2\,dx + \int_\Omega g_0(\varepsilon(u^\delta))\,dx, \end{aligned}$$

we find that

$$\int_\Omega (\sigma_0^\delta : \varepsilon(u^\delta) - g_0(\varepsilon(u^\delta)))\,dx \geqslant 0.$$

Remembering that $\sigma_0^\delta \in K$, we obtain, in accordance with the definition of the function g_0,

$$g_0(\varepsilon(u^\delta)) - \sigma_0^\delta : \varepsilon(u^\delta) \geqslant 0 \quad \text{a.e. in } \Omega$$

and, hence,

$$g_0(\varepsilon(u^\delta)) = \sigma_0^\delta : \varepsilon(u^\delta) \quad \text{a.e. in } \Omega. \tag{5.32}$$

So, it is clear that inequality (5.31) holds as an equality, and we have

$$j^\delta(u^\delta) = \lambda^\delta - \frac{\delta}{2}\int_\Omega \left|\varepsilon(u^\delta)\right|^2 dx. \tag{5.33}$$

Now we can easily prove that u^δ is the solution of the following problem: Find a vector $u^\delta \in \overline{V}_0$ such that

$$j^\delta(u^\delta) = \inf\{\, j^\delta(v) : v \in \overline{V}_0 \,\}. \tag{5.34}$$

Indeed, using (5.32) and (5.28), for $v \in \overline{V}_0$ we have

$$\begin{aligned} j^\delta(v) - j^\delta(u^\delta) &= \frac{\delta}{2}\int_\Omega \left(|\varepsilon(v)|^2 - |\varepsilon(u^\delta)|^2\right) dx + \int_\Omega \left(g_0(\varepsilon(v)) - g_0(\varepsilon(u^\delta))\right) dx \\ &\geqslant \delta \int_\Omega \varepsilon(u^\delta) : \varepsilon(v - u^\delta)\, dx + \frac{\delta}{2}\int_\Omega \left|\varepsilon(v - u^\delta)\right|^2 dx + \int_\Omega \sigma_0^\delta : \varepsilon(v - u^\delta)\, dx \\ &= \frac{\delta}{2}\int_\Omega \left|\varepsilon(v - u^\delta)\right|^2 dx \geqslant 0. \end{aligned}$$

Therefore the inequality $J^\delta(u^\delta) \leqslant J^\delta(u_0) \leqslant J^1(u_0)$ implies (see (5.33)) the estimates

$$0 \leqslant \delta \int_\Omega \left|\varepsilon(u^\delta)\right|^2 dx, \qquad \lambda^\delta \leqslant C_{11}, \tag{5.35}$$

where C_{11} does not depend on δ. Furthermore, applying Lemma 5.1 to the tensor σ_0^δ when $\tau^0 = -\delta\varepsilon(u^\delta)$, $f^0 = \lambda^\delta f$, $F^0 = \lambda^\delta F$, we can establish the inequality

$$\int_\Omega \left|\sigma^\delta\right|^2 dx \leqslant C_{12}, \tag{5.36}$$

where the constant C_{12} also does not depend on δ. Thus we conclude that $\sigma^\delta \rightharpoonup \sigma$ in Σ, where $\sigma \in K$, $\int_\Omega \sigma : \varepsilon(v)\, dx = \lambda^0 M(v)\ \forall v \in V_0$, and

$$\lambda^\delta \to \lambda^0 = \int_\Omega \sigma : \varepsilon(u_0)\, dx = \inf\left\{ \int_\Omega \sigma : \varepsilon(v)\, dx : v \in \overline{V}_0 \right\}.$$

Consequently

$$\lambda^0 = r(\sigma) \leqslant \lambda_s. \tag{5.37}$$

Passing to the limit in (5.31), we find that

$$\lambda_k \leqslant \liminf_{\delta\to 0} j(u^\delta) \leqslant \lim_{\delta\to 0} \lambda^\delta = \lambda_0.$$

This inequality together with (5.37) and (2.18) implies the required relation $\lambda_k = \lambda_s = r(\sigma)$. Theorem 5.1 is proved. □

§6. Appendix 2

LEMMA 6.1. *If* $\sigma \in \Sigma \cap K$ *and* $\operatorname{div} \sigma \in L_2(\Omega; \mathbb{R}^n)$, *then there is a sequence* $\sigma^m \in C^\infty(\overline{\Omega}; \mathbb{M}_s^{n\times n}) \cap K$ *such that*

$$\sigma^m \longrightarrow \sigma \quad \text{in } \Sigma, \qquad \operatorname{div} \sigma^m \longrightarrow \operatorname{div} \sigma \quad \text{in } L_2(\Omega; \mathbb{R}^n).$$

PROOF. Since $\partial\Omega$ is Lipschitz continuous, for any point $x \in \partial\Omega$ we can find a neighborhood O_x such that $O_x \cap \Omega$ is a starshaped set. The set Ω together with all these neighborhoods O_x is an open covering of the compact set $\overline{\Omega}$. Thus, there is a partition of unity such that

$$\varphi_k \in C_0^\infty(\mathbb{R}^n), \qquad \varphi_k \geqslant 0 \quad \text{in } \mathbb{R}^n, \qquad k = 0, 1, 2, \dots, r,$$
$$\operatorname{supp} \varphi_0 \subset \Omega = \Omega^0, \qquad \operatorname{supp} \varphi_k \subset \Omega^k = O_{x_k} \cap \Omega, \qquad k = 1, 2, \dots, r,$$
$$\sum_{k=0}^r \varphi_k \equiv 1 \quad \text{in } \overline{\Omega}.$$

Let x_k^* be the center of the domain Ω^k and $\sigma_k = \varphi_k \sigma$, $k = 0, 1, 2, \dots, r$. We can extend it with the help of a similarity transformation so that Ω^k transforms into

$$\Omega_\lambda^k := \{\, x \in \mathbb{R}^n : x_k^* + \lambda(x - x_k^*) \in \Omega^k \,\}, \qquad 0 < \lambda < 1,$$

and define the function $\sigma_k^\lambda = \sigma(x_k^* + \lambda(x - x_k^*))$ for all $x \in \Omega_\lambda^k$. We also define

$$\sigma_k^{\lambda,\rho}(x) = \int_{\Omega_\lambda^k} \omega_\rho(x - y)\sigma_k^\lambda(y)\,dy, \qquad \sigma_0^\rho(x) = \int_\Omega \omega_\rho(x - y)\sigma_0(y)\,dy,$$

where ω_ρ is the standard averaging kernel.

Using well-known properties of the averaged functions, we can affirm that for any $\varepsilon > 0$ there is a $\delta(\varepsilon) > 0$ such that for $\lambda \in]1 - \delta(\varepsilon), 1[$ it is possible to find a number $\mu(\lambda, \varepsilon) > 0$ satisfying

$$\|\tau^{\lambda,\rho} - \sigma\|_\Sigma < \varepsilon, \qquad \|\operatorname{div} \tau^{\lambda,\rho} - \operatorname{div} \sigma\|_{L_2(\Omega;\mathbb{R}^n)} < \varepsilon,$$
$$\rho \in]0, \mu(\lambda, \varepsilon)[, \qquad \tau^{\lambda,\rho} = \sigma_0^\rho + \sum_{k=1}^r \sigma_k^{\lambda,\rho}.$$

Since $\sigma \in K$, it is easy to deduce the estimate

$$\begin{aligned} &|(\tau^D)^{\lambda,\rho}(x)| + \alpha \operatorname{tr} \tau^{\lambda,\rho} \\ &\quad \leqslant \sqrt{2}k_* \left[\int_\Omega \omega_\rho(x - y)\varphi_0(y)\,dy + \sum_{k=1}^r \int_{\Omega_\lambda^k} \omega_\rho(x - y)\varphi_k^\lambda(y)\,dy \right] \end{aligned}$$

for any $x \in \overline{\Omega}$. Decreasing the values of $\delta(\varepsilon)$ if necessary and $\mu(\lambda, \varepsilon)$, we thus obtain

$$\sup\{\, |(\tau^D)^{\lambda,\rho}(x)| + \alpha \operatorname{tr} \tau^{\lambda,\rho}(x) : x \in \overline{\Omega} \,\} \leqslant \sqrt{2}k_*(1 + \varepsilon).$$

It is evident now that the sequence

$$\sigma^m = \frac{\tau^{\lambda_m,\rho_m}}{1 + \frac{1}{m}}, \quad \text{where } \varepsilon = \tfrac{1}{m},\ \lambda_m = 1 - \tfrac{1}{2}\delta\left(\tfrac{1}{m}\right),\ \rho_m = \tfrac{1}{2}\mu\left(\lambda_m, \tfrac{1}{m}\right)$$

is the required one. □

References

[AG1] G. Anzellotti and M. Giaquinta, *Existence of the displacement field for an elastoplastic body subject to Hencky's law and von Mises' yield condition*, Manuscripta Math. **32** (1980), 101–136.

[AG2] ———, *On the existence of the fields of stresses and displacements for an elasto-perfectly plastic body in static equilibrium*, J. Math. Pures Appl (9) **61** (1982), 219–244.

[AG3] ———, *Convex functionals and partial regularity*, Arch. Rational Mech. Anal. **102** (1988), 243–272.

[DP] D. C. Drucker and W. Prager, *Soil mechanics and plastic analysis or limit design*, Quart. Appl. Math. **10** (1952), 157–165.

[DuL] G. Duvant and J.-L. Lions, *Les inéquations en mécanique et en physique*, Dunod, Paris, 1972; English transl., Springer-Verlag, Berlin, 1976.

[EK] L. C. Evans and D. F. Knerr, *Elastic-plastic plane stress problems*, Appl. Math. Optim. **5** (1979), 331–348.

[G] M. Giaquinta, *Multiple integrals in the calculus of variations and nonlinear elliptic systems*, Ann. of Math. Studies, vol. 105, Princeton Univ. Press, Princeton, NJ, 1983.

[HKi] R. Hardt and D. Kinderlehrer, *Elastic plastic deformation*, Appl. Math. Optim. **10** (1983), 203–246.

[JS] C. Johnson and R. Scott, *A finite element method for problems in perfect plasticity using discontinuous trial functions*, Nonlinear Finite Element Analysis in Structural Mechanics (Proc. European–U.S. Workshop, Bochum, 1980; W. Wunderlich et al., editors), Springer-Verlag, Berlin, 1981, pp. 307–324.

[KT] R. Kohn and R. Temam, *Dual spaces of stresses and strains with Applications to Hencky plasticity*, Appl. Math. Optim. **10** (1983), 1–35.

[LS] O. A. Ladyzhenskaya and V. A. Solonnikov, *Some problems of vector analysis and generalized statements of boundary problems for the Navier-Stokes equations*, Zap. Nauchn. Sem. Leningrad. Otdel. Mat. Inst. Steklov (LOMI) **59** (1976), 81–116; English transl. in J. Soviet Math. **10** (1978), no. 2.

[LU] O. A. Ladyzhenskaya and N. N. Ural′tseva, *Linear and quasilinear elliptic equations*, 2nd rev. ed., "Nauka", Moscow, 1973; English transl. of 1st ed., Academic Press, New York, 1968.

[MO] J. M. M. C. Marquest and D. R. J. Owen, *Some reflections on elastoplastic stress calculation in finite element analysis*, Computers & Structures **18** (1984), 1135–1139.

[MSCh] H. Matthies, G. Strang, and E. Christiansen, *The saddle point of a differential program*, Energy Methods in Finite Element Analysis (Dedicated to the Memory of Professor B. Fraeijs de Veubeke; R. Glowinski et al., editors), Wiley, New York, 1979, pp. 309–318.

[N] A. Nadai, *Theory of flow fracture of solids*, Academic Press, New York, Toronto, and London, 1950.

[R1] S. I. Repin, *On the variational formulations using discontinuous displacements fields for problems of perfect plasticity*, Prikl. Mat. Mekh. **55** (1991), 1026–1034; English transl. in J. Appl. Math. Mech. **55** (1991).

[R2] ———, *Minimization of a class of nondifferentiable functionals by means of relaxation methods*, Zh. Vychisl. Mat. i Mat. Fiz. **27** (1987), 976–983; English transl. in USSR Comput. Math. and Math. Phys. **27** (1987).

[R3] ———, *Variational-difference method for problems of perfect plasticity using a discontinuous conventional finite elements method*, Zh. Vychisl. Mat. i Mat. Fiz. **28** (1988), 449–453; English transl. in USSR Comput. Math. and Math. Phys. **28** (1988).

[R4] ———, *Variational-difference method for solving problems with functionals of linear growth*, Zh. Vychisl. Mat. i Mat. Fiz. **29** (1989), 693–708; English transl. in USSR Comput. Math. and Math. Phys. **29** (1989).

[R5] ———, *Partial relaxation of some mechanical variational problems*, Problems of Pure and Applied Mathematics, Tartu, 1990, pp. 284–287. (Russian)

[R6] ———, *Variational formulations of ideal plasticity problems for discontinuous displacement fields*, Dokl. Akad. Nauk SSSR **320** (1991), 1340–1344; English transl. in, Soviet Phys. Dokl. **36** (1991).

[R7] ———, *Discontinuous solutions of some variational problems in the mathematical theory of plasticity*, Prikl. Mat. Mekh. (to appear); English transl. in J. Appl. Math. Mech. (to appear).

[S1] G. A. Seregin, *Variation-difference scheme for problems in the mechanics of ideally elastoplastic media*, Zh. Vychisl. Mat. i Mat. Fiz. **25** (1985), 237–253; English transl., USSR Comput. Math. and Math. Phys. **25** (1985), 153–165.

[S2] ———, *Differentiability of local extremals of variational problems in the mechanics of perfect elastoplastic media*, Differentsial′nye Uravneniya **28** (1987), 1981–1991; English transl., Differential Equations **23** (1987), 1349–1358.

[S3] ______, *Differentiability properties of the stress tensor in perfect elastic-plastic theory*, Preprint UTM 321-Settembre, Università degli Studi di Trento, Trento, 1990.

[S4] ______, *On differential properties of extremals of variational problems arising in plasticity theory*, Differentsial′nye Uravneniya **26** (1990), 1033–1043; English transl., in Differential Equations **26** (1990).

[S5] ______, *On the regularity of weak solutions of variational problems in plasticity theory*, Algebra i Analiz **2** (1990), no. 2, 121–140; English transl., Leningrad Math. J. **2** (1991), 321–338.

[S6] ______, *Some mathematical problems in the theory of perfect elastic-plastic plates*, LOMI Preprint E-6-91, Leningrad, 1991. (English)

[S7] ______, *On the regularity of the minimizers of some variational problems of plasticity theory*, Algebra i Analiz **4** (1992), no. 5, 181–218; English transl., St. Petersburg Math. J. **4** (1993), 989–1020.

[S8] ______, *On the well-posedness of variational problems of mechanics ideally elastic-plastic media*, Dokl. Akad. Nauk SSSR **276** (1984), no. 1, 71–75; English transl., Soviet Phys. Dokl. **29** (1984), 316–318.

[S9] ______, *On a variational difference scheme for problem of limit equilibrium*, Zh. Vychisl. Mat. i Mat. Fiz. **27** (1987), 83–92; English transl. in USSR Comput. Math. and Math. Phys. **27** (1987), 53–59.

[SiDe] H. J. Siriwardane and C. S. Desai, *Computational procedures for nonlinear three-dimensional analysis with some advanced constitutive laws*, Internat. J. Numer. Anal. Methods Geomech. **7** (1983), 143–171.

[StT] Y. Stephan and R. Temam, *Finite element computation of discontinuous solutions in the perfect plasticity theory*, Computational Plasticity. Part I (Proc. Internat. Conf., Barcelona, 1987; D. R. J. Owen et al., editors), Pineridge Press, Swansea, 1987, pp. 243–256.

[Su] P. Suquet, *Existence et régularité des solutions des équations de la plasticité parfaite*, These de 3 cycle, Université Paris-vi; summary in C. R. Acad. Sci. Paris Sér. D **286** (1978), 1201–1204.

[TStr] R. Temam and G. Strang, *Functions of bounded deformation*, Arch. Rational Mech. Anal. **75** (1980/81), 7–21.

St. Petersburg Technical University

St. Petersburg Branch, Steklov Mathematical Institute, Russian Academy of Sciences

Translated by the authors

Recent Titles in This Series

(*Continued from the front of this publication*)

125 **D. V. Anosov et al.**, Seven Papers in Applied Mathematics
124 **B. P. Allakhverdiev et al.**, Fifteen Papers on Functional Analysis
123 **V. G. Maz′ya et al.**, Elliptic Boundary Value Problems
122 **N. U. Arakelyan et al.**, Ten Papers on Complex Analysis
121 **V. D. Mazurov, Yu. I. Merzlyakov, and V. A. Churkin, Editors,** The Kourovka Notebook: Unsolved Problems in Group Theory
120 **M. G. Kreĭn and V. A. Jakubovič,** Four Papers on Ordinary Differential Equations
119 **V. A. Dem′janenko et al.**, Twelve Papers in Algebra
118 **Ju. V. Egorov et al.**, Sixteen Papers on Differential Equations
117 **S. V. Bočkarev et al.**, Eight Lectures Delivered at the International Congress of Mathematicians in Helsinki, 1978
116 **A. G. Kušnirenko, A. B. Katok, and V. M. Alekseev,** Three Papers on Dynamical Systems
115 **I. S. Belov et al.**, Twelve Papers in Analysis
114 **M. Š. Birman and M. Z. Solomjak,** Quantitative Analysis in Sobolev Imbedding Theorems and Applications to Spectral Theory
113 **A. F. Lavrik et al.**, Twelve Papers in Logic and Algebra
112 **D. A. Gudkov and G. A. Utkin,** Nine Papers on Hilbert's 16th Problem
111 **V. M. Adamjan et al.**, Nine Papers on Analysis
110 **M. S. Budjanu et al.**, Nine Papers on Analysis
109 **D. V. Anosov et al.**, Twenty Lectures Delivered at the International Congress of Mathematicians in Vancouver, 1974
108 **Ja. L. Geronimus and Gábor Szegő,** Two Papers on Special Functions
107 **A. P. Mišina and L. A. Skornjakov,** Abelian Groups and Modules
106 **M. Ja. Antonovskiĭ, V. G. Boltjanskiĭ, and T. A. Sarymsakov,** Topological Semifields and Their Applications to General Topology
105 **R. A. Aleksandrjan et al.**, Partial Differential Equations, Proceedings of a Symposium Dedicated to Academician S. L. Sobolev
104 **L. V. Ahlfors et al.**, Some Problems on Mathematics and Mechanics, On the Occasion of the Seventieth Birthday of Academician M. A. Lavrent′ev
103 **M. S. Brodskiĭ et al.**, Nine Papers in Analysis
102 **M. S. Budjanu et al.**, Ten Papers in Analysis
101 **B. M. Levitan, V. A. Marčenko, and B. L. Roždestvenskiĭ,** Six Papers in Analysis
100 **G. S. Ceĭtin et al.**, Fourteen Papers on Logic, Geometry, Topology and Algebra
99 **G. S. Ceĭtin et al.**, Five Papers on Logic and Foundations
98 **G. S. Ceĭtin et al.**, Five Papers on Logic and Foundations
97 **B. M. Budak et al.**, Eleven Papers on Logic, Algebra, Analysis and Topology
96 **N. D. Filippov et al.**, Ten Papers on Algebra and Functional Analysis
95 **V. M. Adamjan et al.**, Eleven Papers in Analysis
94 **V. A. Baranskiĭ et al.**, Sixteen Papers on Logic and Algebra
93 **Ju. M. Berezanskiĭ et al.**, Nine Papers on Functional Analysis
92 **A. M. Ančikov et al.**, Seventeen Papers on Topology and Differential Geometry
91 **L. I. Barklon et al.**, Eighteen Papers on Analysis and Quantum Mechanics
90 **Z. S. Agranovič et al.**, Thirteen Papers on Functional Analysis
89 **V. M. Alekseev et al.**, Thirteen Papers on Differential Equations
88 **I. I. Eremin et al.**, Twelve Papers on Real and Complex Function Theory
87 **M. A. Aĭzerman et al.**, Sixteen Papers on Differential and Difference Equations, Functional Analysis, Games and Control

(See the AMS catalog for earlier titles)